THE SMARTER WAY TO LEARN HTML & CSS

Hands-On Guide for Beginners and Future Web Developers

Rajender Kumar

For more information, contact: support@JambaAcademy.com.

Published by:
Jamba Academy

Printed in the United States of America
First Printing Edition, 2025

DISCLAIMER

The information provided in *The Smarter Way to Learn HTML & CSS: Hands-On Guide for Beginners and Future Web Developers* is intended solely for educational purposes to help readers learn the fundamentals of HTML and CSS and develop practical web development skills. Every effort has been made to ensure the accuracy and relevance of the content, including code examples, design techniques, and industry practices. However, web development is a dynamic field, and standards, tools, and best practices may evolve over time. The author and publisher do not guarantee that the techniques or examples in this book will produce specific results, such as professional website creation or career success, as outcomes depend on individual effort, prior knowledge, and application of the material.

This book does not constitute professional advice in web development, programming, or related fields. Readers are responsible for evaluating the suitability of the content for their personal or professional projects and for any decisions made based on the book's guidance. The author and publisher shall not be liable for any errors, omissions, or damages arising from the use of this book's content, including but not limited to issues with website functionality, compatibility, or performance. By engaging with this book, you acknowledge that you assume all risks associated with applying its lessons. We encourage readers to supplement their learning with additional resources and stay updated with current web development standards to achieve their goals.

Found Typos & Broken Link

We apologize in advance for any typos or broken links that you may find in this book. We take pride in the quality of our content and strive to provide accurate and useful information to our readers. Please let us know where you found the typos and broken links (if any) so that we can fix them as soon as possible. Again, thank you very much in advance for bringing this to our attention and for your patience.

If you find any typos and broken links in this book, please feel free to email us.

support@JambaAcademy.com

Support

We would love to hear your thoughts and feedback! Could you please take a moment to write a review or share your thoughts on the book? Your feedback helps other readers discover the books and helps authors to improve their work. Thank you for your time and for sharing your thoughts with us!

If there is anything you want to discuss or you have a question about any topic of the book, you can always reach out to us, and we will try to help as much as we can.

support@JambaAcademy.com

To every aspiring web developer and coding newcomer—may this guide open doors to creativity, opportunity, and a brighter digital future. Your journey empowers the web's next chapter, and it's my mission to make coding accessible to all.

TABLE OF CONTENTS

FOREWORD

As technology continues to shape every aspect of our lives, the ability to create and understand the digital world has never been more valuable. Whether you're glimpsing the sleek interface of your favorite app, browsing an online store, or reading a personal blog, you're engaging with products crafted using HTML and CSS—the foundational languages of the web. As a Senior Front-End Developer and lead instructor at one of the world's most respected coding bootcamps, I have witnessed firsthand how mastering these skills can transform not only careers but also lives.

HTML and CSS serve as the backbone of every website and digital experience. HTML (HyperText Markup Language) gives structure to web content, laying out the headlines, paragraphs, images, and links that users interact with daily. CSS (Cascading Style Sheets) breathes life into these structures, providing the colors, layouts, and responsive designs that make websites beautiful and functional across all devices. Together, they empower anyone to bring creative ideas to life on the web, whether for personal projects, business ventures, or ambitious tech careers.

The importance of learning HTML and CSS cannot be overstated. For aspiring web developers, these languages are the essential first step on a journey toward front-end, full-stack, or even UX/UI design roles. Yet their relevance extends far beyond professional programmers. Entrepreneurs, marketers, designers, and educators increasingly depend on a working knowledge of web fundamentals to reach audiences and stay competitive. In today's digital economy, understanding how websites are built is a superpower.

I have seen the doors that open when people learn the language of the web. Consider the story of Priya, a former elementary school teacher who joined my beginner coding class two years ago. With no prior experience, she dove into HTML and CSS, learning to build interactive lesson plans and eventually launching her own educational website. Within a year, Priya transitioned into a full-time role as a junior web developer at an edtech startup, combining her passion for teaching with her new technical skills. Her story is just one among many—each underscoring the transformative potential of web development education.

Another powerful example is the surge in demand for web-savvy professionals across industries. With the rise of global e-commerce, digital portfolios, and remote-first workplaces, employers seek individuals who can create and maintain compelling online presences. Technologies like responsive design, mobile-first layouts, and accessible web standards are not just technical trends, they are now expected skills in the modern workforce.

It is in this context that I am delighted to introduce "*The Smarter Way to Learn HTML & CSS: Hands-On Guide for Beginners and Future Web Developers*." This book stands out in a crowded field for one simple reason: it is relentlessly practical. The author, whose expertise as both a web developer and educator I deeply respect, has crafted a guide that demystifies web development for absolute beginners, providing not only clear explanations but also engaging, project-based exercises that deliver real results.

What sets this book apart is its unwavering focus on learning by doing. Rather than overwhelming you with abstract theory, the author guides you step by step through building real web pages, experimenting with code, and solving practical challenges. Each chapter builds on the last, ensuring you develop both confidence and competence as you progress. You'll learn to design a personal portfolio that showcases your skills, create a blog layout with beautiful typography, and even explore the fundamentals of crafting a simple e-commerce homepage. Along the way, you'll gain insight into global web design trends, accessibility best practices, and responsive design techniques that ensure your websites look great on any device.

The author's passion for teaching shines through every page. Having collaborated with them on several web development workshops, I know firsthand their ability to break down complex concepts into approachable, bite-sized lessons. Their teaching style is patient, encouraging, and rooted in real-world experience. Whether you're a complete novice or someone seeking a structured refresher, you'll find this book to be a supportive companion on your learning journey.

Perhaps what excites me most about this guide is how it empowers you to create and share your own digital projects from the very beginning. The hands-on approach demystifies the coding process, transforming what may seem intimidating into a series of achievable, rewarding milestones. By the time you build your first multi-page site or experiment with advanced CSS styling, you'll not only understand what you're doing but also why it matters—preparing you to tackle more sophisticated projects and even collaborate with other developers.

In an era when technology moves rapidly and opportunities abound, those who take the initiative to learn and apply web development fundamentals position themselves at the forefront of innovation. Whether your dream is to freelance as a web designer, launch a startup, enhance your current profession, or simply gain a deeper appreciation for the digital world, the practical skills in this book will serve as your foundation.

As you turn these pages, I encourage you to embrace a spirit of curiosity and persistence. Code along with each exercise, experiment with your own ideas, and don't be afraid to make mistakes—every error is a step closer to mastery. The web is an open canvas, and with the guidance of "The Smarter Way to Learn HTML & CSS," you are about to unlock its endless creative possibilities.

The journey to becoming a web developer starts with a single line of code. Open your text editor, roll up your sleeves, and dive into the hands-on learning that awaits. Your future in web development begins now.

– Jordan Li

Senior Front-End Developer & Coding Bootcamp Instructor

PREFACE

When I wrote my very first lines of HTML, I had no idea they would set the stage for a lifelong journey into the world of web development. Like many who are just starting out, I remember the thrill of seeing a simple "Hello, world!" appear in my browser — and the frustration that followed as I tried to make sense of unfamiliar tags, cryptic error messages, and endless pages of documentation. Learning to code often felt like learning a foreign language without a phrasebook. Yet, over time, I realized the most powerful lessons came not from reading manuals cover to cover, but from building things, making mistakes, and seeing my skills grow with every project.

My name is Rajender Kumar, and I wrote **The Smarter Way to Learn HTML & CSS: Hands-On Guide for Beginners and Future Web Developers** to make your introduction to web development smoother, more engaging, and, above all, practical. Whether you are a complete beginner, a career changer looking to break into a booming industry, or someone with a creative spark eager to build your own online presence, this book is crafted for you. My mission is simple: to help you become a confident web developer by guiding you through the essential skills and thought processes you need to create real, beautiful websites from scratch.

Why This Book? The Motivation Behind a Smarter Approach

Over the years, I have taught coding to hundreds of students from diverse backgrounds. I've seen the excitement in a graphic designer's eyes as she brings her first portfolio website to life, and I've witnessed the quiet determination of a mid-career professional learning to code after hours, hoping for a fresh start. In each case, a common thread emerged: the need for a friendlier, more hands-on approach to learning HTML and CSS.

Too often, introductory guides drown beginners in abstract theory or expect them to memorize endless syntax before they ever build something meaningful. Traditional computer science courses and even some modern bootcamps can leave students overwhelmed or unprepared for the real challenges of web development. I wanted to change that. I wanted to create a guide that not only teaches the how but the why, empowering you to learn by doing and to understand the bigger picture behind every line of code.

A Personal Anecdote: Learning by Doing

Let me share a moment that shaped my teaching philosophy. Years ago, while mentoring a group of new developers, I noticed one student, Arjun, struggling to grasp the basics of CSS. He'd read the documentation, watched the tutorials, but when it came time to style a webpage, he froze. Instead of assigning more reading, I sat down with him, and we built a simple personal homepage together, step by step. As we worked, he experimented with colors, fonts, and layouts, seeing instant results in his browser. By the end of that session, not only did Arjun understand CSS selectors and properties, but he'd also gained the confidence to keep exploring on his own.

That experience reinforced what I believe is the most effective way to learn coding: through hands-on practice, immediate feedback, and building real projects that matter to you.

Who This Book Is For

If you're picking up this book, you might fit one of these descriptions:

The Absolute Beginner: You've never written code before, but you're curious about how websites work and excited to create your own.

The Career Changer: You're looking to enter the tech industry and want a practical, approachable path into web development.

The Hobbyist or Maker: You have ideas for a blog, portfolio, or e-commerce site and want the skills to bring them to life on your terms.

The Aspiring Professional: You're considering a longer-term career in web design or development and need a solid foundation in HTML and CSS.

No matter what your background, this book requires no prior coding experience. All you need is curiosity, a computer, and the willingness to learn by doing.

What Makes This Book Different

In a world filled with coding resources, what sets this guide apart? **The Smarter Way to Learn HTML & CSS** is not just another reference manual or passive tutorial. It's an active, project-based learning experience designed with you in mind. Here's how:

Practical Projects, Not Just Theory: Each concept is introduced through real-world examples — from creating a personal portfolio page to styling a modern blog or laying out an e-commerce product card. You'll see how the skills you learn apply to websites you use every day.

Step-by-Step Guidance: Every topic is broken down into manageable steps, with clear explanations and hands-on exercises that reinforce your learning. You'll never be left wondering what to do next.

Immediate Application: You'll write code from page one, building your own projects and seeing your progress in real time. This active approach helps concepts "stick" and builds your confidence as a developer.

Modern, Accessible Techniques: You'll learn the latest HTML and CSS standards, including responsive design, flexible layouts, and accessible web practices that prepare you for today's web — not yesterday's.

Beginner-Friendly Language: No confusing jargon or unexplained buzzwords. Complex ideas are demystified with plain language, analogies, and visual examples that anyone can grasp.

Closing the Gap: Beyond Bootcamps and Online Tutorials

The explosion of coding bootcamps and online tutorials has made web development more accessible than ever, but it has also revealed some gaps. Many programs promise rapid results but rush through fundamentals, leaving learners with knowledge that's wide but shallow. Others require a steep upfront investment — in both time and money — that isn't realistic for everyone.

This book is designed as an affordable, self-paced alternative that prioritizes depth over speed and understanding over memorization. It gives you the practical foundation that employers value and that you need to confidently tackle your own projects, whether you're freelancing, job-hunting, or building something for fun.

Empowerment Through Creation

At its heart, web development is about turning ideas into reality. The ability to design and build your own websites is not just a technical skill — it's a form of creative empowerment. With HTML and CSS, you can share your story, showcase your talents, launch a business, or connect with communities around the world.

My hope is that **The Smarter Way to Learn HTML & CSS** will be more than just a book for you. Let it be your companion and cheerleader as you unlock new possibilities, one project at a time. Remember, every expert was once a beginner. With curiosity, persistence, and the right guidance, you too can become a skilled web developer.

Welcome to your journey. Let's start building.

-Rajender Kumar

ACKNOWLEDGEMENTS

Creating **The Smarter Way to Learn HTML & CSS: Hands-On Guide for Beginners and Future Web Developers** has been a transformative journey—one that would not have been possible without the encouragement, expertise, and generosity of a truly inspiring community. As I reflect on the process of bringing this guide to life, I am filled with gratitude for everyone who contributed their time, knowledge, and support.

First and foremost, I want to thank the **web development mentors** whose patient guidance and deep understanding of HTML and CSS have influenced every page of this book. Your willingness to share real-world insights and demystify complex topics ensured that every concept is explained clearly and accessibly. The practical advice and best practices you shared helped shape the hands-on approach that defines this guide, making it a resource that genuinely empowers beginners.

I am also indebted to the vibrant **coding community**, both online and offline, who offered invaluable feedback, shared resources, and fostered a spirit of collaborative learning. Your thoughtful discussions and constructive critiques of early drafts played a pivotal role in refining the book's content and structure. The sense of belonging and shared purpose I found among you continually motivated me to create a guide that is welcoming, practical, and relevant for all aspiring web developers.

A heartfelt thank you goes to the dedicated **beta readers** who took the time to test every tutorial and hands-on exercise. Your honest assessments and candid suggestions were instrumental in ensuring that the instructions were clear, the examples realistic, and the learning curve manageable. The improvements made based on your experiences have made this book more user-friendly for readers taking their first steps into coding.

My sincere appreciation extends to the meticulous **technical reviewers** who scrutinized each HTML and CSS example for accuracy and best practices. Your attention to detail and commitment to high standards enhanced the book's technical integrity and ensured that every code snippet is robust and aligned with modern web development standards.

To my **family and friends**, thank you for your unwavering encouragement and patience throughout the writing process. Your belief in the importance of accessible coding education inspired me to persevere through challenges and approach this project with enthusiasm and care. The countless hours spent drafting, revising, and testing would not have been possible without your steadfast support.

Reflecting on this collaborative journey, I am reminded of the power of community in shaping meaningful educational resources. The collective wisdom, encouragement, and constructive feedback I received have enriched this guide, making it not just a book, but a companion for beginners discovering the world of web development. It is my hope that this spirit of support and shared learning resonates with every reader who picks up **The Smarter Way to Learn HTML & CSS**, and encourages you to contribute to the ever-growing, diverse world of web creators.

With deep gratitude,

Rajender Kumar

INTRODUCTION: YOUR GATEWAY TO THE WORLD OF WEB DEVELOPMENT

Imagine this: In less than a year, Sarah, a graphic designer with no technical background, built her first portfolio website. She learned the basics of coding by dedicating just an hour each evening, transforming her career and unlocking freelance opportunities she never thought possible. Today, Sarah works with clients worldwide, showcasing her work and attracting new business—all because she took the first step to learn **HTML** and **CSS**.

Sarah's story isn't unique. According to a recent report from the U.S. Bureau of Labor Statistics, web development jobs are projected to grow by 16 percent this decade, much faster than the average for all occupations. Every day, millions of people browse news sites, shop online, and connect on social platforms—all powered by the same core technologies: **HTML** and **CSS**. These are the foundational languages that shape the web.

Now, pause for a moment and ask yourself: What if you could build the website, you've been dreaming about, launch a blog to share your ideas, or even pivot into a rewarding career in tech? No matter what your background, learning web development is more accessible than ever. And it all starts with understanding **HTML** and **CSS**.

Welcome to *The Smarter Way to Learn HTML & CSS: Hands-On Guide for Beginners and Future Web Developers*. This book is designed to empower you, the absolute beginner, to confidently create beautiful, professional websites from the ground up. Whether you're a student, a career changer, a creative professional, or simply curious about coding, you'll find a clear, supportive path to becoming a web developer—one practical, engaging project at a time.

Why HTML & CSS Matter

Before diving into code, let's demystify what **HTML** and **CSS** actually are—and why they're so important.

HTML (HyperText Markup Language) is the backbone of every website. It describes the structure and meaning of web content, turning plain text into headings, paragraphs, images, links, and more. Think of **HTML** as the framework or skeleton of a house.

CSS (Cascading Style Sheets) is what makes the web beautiful and engaging. It controls the layout, colors, fonts, and overall visual design. If **HTML** is the skeleton, **CSS** is the paint, furniture, and interior design that bring a website to life.

Together, **HTML** and **CSS** form the essential toolkit for any aspiring web developer or designer. They are the first step in your journey—no matter if your goal is to build a personal blog, launch a business site, or start a new career in tech.

Who This Book Is For

You might be wondering if you need a math degree, prior coding experience, or thousands of dollars in tuition to learn web development. The answer is an emphatic no.

This book is written for:

- **Absolute beginners** who have never written a line of code.
- **Career changers** seeking a flexible, high-demand skill.
- **Creative professionals** (designers, writers, artists) eager to build and control their own online presence.
- **Students** looking to boost their digital literacy and future career prospects.

If you can use a web browser, you can learn to code. All you need is curiosity, motivation, and a willingness to experiment.

The Smarter, Hands-On Approach

Many coding books overwhelm you with jargon or expect you to memorize endless rules before you ever write a line of code. That's not how this book works.

The Smarter Way to Learn HTML & CSS is built on three core principles:

Learn by Doing: You'll be writing code from the very first chapter. Each concept is paired with a practical exercise or mini-project, so you immediately apply what you learn.

Simple Explanations, No Assumptions: Every technical term is explained in plain English, with real-world analogies and visual examples.

Progressive, Project-Based Learning: Each new skill builds on the last, guiding you from simple web pages to responsive, professional-quality sites.

By the end of this book, you'll have created multiple hands-on projects you can proudly share and expand on. More importantly, you'll have the confidence and skills to keep growing as a web developer.

Inspiring Real-World Examples

To motivate you on your journey, let's look at a few people and projects that started just like you:

Alex, the Career Changer: After ten years in hospitality, Alex wanted more flexible work. By learning HTML and CSS, he built a sleek resume website and started freelancing. Within months, he landed a junior web developer job at a local agency, proving that it's never too late to start.

Marie, the Artist and Entrepreneur: Marie wanted to sell her handmade jewelry online. Rather than relying on generic templates, she learned to code her own e-commerce site, showcasing her art with custom galleries and unique branding. Her business grew, and so did her skills.

Global Inspiration: Websites like [Awwwards](https://www.awwwards.com/) and [CSS Design Awards](https://cssdesignawards.com/) highlight cutting-edge designs from around the world. Many featured developers started with simple HTML and CSS, gradually building their portfolios and skills to create award-winning digital experiences.

You don't need to be a genius or have a computer science degree. You just need the drive to learn—and a guide that makes the path clear and enjoyable.

What Makes This Book Different?

There are many resources for learning HTML and CSS. What sets this book apart is its emphasis on **active, hands-on learning** and **real-world application**. Here's what you can expect:

- **Step-by-step tutorials** with clear explanations and code samples.
- **Immediate practice exercises** after each concept, reinforcing your learning.
- **Mini-projects** that let you build real websites, not just toy examples.
- **Modern best practices**, including accessibility, semantic markup, and responsive design.
- **Encouragement and support** at every step, helping you overcome challenges and celebrate successes.

You won't just read about code, you'll write it, test it, and see your work come to life in your own browser.

Your First Step into a Bigger World

Learning HTML and CSS is more than a technical skill. It's a creative superpower, a way to share ideas, launch projects, and connect with people around the globe. With each chapter, you'll unlock new abilities, solve real problems, and gain the confidence to tackle bigger challenges.

Whether your dream is to build a personal blog, launch an online store, or start a new career, this book is your trusted companion. Every page is designed to make learning fun, accessible, and effective—even if you've never coded before.

So, are you ready to start building your first website? Open your computer, roll up your sleeves, and let's code your future—one line at a time.

Welcome to *The Smarter Way to Learn HTML & CSS*. Your web development journey starts now.

ACCESSING THE CODE: COMPANION GITHUB REPOSITORY

To enhance your learning experience with *The Smarter Way to Learn HTML & CSS: Hands-On Guide for Beginners and Future Web Developers*, all code examples, projects, and supplementary resources discussed in this book are available in a dedicated GitHub repository. Visit https://github.com/JambaAcademy/HTML_CSS to access the complete collection of HTML and CSS files, organized by chapter and section, along with images, videos, and other media used in the book's hands-on projects. This repository is designed to complement the book, allowing you to view, edit, and experiment with the code as you progress through each chapter. Whether you're building your first webpage or mastering responsive design, the repository provides a practical way to follow along and apply what you learn. Simply clone the repository or download the files to get started, and refer to the book's instructions for guidance on using each example effectively.

ABOUT THE AUTHOR

Rajender Kumar is a seasoned data professional and the driving force behind The Smarter Way to Learn HTML & CSS: Hands-On Guide for Beginners and Future Web Developers. With over a decade of experience at the crossroads of data science, machine learning, and big data, Rajender has cultivated a reputation for transforming complex technical landscapes into accessible pathways for learning. His illustrious career spans key roles with industry giants such as Walgreens Boot Alliance, Healthcare Retroactive Audits, NEBA, and Loxia Technologies, where he has delivered impactful solutions in data analysis, engineering, and consulting.

Rajender's analytical expertise is matched by his systematic approach to coding education. Years spent designing user-friendly interfaces and compelling data visualizations have given him a deep appreciation for the foundational role of HTML and CSS in the digital world. He understands that clean, well-structured code is not only essential for effective web presentation but also vital for creating intuitive user experiences and meaningful data insights. This unique perspective shapes his teaching philosophy—one that emphasizes logical progression, practical application, and real-world relevance.

What sets Rajender apart as an educator is his remarkable ability to break down intricate coding concepts into simple, digestible steps. Drawing from his consulting background, he excels at demystifying the building blocks of web development, ensuring that even absolute beginners feel confident and supported as they embark on their coding journey. Whether guiding readers through the basics of HTML tags or the intricacies of responsive CSS layouts, Rajender's explanations are always clear, patient, and rooted in best practices.

Rajender's passion for empowering others is at the heart of his work. He recalls his own early days of learning to code, when resources were scarce and guidance was often elusive. Inspired by the transformative power of technology and education, he is deeply committed to making coding accessible and enjoyable for everyone, regardless of background or experience. This commitment is reflected in his hands-on, project-based methodology, which encourages learners to build, experiment, and create from the very first chapter.

Throughout The Smarter Way to Learn HTML & CSS, Rajender draws on real-world examples from diverse fields—such as portfolio websites, blogs, and e-commerce platforms—to illustrate the limitless potential of web development skills. His engaging, practical approach instills not just technical knowledge but also the confidence to tackle real projects and pursue new opportunities in the digital age.

Beyond authorship, Rajender is dedicated to mentoring the next generation of web developers. He regularly contributes to coding communities, leads workshops, and supports initiatives aimed at broadening access to tech education. His unwavering belief in the power of practical, hands-on learning continues to inspire aspiring developers worldwide.

Rajender Kumar's blend of industry expertise, educational vision, and genuine passion for teaching makes him a trusted guide for anyone seeking to master HTML and CSS and unlock a world of creative possibilities online.

"Learning to code unlocks not just the language of technology, but the power to shape your own digital future. Start simple, stay curious, and watch your ideas become reality."

— Reshma Saujani, Founder of Girls Who Code

PREREQUISITES: GETTING READY TO CODE

Before you embark on your journey to learn HTML and CSS with this hands-on guide, it's important to know what you need—and, just as importantly, what you *don't* need. This book is written for absolute beginners, so you don't need any prior programming or web development experience. All you need is a curious mind, a bit of patience, and a willingness to learn by doing.

The Right Mindset: Curiosity, Practice, and Patience

Learning to code can feel intimidating at first, especially if you've never typed a line of code before. It's normal to have questions or even doubts. Will I be able to understand the code? What if I make mistakes? Can I really build a website if I've never coded before?

The answer is yes—you absolutely can. The most important things you can bring to this book are:

Curiosity: Be eager to explore and experiment. Coding is a creative process, and you'll learn best by trying things out for yourself.

Willingness to Practice: Much like learning to play an instrument or a new sport, regular hands-on practice is essential. You'll gain confidence and skill with each small project and exercise.

Patience: Mistakes are part of learning. If something doesn't work right away, don't get discouraged. Debugging and problem-solving are valuable skills that all web developers develop over time.

Imagine you're learning to ride a bicycle. You wouldn't expect to master it on your first try. Coding is similar—each attempt helps you balance a little better, and soon you'll be cruising along.

Basic Computer Skills: What You Really Need

You don't need to be a computer expert, but some basic computer comfort will make your learning experience smoother. Here's what's helpful:

- **Typing and using a keyboard**: You'll be writing code, so being able to type (even slowly) is important.
- **Navigating files and folders**: You should know how to create, open, and save files on your computer.

- **Copying and pasting text**: This will help you move code snippets between the book, your editor, and your browser.
- **Using a web browser**: You should know how to open a website, refresh a page, and use tabs.

If you've ever browsed the web, written an email, or saved a document, you already have the basic skills you need. If you encounter unfamiliar computer tasks, don't worry—this book explains technical steps as they come up.

Technical Setup: Simple Tools for Big Results

To follow along with the hands-on exercises and projects in this book, you'll need just a few free and accessible tools. No complicated setup, no expensive software.

1. A Modern Web Browser

All your HTML and CSS projects will be displayed in a web browser. Browsers are programs like Chrome, Firefox, Edge, or Safari that let you view websites. If your computer is less than 10 years old, you likely already have one installed. If possible, use the latest version of:

Google Chrome: [Download Chrome](https://www.google.com/chrome/)
Mozilla Firefox: [Download Firefox](https://www.mozilla.org/firefox/)
Microsoft Edge: [Download Edge](https://www.microsoft.com/edge)
Apple Safari: (Comes pre-installed on Macs)

Any of these will work for testing your code and seeing your web pages come to life.

2. A Simple Text Editor

To write HTML and CSS, you'll need a text editor—a program for writing plain text. While your computer's default editors like Notepad (Windows) or TextEdit (Mac) will work, a code-friendly editor can make your life much easier by adding helpful features like color-coding and auto-completion.

Recommended beginner-friendly editors include:

Visual Studio Code: [Download VS Code](https://code.visualstudio.com/) (free, highly popular, available for Windows/Mac/Linux)
Sublime Text: [Download Sublime Text](https://www.sublimetext.com/) (free to try, simple interface)
Atom: [Download Atom](https://atom.io/) (free, open-source)

Choose the one you find most comfortable—each is lightweight and easy to install.

3. Optional: Online Coding Playgrounds

If you'd rather not install any software immediately, or want to experiment quickly, online coding playgrounds are a fantastic option. These websites let you write and preview HTML and CSS code directly in your browser, no setup required.

Popular online playgrounds:

CodePen: codepen.io (great for experimenting and sharing code)
JSFiddle: jsfiddle.net (simple, fast, supports HTML/CSS/JavaScript)
repl.it: [replit.com](https://replit.com/~) (supports many programming languages, including web development)

Throughout this book, you'll find exercises and mini-projects you can do in your text editor or on a coding playground.

4. Optional: Beginner-Friendly Tutorials and Resources

If you ever feel stuck or want to reinforce a concept, there are helpful, free tutorials designed for complete newcomers:

MDN Web Docs (HTML Basics): developer.mozilla.org/en-US/docs/Learn/HTML/Introduction_to_HTML
HTML & CSS Crash Course (YouTube, free video tutorials): Search for "HTML crash course" or "CSS for beginners" for guided walkthroughs.
W3Schools: [w3schools.com](https://www.w3schools.com/) (quick reference and interactive examples)

These resources can supplement your learning and provide different explanations if you need more clarity.

Overcoming Common Barriers and Fears

If you're worried about making mistakes or "breaking" your computer, rest assured: writing HTML and CSS is very safe. The worst thing that can happen is your web page doesn't look quite right. You can always undo changes or start fresh.

Here are some common beginner fears, and why you don't need to worry:

"What if I don't understand the code?"

Every new skill takes time. The book breaks down each concept into small, manageable steps, with practical examples and plenty of repetition.

"I'm not good with technology."

If you can open a file, type a few words, and use the internet, you're ready to start coding. You'll build confidence as you go.

"I'm afraid of making mistakes."

Mistakes are not only normal—they're helpful! Every coder, from beginner to expert, learns through trial and error. You can't break anything that can't be fixed by hitting "Undo" or trying again.

"I don't know if I'm creative enough."

Coding is both logical and creative. You'll learn the rules, then use them to design and build sites that reflect your own ideas and style.

Practical Scenarios: Why These Skills Matter

Imagine you're working on your first web page, typing HTML into your text editor. You save your work, open your browser, and reload the page to see the results. Maybe you made a typo and the result isn't perfect. That's okay! You go back, fix the error, and instantly see your improvement.

Or perhaps you're following a project in this book—like building your own personal portfolio or a simple blog layout. You copy a code snippet from an example, paste it into your editor, and tweak a few words to make it your own. Each small change builds your understanding and confidence.

Coding Comfort Checklist

Before you dive into the first chapter, make sure you can:

- Open a web browser and navigate to a website.
- Download and install a simple program (like a text editor).
- Create a new file and save it with a specific name (like `index.html`).
- Type and edit text in a document.
- Copy and paste text between programs.

If you're comfortable with these steps, you're ready to start building web pages and styling them with CSS, no matter your background.

Coding Summary and Hands-On Exercise

Let's put this all together with a practical exercise you can do right now, even if this is your very first day:

Install a free text editor (such as Visual Studio Code).
Open your editor and create a new file named `hello.html`.
Type the following code:

```
<!DOCTYPE html>
<html>
  <head>
    <title>My First Web Page</title>
  </head>
  <body>
    <h1>Hello, world!</h1>
    <p>This is my very first web page.</p>
  </body>
</html>
```

Save the file and double-click it to open in your web browser.
Admire your first website!

You've just practiced the core workflow used by professional developers worldwide: write code in your text editor, preview in your browser, and repeat.

Final Encouragement

Remember: every professional web developer started right where you are now—with curiosity, a blank page, and a desire to create something new. With this guide, you'll move step by step from the basics of HTML and CSS to building real, beautiful web pages. Take your time, practice regularly, and celebrate each small victory. You're about to unlock a whole new world of creativity and opportunity—let's get started!

WHAT YOU'LL LEARN: CODING SKILLS AND CAPABILITIES

Embarking on your web development journey can feel like stepping into a new world—one filled with endless creative possibilities and exciting career paths. **The Smarter Way to Learn HTML & CSS: Hands-On Guide for Beginners and Future Web Developers** is designed to make this journey accessible, rewarding, and truly transformative, no matter your starting point or learning style. Whether you're a total beginner, a self-starter, or someone considering a career change, this guide will equip you with the essential coding skills and practical confidence to build real, modern websites from scratch.

At its core, this book focuses on two foundational languages of the web: **HTML** (HyperText Markup Language) and **CSS** (Cascading Style Sheets). You'll move beyond theory, actively coding your way through each concept with practical exercises and real-world projects that solidify your understanding. Here's what you can expect to master:

- **HTML Fundamentals:** Learn to structure web pages by creating headings, paragraphs, lists, links, images, tables, forms, and more using HTML tags.
- **CSS Styling:** Discover how to style your websites with colors, fonts, spacing, borders, backgrounds, and more using powerful CSS properties.
- **Responsive Design:** Build websites that look great on any device—desktop, tablet, or smartphone—using modern responsive design techniques like media queries and flexible layouts.
- **Web Page Layouts:** Master the art of arranging content with CSS Grid and Flexbox, two essential tools for creating visually appealing layouts.
- **Typography and Visual Design:** Explore the basics of web typography, working with fonts, font sizes, and spacing to create attractive, readable content.
- **Practical Web Projects:** Apply your skills to build complete, real-world web projects, such as a personal portfolio, a blog homepage, or a simple e-commerce product page.

As you progress, you'll not only write code but understand how each element and style comes together to create a seamless web experience. You'll become fluent in the process of planning, building, and styling professional-quality web pages—skills that form the backbone of all modern websites.

Hands-On Projects: Building Your Portfolio

Learning is most effective when it's applied. Throughout this book, you'll complete hands-on projects that mirror real-world web development challenges. These projects are designed to reinforce each concept while giving you tangible pieces for your coding portfolio:

Personal Portfolio Website: Showcase your skills and creativity by building a custom website to present your work, hobbies, or resume—an essential asset for freelancers and job seekers.

Simple Blog Homepage: Create a blog landing page featuring articles, images, and navigation menus, practicing content structure and visual hierarchy.

E-commerce Product Page: Design a clean, attractive product page with images, descriptions, and pricing, learning how to present information clearly and responsively.

Each project introduces new HTML elements and CSS techniques, helping you see how individual skills combine to create polished, user-friendly websites.

Real-World Applications: From Personal Projects to Professional Work

Mastering HTML and CSS opens the door to a variety of opportunities. Here's how you can put your new skills to work:

Freelance Work and Side Hustles

Many businesses and individuals need simple, attractive websites. As a freelancer, you'll use your HTML and CSS expertise to:

Build landing pages for local businesses, artists, or non-profits.

Update or restyle existing websites to be mobile-friendly and visually modern.

Offer custom website creation services for clients needing personal portfolios or event pages.

Career Transitions and Professional Growth

If you're considering a career change, HTML and CSS are the perfect starting point for roles like:

Junior Web Developer: Many entry-level web developer positions require a solid foundation in HTML and CSS before moving on to advanced languages.

Web Designer: Combine your design sense with coding skills to create visually compelling websites, collaborating with clients or agencies.

Content/Marketing Specialist: Improve digital marketing materials by creating or editing web pages, emails, and landing pages.

Many professionals in marketing, design, or communications discover that even basic coding skills can make them invaluable team members.

Personal and Creative Projects

Your skills aren't limited to the workplace! Use HTML and CSS to:

Launch a personal blog, art gallery, or travel journal.
Create event pages for weddings, fundraisers, or clubs.
Design custom online resumes or digital business cards.

The ability to bring your ideas to life on the web is empowering, allowing you to share your passions and accomplishments with the world.

Learning for Everyone: Adapting to Your Pace and Background

This book is built for flexibility and inclusivity. Each chapter is structured to guide you step by step, with clear explanations, coding examples, and hands-on exercises. Whether you're a:

- **Complete Beginner:** No prior coding knowledge required; every concept is introduced from the ground up, using simple language and relatable examples.
- **Career Changer:** Practical, project-based learning helps you build a portfolio and develop job-ready skills at your own pace.
- **Self-Taught Learner:** Fill in gaps, reinforce core concepts, and discover best practices for modern web development, even if you've dabbled in HTML or CSS before.

Every exercise is designed to be achievable, giving you instant feedback and a sense of accomplishment. You'll never feel rushed—progress at your own speed, revisit sections as needed, and watch your coding confidence grow.

Staying Current: Modern Web Development and Responsive Design

The web evolves rapidly, and this book is crafted with today's standards in mind. You'll learn not only traditional HTML and CSS but also the latest best practices, including:

- Writing clean, semantic HTML for accessibility and search engine optimization (SEO).
- Using CSS Grid and Flexbox for flexible, modern layouts.
- Creating mobile-first designs to ensure your sites look great on all devices.
- Understanding web design trends such as minimalism, bold typography, and dark mode themes.

You'll build skills that are immediately relevant in today's web development workforce and adaptable as new technologies emerge.

Bringing the Web to Life: Your Next Steps

By the end of this book, you'll have the skills to design and code professional-quality websites, build a portfolio to showcase your work, and take the first steps toward freelance gigs, new careers, or personal projects. The hands-on, project-driven approach ensures you don't just read about web development—you experience it, one line of code at a time.

Whether you dream of launching your own business, becoming a sought-after developer, or simply expressing yourself online, learning HTML and CSS is the gateway. With commitment, creativity, and the practical guidance in **The Smarter Way to Learn HTML & CSS**, you'll unlock the ability to shape your digital future and join a global community of creators, innovators, and problem-solvers.

The web is waiting—let's build it together.

Rajender Kumar

The Smarter Way to Learn HTML & CSS: Hands-On Guide for Beginners and Future Web Developers

1. GETTING STARTED WITH HTML AND CSS

Welcome to the exciting first step on your journey to becoming a web developer. This chapter lays the foundation for everything you'll build in the world of websites, offering the tools and knowledge to transform your ideas into interactive, visually appealing web pages. Whether you dream of launching a personal portfolio, starting a blog, or eventually building e-commerce sites, mastering the basics of HTML and CSS is where it all begins.

Here, you'll discover what HTML is and how it provides the essential structure for every site on the internet. We'll break down the anatomy of a web page, showing you how simple building blocks—known as HTML elements—combine to organize content, headings, images, and more. But the structure is only half the story. With an introduction to CSS, you'll see how styling turns plain web pages into polished, professional designs using colors, fonts, and layouts.

You don't need any prior experience, just curiosity and a willingness to experiment. We'll guide you step by step, from setting up your code editor to writing your very first lines of code. You'll even create a "Hello World!" page, marking your official entry into coding. Along the way, you'll see how web browsers and the internet bring your creations to life, connecting your work to a global audience.

By the end of this chapter, you'll have built your first web page and gained the confidence to take on more ambitious projects. This hands-on approach ensures that you're not just learning concepts, but actively applying them, setting you up for success as you explore the limitless possibilities of web development.

1.1. What is HTML? Understanding Web Page Structure

HTML, or **HyperText Markup Language**, is the backbone of every website you visit. Think of HTML as a set of instructions that tell your web browser how to display text, images, links, and other content on a page. Understanding the structure of HTML is the first crucial step in building your own web pages from scratch.

The Building Blocks of HTML

Web pages are made up of **HTML elements**. Each element has a specific purpose, like displaying a heading, a paragraph, or an image. Elements are defined using **tags**, which are keywords wrapped in angle brackets (`< >`). Most HTML elements come in pairs: an **opening tag** to start the element and a **closing tag** to end it.

For example, to create a paragraph, you use:

```
<p>This is a paragraph of text.</p>
```

`<p>` is the **opening tag**
`</p>` is the **closing tag**
The text in between is the **content** of the element

The Basic Structure of an HTML Document

Every HTML page follows a basic structure. This structure ensures that browsers can interpret and display your content correctly. Here's a simple template you'll use often:

```
<!DOCTYPE html>
<html>
  <head>
    <title>My First Web Page</title>
  </head>
  <body>
    <h1>Welcome to My Web Page!</h1>
    <p>This is my first step into web development.</p>
  </body>
</html>
```

Let's break down what each part does:

`<!DOCTYPE html>`: Tells the browser to expect an HTML5 document.
`<html>`: The root element that wraps all the content.
`<head>`: Contains meta-information about the page, like the title shown in your browser tab.
`<title>`: Sets the text for the browser tab or window.
`<body>`: Holds everything you want to display on the page, like text, images, and links.

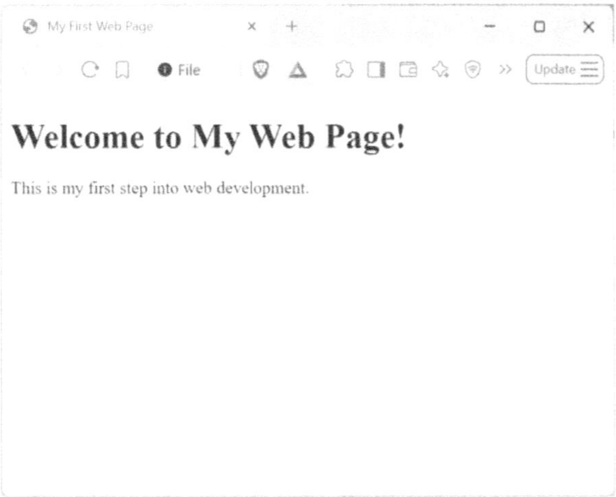

Hands-On Practice: Creating Your First Web Page

Let’s build your very first web page together! This exercise will help you understand how HTML elements fit together to form the structure of a real website.

Open a text editor (like Notepad, VS Code, or Sublime Text).
Type the following code:

```
<!DOCTYPE html>
<html>
  <head>
    <title>Beginner Portfolio</title>
  </head>
  <body>
    <h1>Hello! I’m Learning HTML</h1>
    <p>This is my first portfolio page. I created it from scratch using
HTML.</p>
    <h2>About Me</h2>
    <p>I’m excited to learn web development and build amazing
websites!</p>
    <ul>
      <li>Learning HTML</li>
      <li>Exploring CSS</li>
      <li>Building Projects</li>
    </ul>
  </body>
</html>
```

Save the file as `portfolio.html` on your computer.
Double-click the file to open it in your web browser.

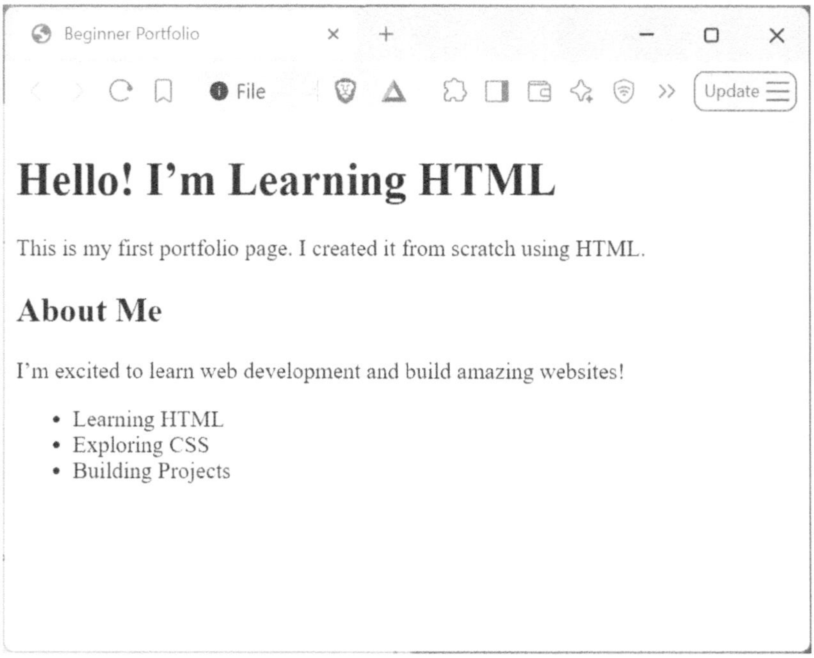

You've just written your first HTML page! Notice how the headings (`<h1>`, `<h2>`), paragraphs (`<p>`), and list (`<ul>`, `<li>`) create a clear structure for your content. This is the same way professional websites are organized, just with more advanced features as you progress.

Real-World Example: Blog Post Layout

Many websites, like blogs and news sites, rely on structured HTML to present articles and information. Let's look at a simplified example of a blog post layout:

```
<!DOCTYPE html>
<html>
  <head>
    <title>My First Blog Post</title>
  </head>
  <body>
    <h1>How I Started Learning Web Development</h1>
    <p>Posted on <strong>June 1, 2024</strong> by <em>Alex</em></p>
    <p>Today, I wrote my first lines of HTML and built a simple web
page. Here's what I learned:</p>
    <ol>
      <li>Every web page needs a solid structure.</li>
      <li>HTML uses tags to organize content.</li>
      <li>Practice makes perfect!</li>
    </ol>
  </body>
</html>
```

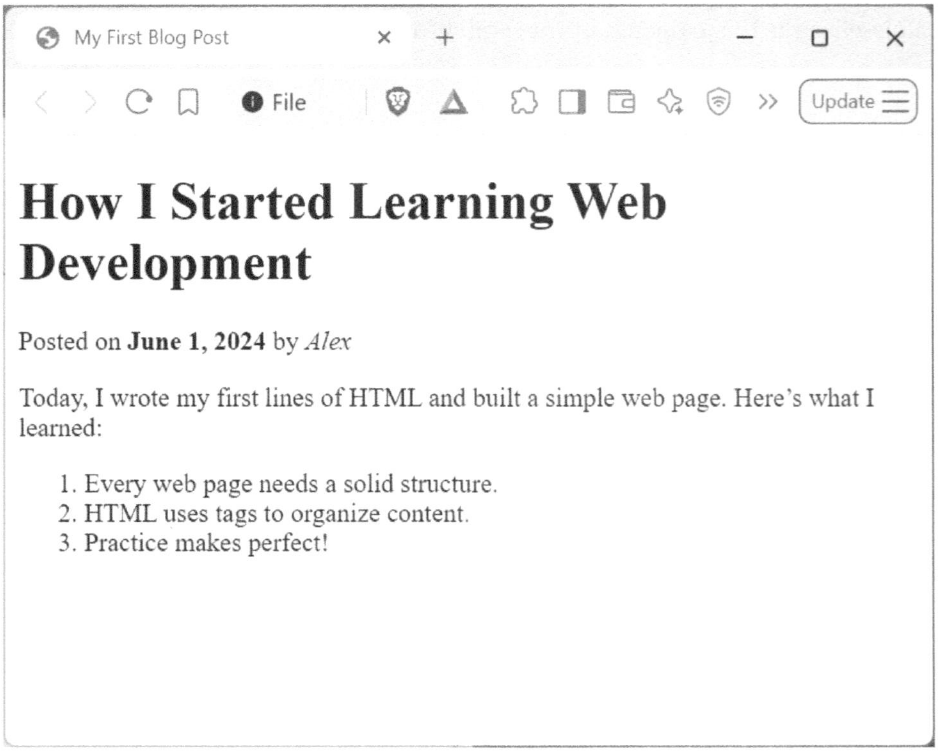

In this example, you see the use of headings, paragraphs, bold (`<strong>`) and italic (`<em>`) text, and an ordered list (`<ol>`). These elements help organize and emphasize different parts of the content, making it easier for readers to follow.

Key HTML Elements for Beginners

As you start building more complex pages, you'll use a variety of HTML elements. Some of the most important ones to remember are:

Headings (`<h1>` to `<h6>`) for titles and section headings
Paragraphs (`<p>`) for blocks of text
Lists: Unordered (`<ul>`, `<li>`) and ordered (`<ol>`, `<li>`)
Links (`<a href="URL">`) for navigation
Images (`<img src="URL" alt="Description">`) for photos and graphics

Each of these plays a role in organizing your web page and making it easy for users to read and navigate.

Quick Coding Exercise: Build a Simple Product Card

Try creating a basic product card, like you might see on an e-commerce site:

```
<!DOCTYPE html>
<html>
  <head>
    <title>Product Card Example</title>
  </head>
  <body>
    <h2>Wireless Headphones</h2>
    <p><strong>Price:</strong> $59.99</p>
    <p>Enjoy high-quality sound without the wires. Perfect for music
lovers on the go.</p>
    <ul>
      <li>Bluetooth connectivity</li>
      <li>12-hour battery life</li>
      <li>Lightweight design</li>
    </ul>
  </body>
</html>
```

This exercise reinforces how HTML elements work together to create structured, meaningful content. If you have active internet connection, you can use the online code editor (e.g. https://htmledit.squarefree.com/) to instantly see your webpage side by side with your code.

By practicing these examples, you've learned how to use HTML elements to build the structure of a real web page. Remember, every website starts with a solid HTML foundation. Try modifying these examples or creating your own simple web pages to strengthen your skills.

Next up, you'll discover how to give your web pages personality and style. In the next section, get ready to learn the basics of **CSS**, the language that lets you add color, layout, and flair to your HTML creations!

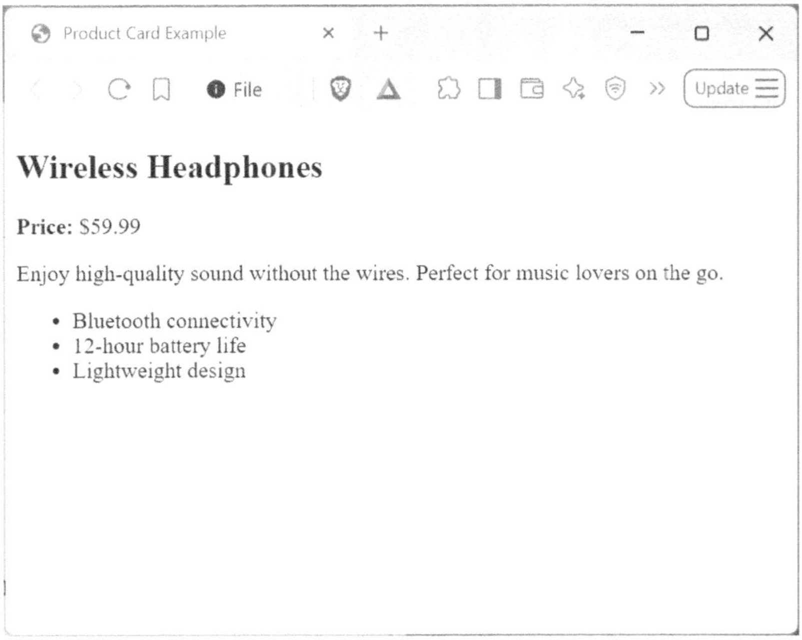


```
<!DOCTYPE html>
<html>
  <head>
    <title>Product Card Example</title>
  </head>
  <body>
    <h2>Wireless Headphones</h2>
    <p><strong>Price:</strong> $59.99</p>
    <p>Enjoy high-quality sound without the wires. Perfect for music lovers on the go.</p>
    <ul>
      <li>Bluetooth connectivity</li>
      <li>12-hour battery life</li>
      <li>Lightweight design</li>
    </ul>
  </body>
</html>
```

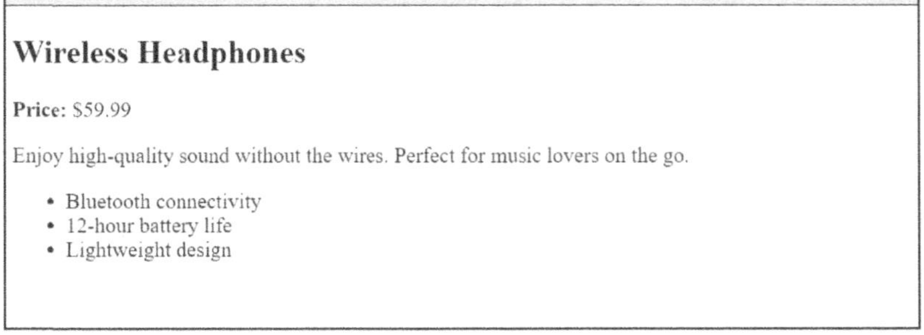

1.2. Introduction to CSS: Styling Your Web Pages

So far, you've learned how HTML provides the structure for your web pages. But what if you want to make your site visually appealing, with colors, custom fonts, and unique layouts? That's where **CSS** comes in. CSS stands for **Cascading Style Sheets**, and it's the language that transforms your plain HTML into eye-catching, professional-looking websites.

What Is CSS and Why Do You Need It?

CSS allows you to control the appearance of every aspect of your web page, from background colors and text sizes to spacing and alignment. Think of your HTML as the blueprint of a house, and CSS as the interior decorating. With just a few lines of CSS, you can turn a simple layout into a site that stands out.

For example, consider a personal portfolio website. HTML defines the structure—your name, photo, and projects. CSS adds the color scheme, stylish fonts, and smooth spacing that make your portfolio memorable and unique.

The Basics: How CSS Works with HTML

To use CSS, you write **rules** that describe how HTML elements should look. Each rule targets an HTML element and applies styles to it. A basic CSS rule has a **selector** (which element to style) and a **declaration block** (what styles to apply).

Here's the structure:

```
selector {
  property: value;
}
```

Let's see a practical example. Suppose you have this HTML for a simple blog heading:

```
<h1>My Travel Blog</h1>
```

You can change the color and font size of this heading with CSS:

```
h1 {
  color: teal;
  font-size: 36px;
}
```

Three Ways to Add CSS to Your Web Page

You can use CSS in your HTML documents in three main ways. For now, let's try the easiest: **inline styles** and **internal styles**.

Inline styles: Add the `style` attribute directly to an HTML tag.
Internal styles: Place your CSS rules inside a `<style>` tag in the `<head>` section of your HTML file.

Let's practice both methods.

Inline CSS Example

Try adding some immediate style to a paragraph:

```
<p style= "color: darkblue; background-color: #e0f7fa; padding: 10px;">
  Welcome to my first styled web page!
</p>
```

This method is great for quick tests, but not ideal for larger projects because it mixes content and style.

Internal CSS Example

A better, more organized approach is using an internal style sheet. Place your CSS rules inside a `<style>` tag in the `<head>` of your HTML document.

```
<!DOCTYPE html>
<html>
  <head>
    <title>Practice: My First Styled Page</title>
    <style>
      body {
        background-color: #f9f9f9;
      }
      h1 {
        color: #2c3e50;
        text-align: center;
      }
      p {
        font-family: Arial, sans-serif;
        font-size: 18px;
      }
    </style>
  </head>
  <body>
    <h1>My Styled Portfolio</h1>
    <p>This is my very first attempt at web design using HTML and
CSS!</p>
  </body>
</html>
```

Try copying this example and see how each CSS rule changes your web page's appearance. Experiment by changing colors or font sizes to see instant results.

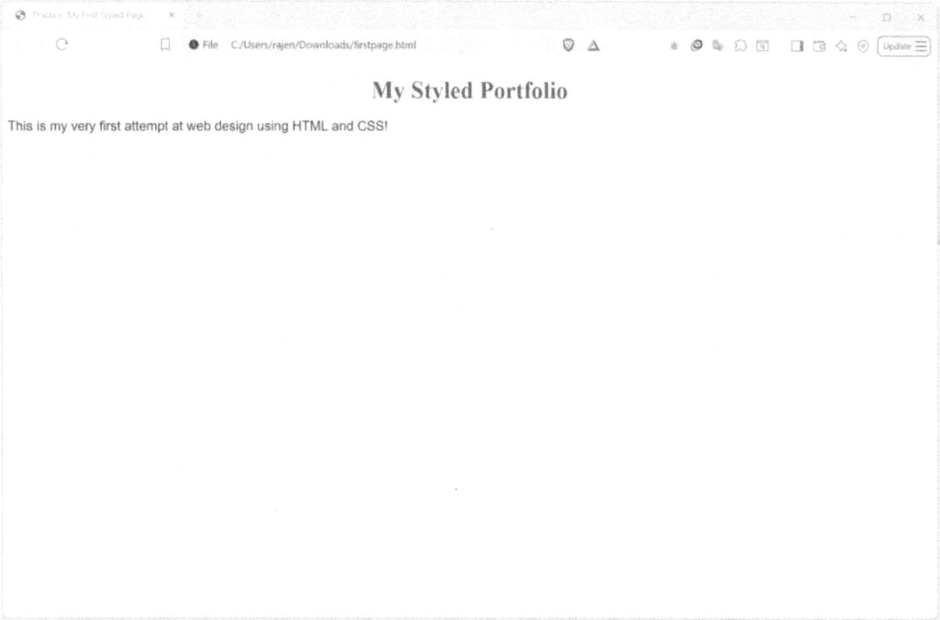

Real-World Practice: A Simple Product Card

Let's see how CSS is used in a real-world scenario, like a basic e-commerce website displaying a product. Here's a hands-on exercise:

Create a simple HTML structure for your product card:

```
<div class= "product-card">
  <h2>Wireless Headphones</h2>
  <p>High-quality sound, noise-cancelling, and 20-hour battery
life.</p>
  <span class= "price">$59.99</span>
</div>
```

Add internal CSS to style your card:

```
<style>
  .product-card {
    border: 1px solid #ddd;
    border-radius: 8px;
    padding: 20px;
    background-color: #fff;
    width: 300px;
    margin: 30px auto;
    box-shadow: 0 2px 8px rgba(0,0,0,0.07);
  }
  .product-card h2 {
```

```
    color: #27ae60;
    font-size: 24px;
    margin-bottom: 8px;
  }
  .product-card .price {
    color: #c0392b;
    font-weight: bold;
    font-size: 20px;
  }
</style>
```

Place both the HTML and CSS inside your HTML file and open it in your browser. You'll see a neatly styled product card, just like those on modern online stores.

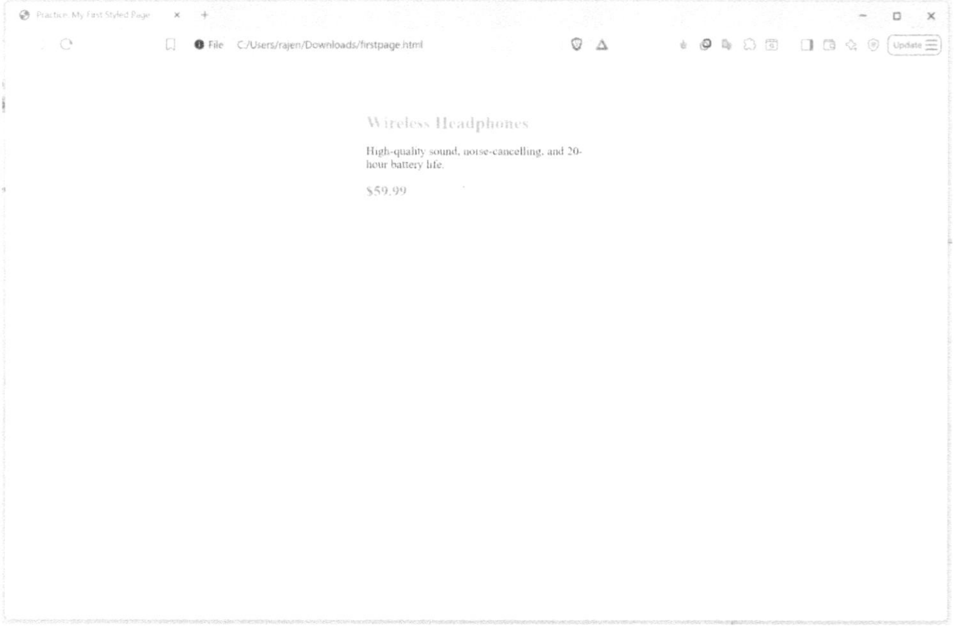

Key CSS Properties for Beginners

As you experiment, here are some essential CSS properties you'll use often:

- **color:** Sets the text color.
- **background-color:** Changes the background color.
- **font-size:** Adjusts text size.
- **font-family:** Sets the typeface.
- **padding:** Adds space inside an element.
- **margin:** Adds space outside an element.
- **border:** Draws a border around an element.
- **text-align:** Aligns text (like `center` or `right`).

Mix and match these properties to see how they affect your page.

Quick Coding Recap

You've now seen how CSS can instantly transform your HTML from basic to beautiful. Practice by changing colors, fonts, and spacing in your own sample pages. The more you experiment, the more confident you'll become in styling your web creations.

Up next, you'll learn how to set up your own coding environment, making it even easier to write, save, and preview your HTML and CSS. Get ready to discover the tools and tips that professional web developers use every day!

1.3. Setting Up Your Coding Environment

Starting your journey as a web developer means creating a comfortable space where you can write code, test ideas, and see your progress instantly. This space is called your **coding environment**. Having the right tools from the beginning helps you work faster, avoid frustration, and develop real-world skills. Let's set up everything you need so you can start writing and previewing your own HTML and CSS code right away.

Why You Need a Coding Environment

A coding environment is where you create, edit, and test your code. Professional web developers often use specialized software, but beginners can get started with simple, free tools on any computer. Setting up your environment involves choosing a **code editor**, organizing your files, and using a web browser to preview your work.

Imagine you want to build a simple **portfolio website** to showcase your projects or a **blog homepage** to share your ideas. Both require editing HTML and CSS files and viewing the results in a browser. A well-organized environment lets you do this easily.

Choosing Your Code Editor

A **code editor** is a program where you write your HTML and CSS. While you can use basic editors like Notepad (Windows) or TextEdit (Mac), using a code editor designed for web development makes coding easier and more enjoyable. Here are some excellent, free options:

- **Visual Studio Code (VS Code):** Popular, beginner-friendly, and packed with features.
- **Sublime Text:** Lightweight and fast, with a clean interface.
- **Atom:** Open-source and customizable.

For this guide, we'll use **VS Code** as an example, but you can choose any editor you prefer.

Installing Visual Studio Code (VS Code)

Let's quickly install VS Code so you can start coding:

Go to the [Visual Studio Code website](https://code.visualstudio.com/).
Click the **Download** button for your operating system (Windows, Mac, or Linux).
Run the downloaded file and follow the installation instructions.

Once installed, open VS Code. You'll see a welcome screen—this is your starting point for creating web pages.

Organizing Your Project Files

Before writing any code, create a folder to keep your project organized. This is where you'll store all your HTML and CSS files.

Create a new folder on your desktop and name it `my-first-website`.
Open VS Code.
In VS Code, go to **File > Open Folder** and select the folder you just created.

Now you're ready to create your first files.

Creating Your First HTML and CSS Files

Let's set up the most basic structure for your website, using a **portfolio** as an example. You'll create two files: one for HTML and one for CSS.

In VS Code, right-click in the file explorer and select **New File**.
Name the file `index.html`.
Create another file and name it `styles.css`.

Your folder should now look like this:

```
my-first-website/
  ├─ index.html
  └─ styles.css
```

Writing Basic HTML Markup

Open `index.html` and type the following code. This is the basic structure for every web page:

```
<!DOCTYPE html>
<html lang= "en">
  <head>
    <meta charset= "UTF-8">
    <title>My Portfolio</title>
    <link rel= "stylesheet"href= "styles.css">
  </head>
  <body>
```

```
    <h1>Welcome to My Portfolio</h1>
    <p>This is the first step in my web development journey.</p>
  </body>
</html>
```

The `<!DOCTYPE html>` line tells the browser you're using HTML5.
`<html>`, `<head>`, and `<body>` organize your page.
`<link rel="stylesheet" href="styles.css">` connects your HTML to your CSS file, so you can add styles.

Writing Your First CSS

Open `styles.css` and add the following code:

```
body {
  background-color: #f0f4f8;
  font-family: Arial, sans-serif;
  color: #333;
}

h1 {
  color: #3366cc;
}
```

This CSS changes the background color of your page, sets a clean font, and styles your main heading. Feel free to adjust the colors or fonts to see how the page changes.

Previewing Your Website in a Browser

Now let's see your work:

In your project folder, double-click `index.html`.
Your default web browser will open, and you'll see your styled portfolio page.

Try changing the text in `<h1>` or the color in your CSS, save the files, and refresh the browser. Notice how your changes appear instantly. This feedback loop is how you'll learn and improve your web development skills.

Real-World Scenario: Creating a Blog Homepage

Suppose you want to experiment with a **blog homepage**. Duplicate the steps above, but in your `index.html`, try changing the content:

```
<h1>My Coding Blog</h1>
<p>Sharing tips and projects as I learn HTML and CSS!</p>
```

Customize your CSS to give your blog a unique look. This hands-on practice prepares you for building real websites.

You've set up your own coding environment, created your first HTML and CSS files, and previewed your work in a browser. Practicing these steps helps you build confidence in managing web projects and making changes you can see instantly. Experiment with different texts and colors to make the space truly your own.

Next, you'll dive into the exciting process of building your very first web page from scratch in the section 'Your First Web Page: Hello World!'. Get ready to write real HTML, add new elements, and watch your ideas come to life—one line of code at a time!

1.4. Your First Web Page: Hello World!

Imagine opening your browser and seeing a web page you built with your own code. In this section, you'll do just that by creating a simple "Hello World!" web page, a classic first project for every new coder. This hands-on exercise introduces you to writing and saving HTML code, previewing your results instantly, and understanding how each element builds your web page from the ground up. No special tools or complicated software are required—just your computer, a text editor, and a browser.

Setting Up: Your Coding Toolkit

Before you begin, let's make sure you have everything you need to write and view HTML code:

Text Editor: Use a simple program like Notepad (Windows), TextEdit (Mac, set to plain text), or a free code editor such as VS Code or Sublime Text.
Web Browser: Chrome, Firefox, Edge, or Safari will all work perfectly.

Writing Your First HTML Code

Every web page is built from an HTML file. Let's walk through the process of creating your very first one, step by step.

Open your text editor and start a new file.
Copy and paste the following HTML code:

```
<!DOCTYPE html>
<html>
  <head>
    <title>Hello World!</title>
  </head>
  <body>
    <h1>Hello World!</h1>
    <p>Welcome to your first web page. You're officially a web
developer!</p>
  </body>
</html>
```

Save the file as `hello.html`. Remember to use the `.html` extension so browsers recognize it as a web page.

Double-click your file or open it with your browser. You should see a big heading that says "Hello World!" and a greeting underneath.

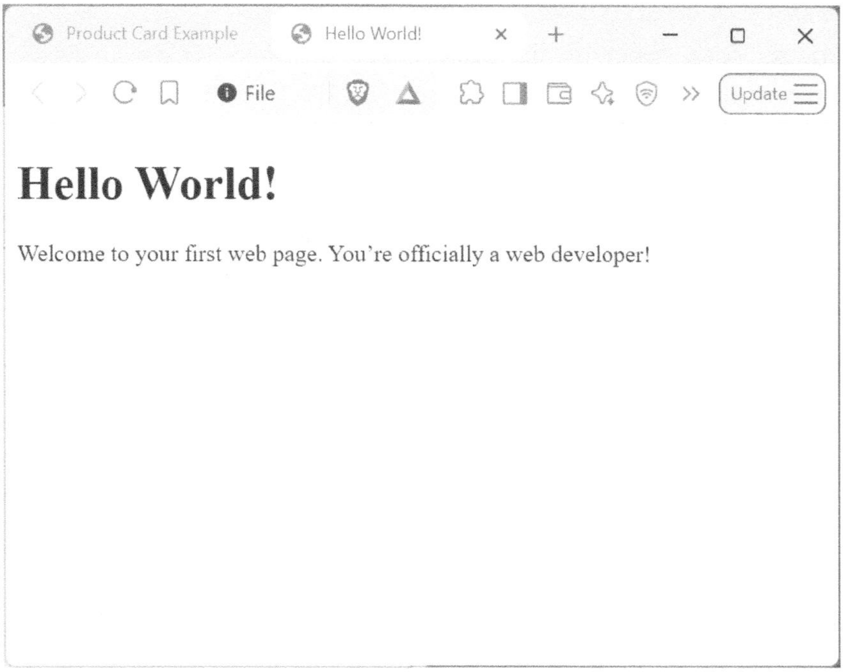

Let's break down what each part of this code does, so you can start building your understanding from the ground up.

Exploring the Structure: What Does Each Line Mean?

`<!DOCTYPE html>` tells the browser that this file uses modern HTML5.

`<html>` and `</html>` wrap everything, marking the start and end of your web page.

`<head>` contains information about your page (not visible to visitors), such as the `<title>`, which appears in the browser tab.

`<body>` is where the content of your page lives—what visitors see.

`<h1>` is a heading. It's usually the biggest, most important title on the page.

`<p>` stands for paragraph and holds regular text.

Customizing Your First Page

Let's make your page unique. Change the text inside the `<h1>` and `<p>` elements:

```
<h1>My First Portfolio</h1>
<p>This is the start of my journey as a web developer. Stay tuned for
more projects!</p>
```

Save and refresh your browser to see your changes. Congratulations—you're already personalizing your site, just like real portfolio pages used by web designers worldwide.

Adding Simple CSS: Making It Stylish

HTML provides structure, but CSS (Cascading Style Sheets) makes your page attractive. You can add CSS directly inside your HTML file for quick experiments.

Inside the `<head>` section, add the following `<style>` block:

```
<style>
  body {
    background-color: #f0f8ff;
    font-family: Arial, sans-serif;
    color: #333333;
  }
  h1 {
    color: #008080;
  }
  p {
    font-size: 1.2em;
  }
</style>
```

Save and refresh your browser. Notice the background is now a soft blue, your heading is teal, and the text is easier to read.

This is your first taste of how CSS changes the look and feel of your web page instantly—just like the sleek, modern designs seen on tech blogs and personal portfolios.

Real-World Example: Your Own Introduction Page

Imagine you want to introduce yourself to future employers or friends. You can easily adapt your "Hello World!" page to serve as a personal intro website.

```
<!DOCTYPE html>
<html>
  <head>
    <title>About Me</title>
    <style>
      body {
        background-color: #e6f7ff;
        font-family: Verdana, sans-serif;
        margin: 40px;
      }
      h1 {
        color: #005f73;
      }
      p {
        line-height: 1.6;
```

```
        font-size: 1.1em;
      }
    </style>
  </head>
  <body>
    <h1>Hi, I'm Alex!</h1>
    <p>
      Welcome to my first web page. I'm excited to learn HTML & CSS and
build amazing websites.
      This page is where my coding adventure begins!
    </p>
  </body>
</html>
```

Try editing your own name and message. With just a few lines of HTML and CSS, you've created a personal web page—a foundation for a future portfolio or resume site.

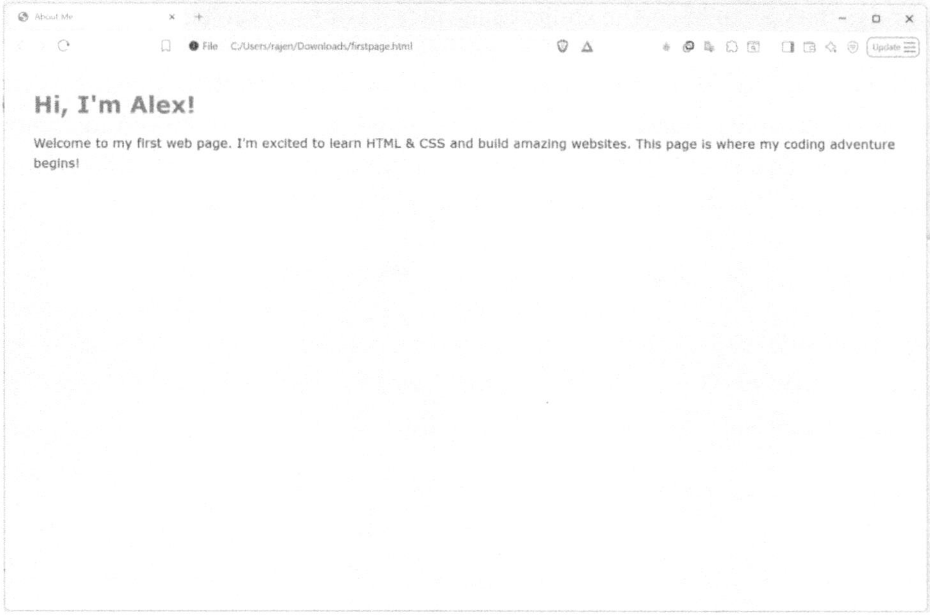

You've just built your first web page using HTML and CSS, learned how to structure content, and experimented with styling. Each change you make gives immediate feedback, helping you understand how code shapes real web pages. Practice by changing text, colors, and fonts to see how your site transforms.

Next, you'll discover how individual web pages connect to form the worldwide web. Get ready to explore how the internet links pages together and how you can build multi-page websites that anyone, anywhere, can visit!

1.5. How the Internet Connects Web Pages

When you visit a website, it often feels like magic: you type a web address into your browser, and suddenly, a rich, interactive page loads before your eyes. But what's happening behind the scenes? The answer lies in how the internet connects individual web pages, weaving them together to form the vast "web" we use every day. As a new web developer, understanding these connections is key to building websites that users can navigate with ease.

The Web: A Network of Linked Pages

Websites are made up of separate pages, each written in HTML. These pages live on web servers all over the world. When you use a browser to visit a website, your computer requests these pages, and the server sends them back for you to view. What truly brings the web to life is the ability to move from one page to another using **links**.

A link is a clickable element, often underlined or highlighted, that points to another web page. Links can connect pages within the same website or take you to completely different sites. This simple feature allows users to explore online portfolios, shop in e-commerce stores, or read articles on blogs from anywhere in the world.

Let's see how you can create these connections with HTML.

Creating Links with the `<a>` Tag

The **anchor tag** (`<a>`) is the primary way to create links between web pages. The **href** attribute specifies the destination of the link.

Here's a basic example:

```
<a href= "about.html">About Me</a>
```

This line creates a link labeled "About Me." When clicked, it will take the user to a page called `about.html`. If that file exists in your project folder, the browser will load it instantly.

Step-by-Step: Building a Simple Multi-Page Website

Let's put this into practice by connecting two pages: a **home page** and an **about page**. This scenario is common for personal portfolios or small business websites.

Create the Home Page

Open your code editor and create a new file called `index.html`. Add this HTML structure:

```
<!DOCTYPE html>
<html lang= "en">
<head>
  <meta charset= "UTF-8">
  <title>My Portfolio</title>
</head>
<body>
  <h1>Welcome to My Portfolio!</h1>
  <p>This is my homepage. Learn more <a href= "about.html">about
me</a>.</p>
</body>
</html>
```

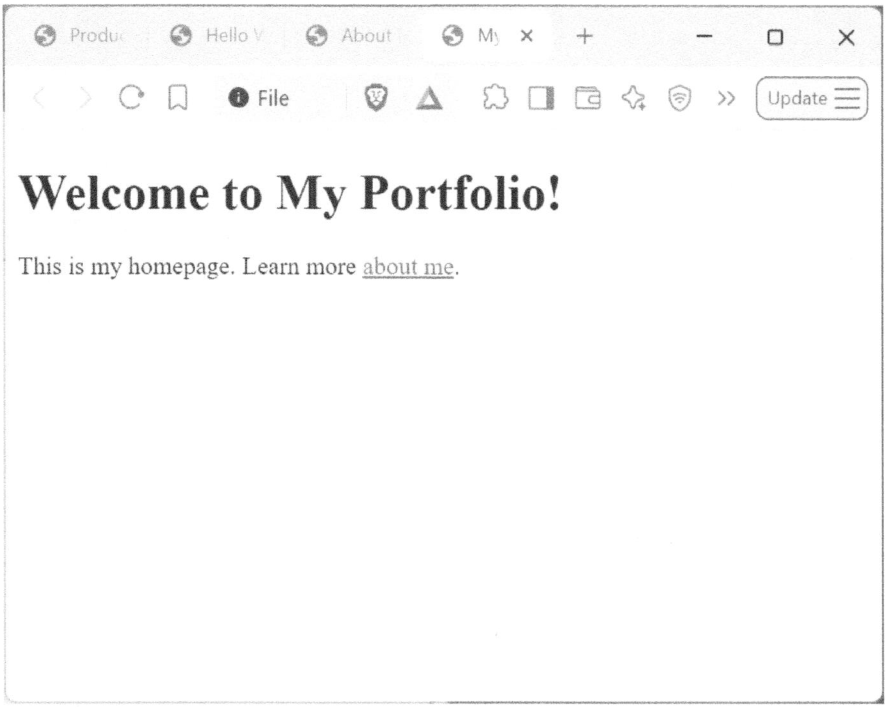

Create the About Page

Now, make a new file named `about.html` in the same folder:

```
<!DOCTYPE html>
<html lang= "en">
<head>
  <meta charset= "UTF-8">
  <title>About Me</title>
</head>
<body>
  <h1>About Me</h1>
  <p>Hello! I'm an aspiring web developer. Go <a href=
"index.html">back to the homepage</a>.</p>
</body>
</html>
```

Now, when you open `index.html` in your browser and click the "about me" link, it takes you to `about.html`. The "back to the homepage" link on that page returns you to `index.html`. You've just built a basic website with two connected pages!

Connecting Pages: Real-World Example

Many websites use navigation menus to help visitors move between sections. Take an online portfolio as an example. It often includes links like "Home," "Projects," "Contact," and "About." You can build a simple navigation menu using the `<a>` tag.

```
<nav>
  <a href= "index.html">Home</a> |
  <a href= "projects.html">Projects</a> |
  <a href= "contact.html">Contact</a>
</nav>
```

This code creates a horizontal menu. If you add `projects.html` and `contact.html` to your project folder, the links will connect all those pages together, just like professional websites.

Styling Links with CSS

Links don't just have to be blue and underlined. You can use CSS to style them and fit your site's look. Here's a quick example:

```
a {
  color: #3498db;
  text-decoration: none;
  font-weight: bold;
}

a:hover {
  color: #e67e22;
  text-decoration: underline;
}
```

To apply this styling, save it in a file called `styles.css` and link it in your HTML's `<head>` section:

```
<link rel= "stylesheet"href= "styles.css">
```

Now your links will appear in a custom color and change when hovered.

You've just learned how the internet connects web pages using HTML links. By creating and linking multiple HTML files, you can build a simple, navigable website. With a touch of CSS, you can make your links visually appealing and interactive.

You're now ready to put your new knowledge into action! In the next section, we'll dive right into a hands-on exercise where you'll build your very first simple web page from scratch. Get ready to start coding, see instant results, and take your first real step toward becoming a web developer.

1.6. Hands-On Exercise: Building Your First Simple Web Page

Ready to build your very first web page? Let's dive in and start coding! In this hands-on exercise, you'll learn how to create a simple web page from scratch using HTML and a touch of CSS for basic styling. This practical approach will help you see immediate results and give you the confidence to explore more complex projects later.

Step 1: Create a New HTML File

Every web page starts as a plain text file. HTML files use the `.html` extension.

Open your text editor.
Save a new file as `my-first-webpage.html`.

You're ready to write your first lines of HTML!

Step 2: Writing the Basic HTML Structure

A web page needs a structure so browsers know how to display the content. Let's start with a simple HTML template.

Type the following code into your file:

```
<!DOCTYPE html>
<html lang="en">
  <head>
    <meta charset="UTF-8">
    <title>My First Web Page</title>
  </head>
  <body>
    <!-- Content will go here -->
  </body>
</html>
```

The `<!DOCTYPE html>` tells browsers this is an HTML5 document.
The `<html>` element wraps everything on the page.
Inside `<head>`, you add information about the page.
The `<body>` contains the visible content.

Step 3: Adding Headings and Paragraphs

Let's make your web page look like a personal introduction, similar to a simple online portfolio or a welcome note.

Inside the `<body>` tags, add:

```
    <h1>Welcome to My First Web Page!</h1>
    <p>Hello! My name is Alex. I'm learning how to build websites using
HTML and CSS.</p>
    <p>This is my very first web page. I can't wait to add more
features as I learn!</p>
```

`<h1>` creates a large heading, perfect for your page title.
`<p>` adds paragraphs of text, making your content easy to read.

Step 4: Inserting an Image

Images make web pages more interesting. Let's add a profile picture or a favorite photo.

Find a small image on your computer, or use a placeholder image from the web (for practice).
Add this code below your paragraphs:

```
    <img src="https://via.placeholder.com/150" alt="Profile photo">
```

The `src` attribute points to your image's location.
The `alt` attribute describes the image for accessibility.

Step 5: Creating a Link

Links connect your page to others. Let's add a link to your favorite website or an online portfolio.

Below the image, add:

```
    <p>Check out my favorite website: <a
href="https://www.wikipedia.org/">Wikipedia</a></p>
```

The `<a>` tag creates a clickable link.
The `href` attribute sets the web address.

Step 6: Saving and Viewing Your Web Page

Save your file (`my-first-webpage.html`).
Open the file in your web browser (double-click it or right-click and select "Open with" your browser).

You'll see your content styled by the browser's default settings. It already looks like a small personal profile page, a common type of site for portfolios or blog introductions.

Step 7: Adding Basic CSS Styling

Now, let's add some simple styling to make your page more visually appealing. We'll do this right inside the HTML file using a `<style>` tag in the `<head>` section.

Inside the `<head>` tag, add:

```
<style>
  body {
    background-color: #f9f9f9;
    font-family: Arial, sans-serif;
    margin: 40px;
  }
  h1 {
    color: #2c3e50;
  }
  p {
    color: #555;
    font-size: 1.1em;
  }
  img {
    border-radius: 50%;
    margin-top: 16px;
  }
  a {
    color: #2980b9;
    text-decoration: none;
  }
  a:hover {
    text-decoration: underline;
  }
</style>
```

The `body` rule changes the background and font.
`h1` sets the heading color.
`p` adjusts paragraph text.
`img` makes your image round.
`a` and `a:hover` style your links.
Save your file and refresh your browser to see the changes.

Your page now has a clean, modern look, similar to many simple portfolio or personal introduction sites you see on the web.

You've just created your first HTML web page and added basic CSS for style. You learned how to set up the page structure, add headings, paragraphs, images, and links, and apply your own colors and fonts. Practicing these steps helps you build confidence and prepares you for more advanced web development.

2. MASTERING HTML PAGE STRUCTURE

Every memorable website begins with a well-structured foundation. In this chapter, you'll unlock the secrets behind organizing web pages using HTML's essential building blocks. Whether you dream of designing a personal blog, an online portfolio, or even the next big e-commerce platform, knowing how to structure your content is the first step toward bringing your ideas to life online.

Here, you'll discover how the **HTML document skeleton** forms the backbone of every page, providing both browsers and visitors with a clear framework for your content. You'll learn to craft meaningful **page titles** and add **metadata** that boosts your site's accessibility and searchability. As you progress, you'll practice using **headings** to establish content hierarchy, making your pages easier to read and navigate for everyone.

We'll also explore essential tools for presenting and emphasizing text, such as creating **paragraphs**, inserting **line breaks**, and applying **bold** or **italic** styles for added emphasis. Alongside these skills, you'll adopt best practices for commenting and organizing your HTML code, setting yourself up for smooth collaboration and future updates.

By the end of this chapter, you'll apply your new abilities in a hands-on exercise, building the structure for a simple blog post—a practical project that mirrors real-world website creation. Not only will these skills empower you to design professional-quality pages, but they'll also give you the confidence to continue exploring web development with clarity and creativity. Let's begin shaping your first web page, step by step.

2.1. The Essential HTML Document Skeleton

Before you can build beautiful and interactive websites, you need to understand the blueprint that every webpage follows. This blueprint is called the **HTML document skeleton**. Think of it as the foundation and framework of your website, much like the walls and beams of a house. Every webpage starts with this essential structure, ensuring browsers know how to interpret and display your content.

Why the HTML Skeleton Matters

The HTML skeleton provides a standardized way for browsers to read and render your site. It tells the browser, "This is an HTML document," and organizes everything on your page, from visible text to hidden metadata. By mastering this structure, you'll write code that works reliably across different browsers and devices, whether you're building a portfolio or a blog.

Breaking Down the Skeleton: Key Elements

Let's look at the essential components that make up every HTML document. Here's a quick overview:

`**<!DOCTYPE html>**`: Declares the document type and version of HTML.
`**<html>**`: The root element that contains all HTML content.
`**<head>**`: Holds metadata, links, and settings (not shown on the page).
`**<body>**`: Contains everything visible on the page (text, images, links).

Building Your First HTML Skeleton: Step-by-Step

Let's create the most basic HTML page using the skeleton. Open your favorite code editor (like VS Code or Notepad), and follow these steps:

Start with the document declaration.
Add the root `<html>` element.
Inside `<html>`, create a `<head>` section.
Add a `<body>` section for your visible content.

Here's what the code looks like:

```
<!DOCTYPE html>
<html>
  <head>
    <!-- Metadata and links will go here -->
  </head>
  <body>
    <!-- Your visible content will go here -->
  </body>
</html>
```

Each part of this skeleton plays a specific role. The `<!DOCTYPE html>` line at the top tells the browser to use the latest HTML standards. The `<html>` tags surround everything on your page, while the `<head>` is where you'll soon add important settings like the page title, and the `<body>` is where all the fun happens—your text, images, and more.

Real-World Example: Personal Portfolio Homepage

Imagine you want to create a simple portfolio homepage. Using what you've just learned, here's how you'd set up the skeleton:

```
<!DOCTYPE html>
<html>
```

```
  <head>
    <!-- We'll add a title and metadata soon! -->
  </head>
  <body>
    <h1>Welcome to My Portfolio</h1>
    <p>Hi, I'm Alex! Explore my work and projects below.</p>
  </body>
</html>
```

With just this structure, you have a valid webpage! You can open the file in your browser and see the heading and paragraph displayed.

Quick CSS Touch: Making It Look a Bit Better

While CSS will be explored in depth later, here's a tiny preview of how you'd add some style. Place this inside your `<head>` within `<style>` tags:

```
<head>
  <style>
    body {
      font-family: Arial, sans-serif;
      background-color: #f3f3f3;
      margin: 40px;
    }
    h1 {
      color: #3366cc;
    }
  </style>
</head>
```

Now your page uses a clean font, a gentle background color, and a colored heading—demonstrating how structure and style work together.

Practice: Create Your Own Blog Skeleton

Try setting up a basic HTML skeleton for a personal blog homepage. Replace the content inside `<body>` with your own name and a welcoming message:

```
<!DOCTYPE html>
<html>
  <head>
    <!-- Blog metadata will go here soon -->
  </head>
  <body>
    <h1>Jane's Blog</h1>
    <p>Welcome to my world of ideas and stories!</p>
  </body>
</html>
```

Open the file in your browser to see your blog's starter page.

You've just learned how every webpage starts with the same reliable HTML skeleton, giving your code structure and clarity. Practice building skeletons for different types of sites, like portfolios or blogs, to reinforce this foundational skill. Remember, mastering the basics makes advanced web development much easier!

Next, we'll make your webpages smarter by adding titles and metadata inside the `<head>` section. Get ready to discover how to name your pages and help browsers (and search engines) understand your site even better!

2.2. Adding Page Titles and Metadata

Before a website even displays its content, the browser checks for vital information inside a special hidden area called the **head**. This section is where you set the identity of your page, making it smarter and more discoverable. In this part of your HTML journey, you'll learn how to add a page title and essential metadata, which play key roles in how your site appears in browser tabs and search engines.

Understanding the `<head>` Section

The **`<head>`** element sits at the very top of your HTML document, right after the opening **`<html>`** tag. While users don't see what's inside the head directly on the page, everything you put here influences the way your website is presented and understood by both browsers and search engines.

Here's a basic structure of a web page with a head section:

```
<!DOCTYPE html>
<html>
  <head>
    <!-- Information about your page goes here -->
  </head>
  <body>
    <!-- Visible content for users goes here -->
  </body>
</html>
```

Let's dive into two of the most important things you'll add to the head: the page title and metadata.

Adding a Page Title

The **`<title>`** tag gives your web page a name. This title shows up in the browser tab and when someone bookmarks your page. It's also what search engines display in search results, so it's important to make it clear and descriptive.

Let's code your first page title:

Open your `index.html` file or create a new HTML file.
Locate the `<head>` section near the top.

Add a `<title>` tag with a descriptive name between the opening and closing tags.

Example:

```
<!DOCTYPE html>
<html>
  <head>
    <title>My Personal Portfolio</title>
  </head>
  <body>
    <h1>Welcome to My Portfolio!</h1>
    <p>Explore my projects and skills.</p>
  </body>
</html>
```

Try changing the title to something unique for your project, like “Delicious Cupcake Recipes” or “Learn HTML the Smart Way!”

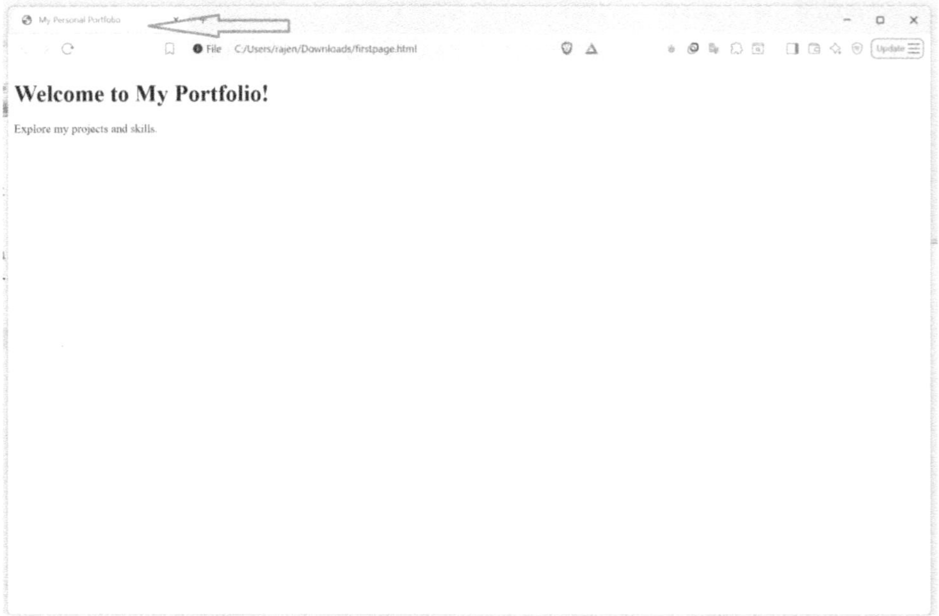

Setting Up Metadata for Your Website

Metadata is extra information about your webpage that helps browsers, search engines, and social media platforms understand what your site is about. Metadata is added using **`<meta>`** tags inside the head.

Some common uses for meta tags include:

- **Describing your page** for search engines.
- **Setting the character encoding** so special characters display correctly.
- **Making your site mobile-friendly**.

Let's add some important meta tags step by step.

1. Setting the Character Encoding

To ensure your site displays all characters (including symbols and emojis) correctly, add this line first in your head:

```
<meta charset="UTF-8">
```

2. Adding a Description for Search Engines

A description helps search engines summarize your site in search results. Add this meta tag to describe your page:

```
<meta name="description" content="A showcase of my web development
projects and coding skills.">
```

3. Making Your Site Mobile-Friendly

With so many people browsing on phones and tablets, it's smart to add this tag to make your site responsive:

```
<meta name="viewport" content="width=device-width, initial-scale=1.0">
```

Putting It All Together

Here's a real-world example for a beginner's portfolio site:

```
<!DOCTYPE html>
<html>
  <head>
    <meta charset="UTF-8">
    <title>My Web Developer Portfolio</title>
    <meta name="description" content="A modern portfolio showcasing my
HTML and CSS projects.">
    <meta name="viewport" content="width=device-width, initial-
scale=1.0">
  </head>
  <body>
    <h1>Welcome!</h1>
    <p>Check out my latest website builds and design experiments.</p>
  </body>
</html>
```

Or, if you're building a simple blog, try this:

```
<!DOCTYPE html>
<html>
  <head>
    <meta charset="UTF-8">
    <title>Smart Learner Blog</title>
    <meta name="description" content="Tips and tutorials for learning
HTML and CSS the smart way.">
    <meta name="viewport" content="width=device-width, initial-
scale=1.0">
```

```
  </head>
  <body>
    <h1>Welcome to the Smart Learner Blog!</h1>
    <p>Begin your web development journey with practical tips and real
examples.</p>
  </body>
</html>
```

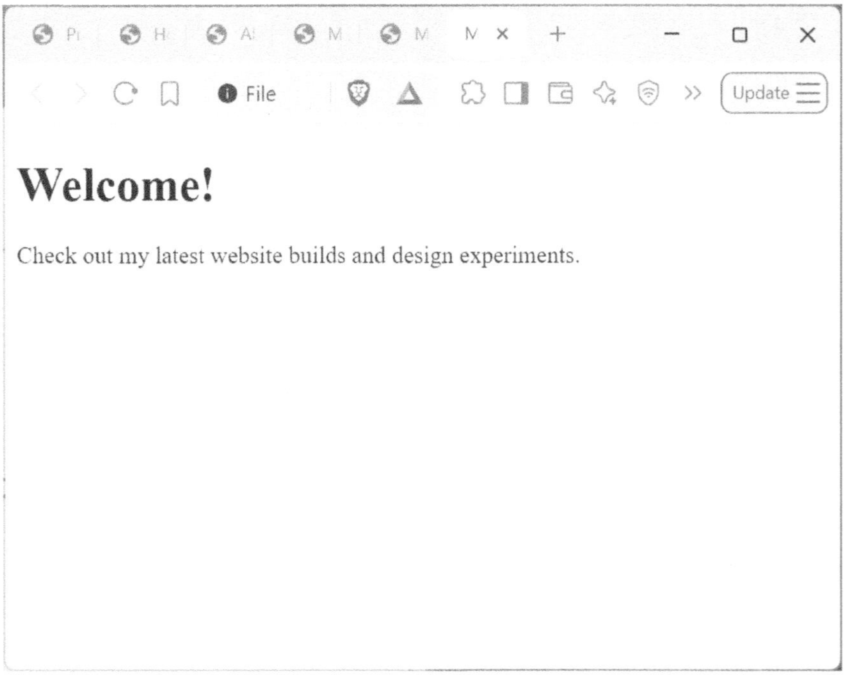

Open your own HTML file and try adding a unique title and at least two meta tags (character encoding, description, or viewport). Refresh your browser to see how the title appears in the tab, and use your text editor's preview to check for errors.

You've now given your web pages a smart identity with titles and metadata, making them easier to find and friendlier for users and search engines. Remember, the head section is your site's backstage pass to better visibility and usability. Keep practicing by customizing titles and metadata for each new page you build.

Next up, you'll learn how to organize your web content with headings—an essential skill for creating readable, well-structured pages. Get ready to dive into the world of HTML headings and discover how they make your websites more user-friendly and accessible!

2.3. Structuring Content with Headings

Headings are like signposts on your webpage, guiding visitors through your content and making everything easier to read. Search engines and assistive technologies also rely on clear headings, which is why mastering their structure is a core skill for every web developer. Let's get hands-on by learning how to create and style headings in your HTML pages.

What Are HTML Headings?

HTML provides six levels of headings, from **<h1>** to **<h6>**. Each level represents a level of importance:

<h1> is the most important heading (usually the main title).
<h2> to **<h6>** are used for subheadings, each one less prominent than the previous.

Headings not only make your content easy to scan, but they also help organize information for users and search engines.

Creating Headings: Hands-On Coding Steps

Let's build a simple homepage for a **personal blog**. We'll use headings to organize the title, sections, and subtopics.

Open your favorite code editor and create a new file named `index.html`.
Add the following HTML structure:

```
<!DOCTYPE html>
<html>
  <head>
    <title>My Personal Blog</title>
  </head>
  <body>
    <h1>Welcome to My Personal Blog</h1>
    <h2>About Me</h2>
    <h3>My Background</h3>
    <h2>Latest Posts</h2>
    <h3>Why I Love Web Development</h3>
    <h3>Traveling the World</h3>
  </body>
</html>
```

Notice how **<h1>** introduces the page, and **<h2>** breaks content into key sections. Subtopics use **<h3>**. This hierarchy helps readers and search engines understand your content's structure.

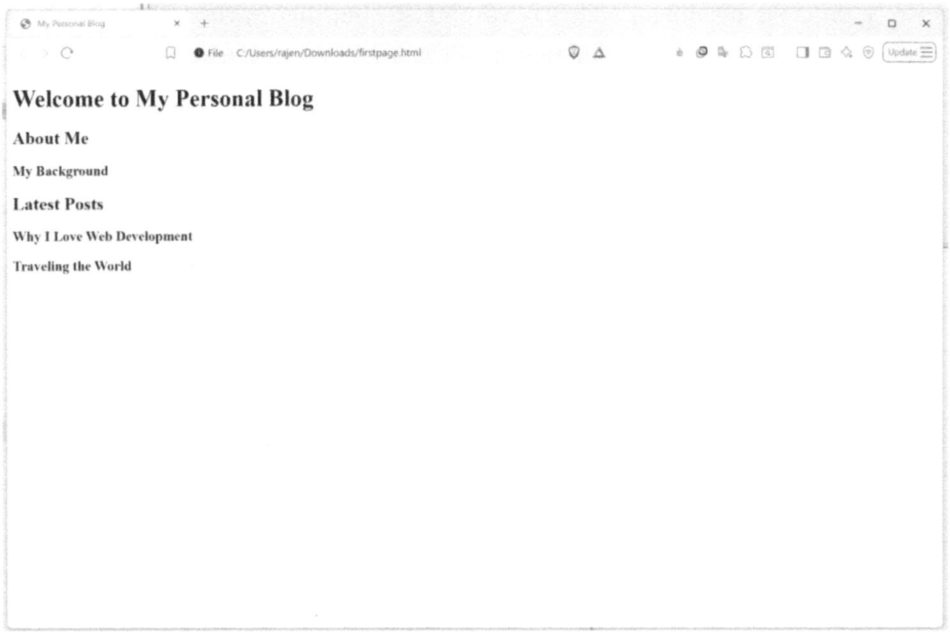

Real-World Example: E-Commerce Product Page

Now, imagine you're building a product page for an online store. Headings can organize product information clearly:

```
<!DOCTYPE html>
<html>
  <head>
    <title>Classic Sneakers - ShopNow</title>
  </head>
  <body>
    <h1>Classic Sneakers</h1>
    <h2>Product Details</h2>
    <h3>Description</h3>
    <h3>Materials</h3>
    <h2>Customer Reviews</h2>
    <h3>Top Rated Review</h3>
    <h3>Write a Review</h3>
  </body>
</html>
```

Using headings like this not only improves readability but also makes your site look more professional.

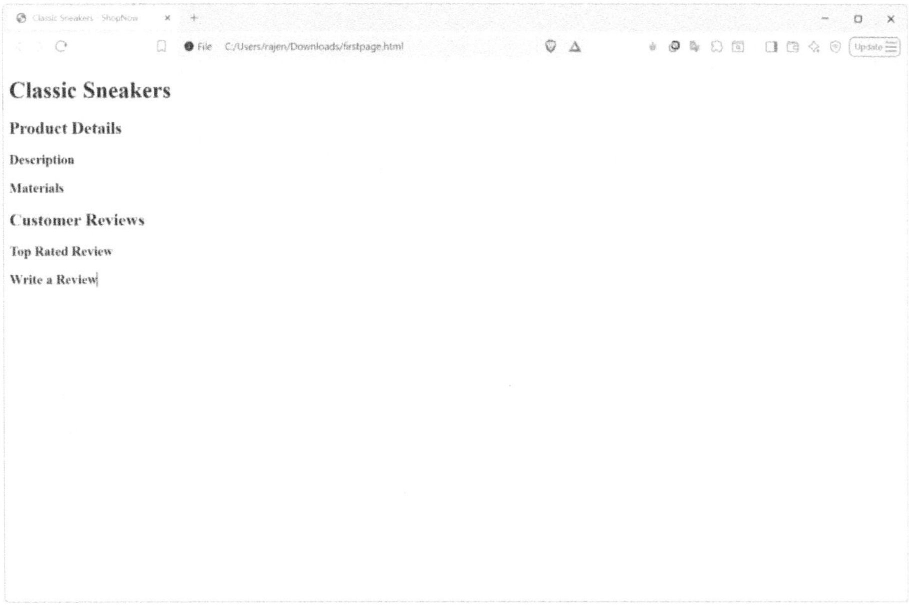

Styling Headings with CSS

By default, browsers display headings in bold and with varying sizes. You can customize their appearance with CSS for a unique style. Let's give our blog headings a fresh look:

Add a **<style>** tag inside the **<head>** section:

```
<head>
  <title>My Personal Blog</title>
  <style>
    h1 {
      color: #2c3e50;
      font-size: 2.5em;
      text-align: center;
    }
    h2 {
      color: #2980b9;
      font-size: 2em;
      margin-top: 30px;
    }
    h3 {
      color: #16a085;
      font-size: 1.5em;
      margin-left: 20px;
    }
  </style>
</head>
```

Refresh your page in the browser to see how each heading stands out with its own color and size.

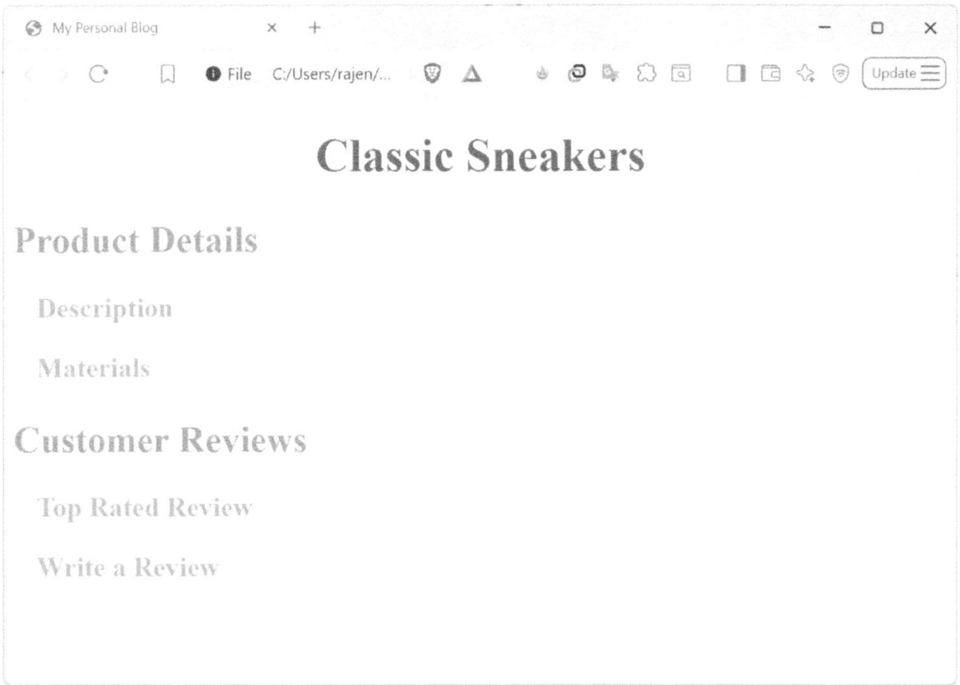

Best Practices for Using Headings

Use only one **<h1>** per page for the main title.

Always keep heading levels in order. For example, don't jump from **<h1>** to **<h4>** without using **<h2>** and **<h3>**.

Headings should describe the content that follows, making navigation logical for all users.

Try reorganizing the blog homepage to add a new section called "Contact" using appropriate heading levels. Experiment with different colors or font sizes in your CSS to see how styling affects readability.

You've just learned how to use HTML headings to structure your web content clearly and professionally. By practicing with real-world examples like blogs and product pages, you're building the foundation for accessible, organized websites. Don't forget to style your headings with CSS for extra visual impact!

Next up, we'll explore how to add paragraphs and line breaks to your web pages. Get ready to make your content even more engaging and readable with these essential HTML tools!

2.4. Creating Paragraphs and Line Breaks

Writing content on the web isn't just about displaying information; it's about making that information easy to read and visually inviting. Two of the most fundamental tools in your HTML toolkit for creating readable web pages are **paragraphs** and **line breaks**. Mastering these will instantly give your web pages a professional, organized look and lay the groundwork for more advanced layouts.

What Are Paragraphs and Why Use Them?

A **paragraph** in HTML is defined using the `<p>` tag. Just like in a book or article, paragraphs help break up text into manageable, readable chunks. Without paragraphs, your content would appear as one long, hard-to-read block. Using `<p>` tags not only improves clarity but also helps search engines and web browsers understand the structure of your content.

Let's see how to create paragraphs in HTML:

```
<p>This is my very first paragraph on the web! HTML makes it easy to
organize text for readers.</p>
<p>Here is another paragraph. Notice how the browser automatically adds
space between them for better readability.</p>
```

When you open this code in your browser, you'll notice each paragraph starts on a new line, with a visible gap separating them. This spacing is handled by the browser's default styling, making your content much easier to scan and enjoy.

Real-World Example: Writing an "About Me" Section

Imagine you're building a simple portfolio website. An "About Me" section is a perfect place to practice paragraphs. Here's how you could structure it:

```
<p>Hello! My name is Sam, and I'm learning to build websites with HTML
and CSS. I enjoy solving problems and bringing creative ideas to life
on the web.</p>
<p>Outside of coding, I love photography, hiking, and exploring new
technologies. I'm excited to share my journey as I grow my web
development skills!</p>
```

Try typing your own introduction in two or three paragraphs. Open it in your browser and see how much more inviting it looks compared to a single block of text.

Creating Line Breaks with `<br>`

Sometimes you want to add a new line within a paragraph, but you don't want to start a whole new paragraph. This is where the **line break** tag `<br>` comes in handy.

Unlike `<p>`, the `<br>` tag does not add extra space; it simply moves the following text to a new line.

Let's look at an example:

```
<p>My favorite hobbies
include:<br>Photography<br>Coding<br>Traveling</p>
```

This will display as:

My favorite hobbies include:

Photography

Coding

Traveling

Each item appears on its own line, but they're all part of the same paragraph. The `<br>` tag is a self-closing tag, which means it doesn't need a closing tag.

Applying Simple CSS for Visual Impact

You can use CSS to style your paragraphs and make them stand out. For example, let's add some color and spacing to your paragraphs:

```
<!DOCTYPE html>
<html>
<head>
  <style>
    p {
      color: #333399;
      font-size: 1.1em;
      line-height: 1.6;
      margin-bottom: 20px;
    }
  </style>
</head>
<body>
  <p>Welcome to my travel blog! Here, I share stories from my
adventures around the world.</p>
  <p>From the mountains of Peru to the beaches of Thailand,<br>each
journey teaches me something new.</p>
</body>
</html>
```

This simple CSS makes the text more readable and visually appealing. The `margin-bottom` property creates extra space between paragraphs, while `color` and `font-size` add personality.

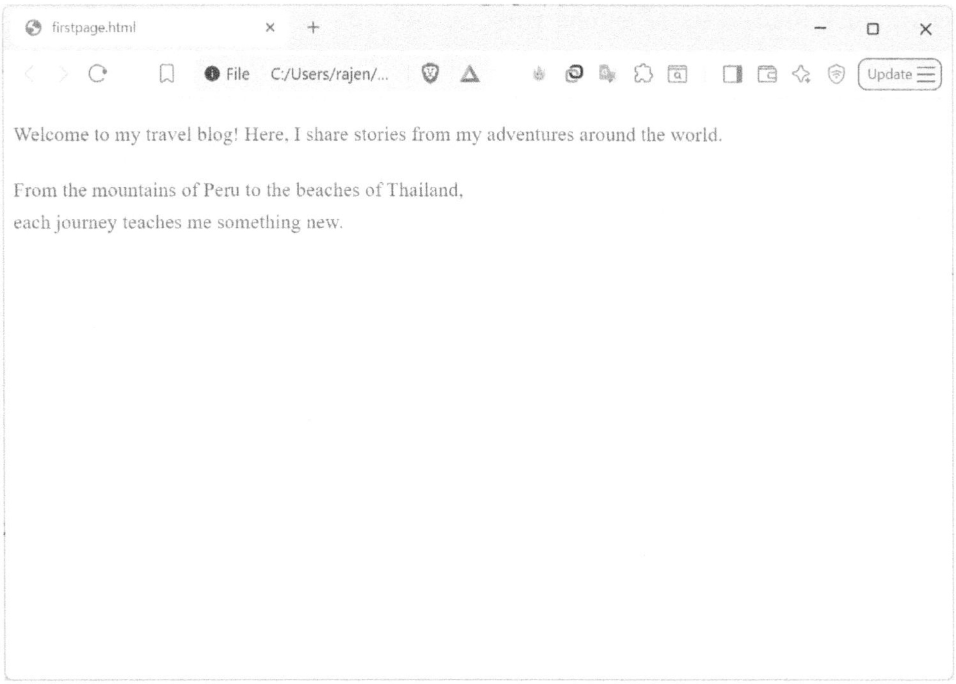

Practice Scenario: E-Commerce Product Descriptions

Many online shops use paragraphs and line breaks to describe products. Try this exercise:

```
<p><strong>Wireless Headphones</strong><br>
Experience high-quality sound, all-day comfort, and a sleek design.<br>
Includes: Carry case, charging cable, and user manual.</p>
```

This approach keeps the product details organized and easy to scan for customers.

You've just learned how to use the `<p>` tag to organize your content and the `<br>` tag to create line breaks inside paragraphs. Practice by writing your own short web page with three paragraphs, adding line breaks where it makes sense. Experiment with simple CSS to see how small changes can enhance your page's look and readability.

Next up, we'll explore how to make specific words stand out using bold, italic, and other text-emphasizing tags. Get ready to add some flair and personality to your web pages as you master the art of highlighting important content!

2.5. Emphasizing Text: Bold, Italic, and More

Have you ever noticed how some words and phrases on websites instantly catch your eye? This is not by accident. Web developers use special HTML tags to emphasize important information, draw attention to key ideas, and make content more readable.

Let's unlock the secrets of emphasizing text, so you can start making your own web pages lively, engaging, and easy to understand.

Before you start, open your favorite code editor and create a file called `emphasis.html`. As you work through the examples, type the code yourself—this hands-on approach is the smartest way to remember what you learn!

Bold Text: Making Words Stand Out

Sometimes you want to make certain words jump off the page, like section headings or important instructions. In HTML, the **`<strong>`** tag is used for strong importance, and **`<b>`** is used for simple bold styling.

Open your `emphasis.html` file.
Add the following code inside the `<body>` tag:

```
<p>This is a <strong>very important</strong> announcement.</p>
<p>Click the <b>Buy Now</b> button to get started.</p>
```

`<strong>` not only makes text bold, it also tells search engines and screen readers that this text is important.
`<b>` is mostly for visual boldness, without extra meaning.

View your file in a web browser. You'll see that both tags make words bold, but remember: **`<strong>`** is preferred for highlighting meaningful information.

Italic Text: Adding Subtle Emphasis

You might want to add a softer emphasis, like highlighting a quote or a note. That's where **`<em>`** for emphasis and **`<i>`** for italic styling come in.

Add these lines to your HTML:

```
<p>Please <em>read carefully</em> before proceeding.</p>
<p>The word <i>café</i> comes from French.</p>
```

`<em>` is used for emphasis and is detected by assistive technologies.
`<i>` simply makes text italic.

Take a look in your browser. Notice how `<em>` and `<i>` both tilt the text, but `<em>` carries extra meaning for accessibility.

Combining Bold and Italic for Extra Emphasis

Sometimes, you need to really make something stand out—think about error messages in a portfolio or a product highlight on an e-commerce site.

Try this combination:

```
<p><strong><em>Warning:</em></strong> Your password is about to
expire.</p>
```

This line appears both bold and italic, showing its importance to users and screen readers alike.

Practical Example: Creating an Attention-Grabbing Blog Post

Let's imagine you're building a blog. Headlines and key points need to catch your reader's eye. Here's a simple example:

```
<h2>Top Tips for Learning Web Development</h2>
<p><strong>Stay consistent:</strong> Practice coding every day.</p>
<p><em>Remember:</em> Everyone starts as a beginner!</p>
```

You can use these tags in your own project—a travel blog, personal website, or even a product description page. Notice how the `<strong>` tag emphasizes the tip, while `<em>` gives a friendly reminder.

Customizing Emphasized Text with CSS

HTML tags are just the beginning. You can style your bold and italic text further with CSS. Let's add a splash of color:

In the `<head>` of your HTML, add:

```
<style>
  strong {
    color: #e67e22; /* Bright orange for bold text */
  }
  em {
    color: #2980b9; /* Deep blue for italic text */
  }
</style>
```

Reload your browser. Now, any text inside `<strong>` is orange, and `<em>` is blue.

This technique is especially useful on personal portfolios or online shops, where you want emphasized text to fit your brand's style.

Practice by writing a short "About Me" paragraph for your portfolio, using `<strong>`, `<em>`, `<b>`, and `<i>` tags to highlight your name, your passion, or a favorite quote. Experiment with CSS colors to reflect your personality.

By mastering these simple tags, you give your web pages personality and clarity, making sure your important messages never get missed.

Ready to keep your code clean and easy to read? In the next section, you'll discover how to add comments and organize your HTML code, making your projects easier to manage as they grow. Let's make your code as readable as your content!

2.6. Commenting and Organizing Your HTML Code

Keeping your HTML code clean and well-organized is just as important as making your website look great. As your projects grow from a simple page to a full portfolio or blog, a messy codebase can slow you down and make it harder to find and fix mistakes. Smart web developers use **comments** and thoughtful organization to make their code easy to read, understand, and update. In this section, you'll learn how to use HTML comments and proven organization techniques, giving you confidence to manage even bigger projects in the future.

Why Comment and Organize Your Code?

Imagine you're building an online portfolio to showcase your art or a product catalog for a small business. As new features are added, keeping track of what each section does can become challenging. **Comments** let you leave notes for yourself (or teammates), while good organization keeps related elements together. This practice:

Makes your code easier to understand later—even months after you've written it.
Helps others collaborate with you more smoothly.
Reduces errors by making it easier to spot problems.

Adding Comments in HTML

A **comment** in HTML is a line or block of text ignored by browsers, meant only for people reading the code. Comments are perfect for explaining sections, marking important notes, or temporarily hiding code without deleting it.

To add a comment, wrap your note like this:

```
<!-- This is a comment. The browser won't display it. -->
```

Let's see a real-world example. Suppose you're designing a personal blog homepage. Here's how you might use comments to explain the structure:

```
<!DOCTYPE html>
<html lang="en">
  <head>
    <meta charset="UTF-8">
    <title>My Personal Blog</title>
  </head>
  <body>
    <!-- Header section starts -->
    <header>
```

```
      <h1>Welcome to My Blog!</h1>
    </header>
    <!-- Header section ends -->

    <!-- Main content starts -->
    <main>
      <!-- Latest blog post -->
      <article>
        <h2>My First Post</h2>
        <p>Today I'm starting my web development journey!</p>
      </article>
    </main>
    <!-- Main content ends -->

    <!-- Footer section starts -->
    <footer>
      <p>© 2024 My Personal Blog</p>
    </footer>
    <!-- Footer section ends -->
  </body>
</html>
```

Notice how each section is clearly labeled with comments. This makes it much easier to scan the code and find what you're looking for.

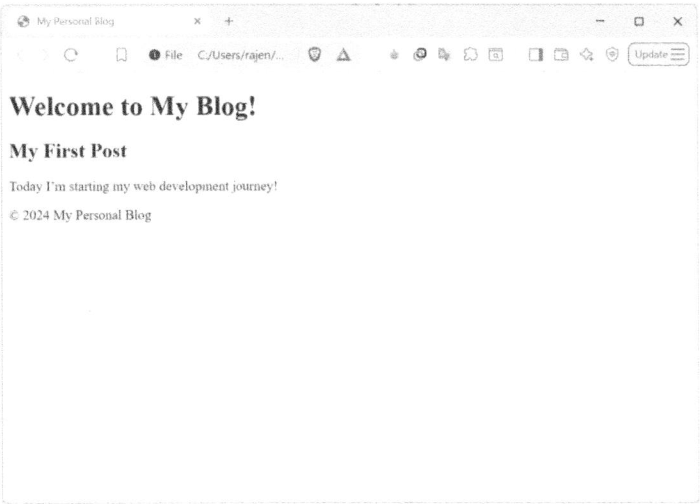

Organizing HTML for Readability

Organization goes beyond just adding comments. You'll want to group related elements together and use indentation consistently so your code is visually structured.

Indent Nested Elements: Each time you nest elements (for example, placing a `<p>` inside a `<section>`), add two spaces or a tab to make the hierarchy clear.

Group Related Sections: Keep elements that belong together (like navigation links or a set of product cards) in the same container, such as a `<nav>` or `<section>`.

Use Comments to Separate Large Sections: Clearly mark where sections begin and end.

Let's organize a small e-commerce product listing:

```
<!DOCTYPE html>
<html lang="en">
  <head>
    <meta charset="UTF-8">
    <title>Shop Easy - Product Catalog</title>
  </head>
  <body>
    <!-- Navigation menu -->
    <nav>
      <ul>
        <li><a href="#">Home</a></li>
        <li><a href="#">Products</a></li>
        <li><a href="#">Contact</a></li>
      </ul>
    </nav>

    <!-- Products section starts -->
    <section>
      <!-- Product 1 -->
      <div class="product-card">
        <h2>Eco Water Bottle</h2>
        <p>Reusable bottle for everyday use.</p>
      </div>
      <!-- Product 2 -->
      <div class="product-card">
        <h2>Yoga Mat</h2>
        <p>Comfort and grip for all levels.</p>
      </div>
    </section>
    <!-- Products section ends -->
  </body>
</html>
```

Try adding your own comments to mark where the product cards begin and end, or to remind yourself to add more products later.

Practical Coding Exercise

Let's practice:

Open your code editor and create a new HTML file.
Write a simple page with a header, main content, and footer.
Add comments to label each section.

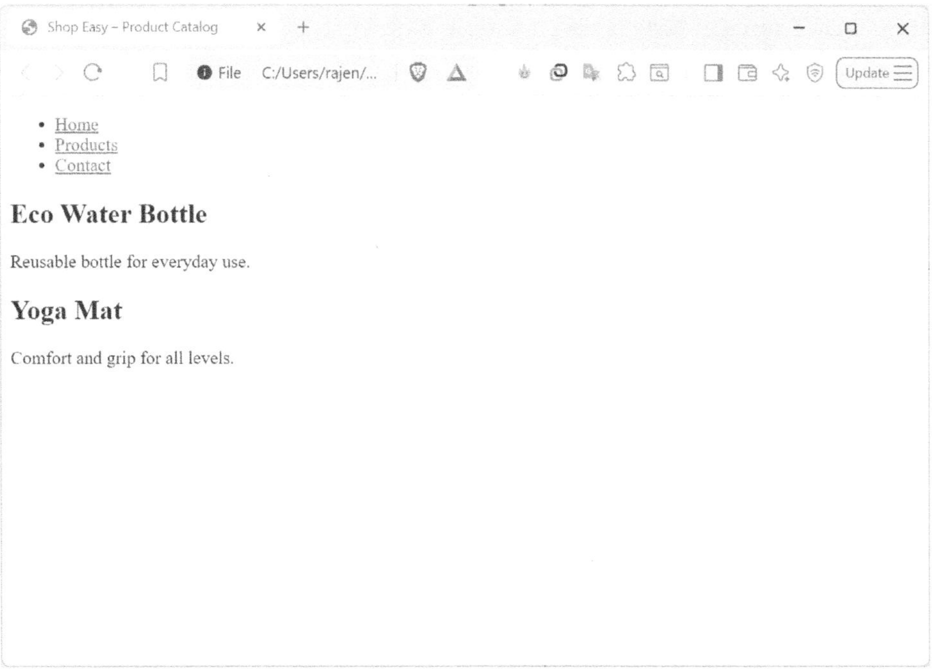

Use indentation to organize nested elements.

For example:

```
<!-- Header begins -->
<header>
  <h1>My Web Portfolio</h1>
</header>
<!-- Header ends -->

<!-- Main content begins -->
<main>
  <p>Check out my latest projects below.</p>
</main>
<!-- Main content ends -->

<!-- Footer begins -->
<footer>
  <p>Contact me: email@example.com</p>
</footer>
<!-- Footer ends -->
```

By using comments and organizing your HTML with clear structure and indentation, you make your code easier to read and maintain. This is a professional habit that will pay off as your projects become more complex and collaborative.

Up next, you'll put your new skills into action by structuring a simple blog post layout. Get ready to dive deeper and build a web page section by section, just like the pros!

2.7. Hands-On Exercise: Structuring a Simple Blog Post

Building your own blog post is a fantastic way to practice the essential skills of structuring HTML content. Blog layouts are a cornerstone of the web, used by everyone from travel bloggers to tech news sites. In this exercise, you'll create a simple blog post using fundamental HTML elements. This will help you understand how real-world sites organize content and prepare you for more advanced layouts in the future.

Planning Your Blog Post Layout

Before writing any code, it's important to visualize the structure of your blog post. Most blog articles share a similar layout:

- A **main heading** for the post title
- An **author section** with a name and publication date
- The **main content**, split into paragraphs
- One or more **subheadings** to break up the text
- An **image** to support the content
- A **conclusion** or call to action

These elements make your page readable and visually appealing. Let's see how to turn this plan into HTML.

Step-by-Step: Building the Blog Post

Follow these instructions to build your first structured HTML page. Open your favorite code editor and create a new file called `blog.html`.

Start with the HTML Skeleton

Every HTML document begins with the following structure:

```
<!DOCTYPE html>
<html lang="en">
  <head>
    <meta charset="UTF-8">
    <title>My First Blog Post</title>
  </head>
  <body>
    <!-- Blog content will go here -->
  </body>
</html>
```

Add the Blog Title

Use the **<h1>** tag for the main heading. This tells browsers and search engines what your post is about.

```
<h1>Discovering Local Cafés: A Coffee Lover's Guide</h1>
```

Include Author and Date

Use a **<p>** tag to display the author and publication date. Add a **<strong>** tag to highlight the author's name.

```
<p>By <strong>Alex Morgan</strong> | Published June 6, 2024</p>
```

Insert an Image

Images make blog posts more engaging. Use the **<img>** tag and include a description with the `alt` attribute for accessibility.

```
<img src="coffee-cup.jpg" alt="A steaming cup of coffee on a café
table">
```

Write the Introduction Paragraph

Use the **<p>** tag for your opening paragraph.

```
<p>If you love coffee as much as I do, exploring local cafés is the
perfect weekend adventure. Here's my guide to finding the best brews in
town.</p>
```

Add a Subheading and More Content

Use the **<h2>** tag for subheadings and more **<p>** tags for supporting paragraphs.

```
<h2>Finding the Perfect Spot</h2>
<p>Look for cafés with cozy atmospheres, friendly staff, and a focus
on quality beans. Don't be afraid to ask for recommendations!</p>
```

Conclude Your Post

Finish with a closing paragraph and a call to action.

```
<p>Ready to start your coffee adventure? Share your favorite café in
the comments below!</p>
```

Your Complete Blog Post

Here's how your full blog post structure should look:

```
<!DOCTYPE html>
<html lang="en">
  <head>
    <meta charset="UTF-8">
    <title>My First Blog Post</title>
  </head>
```

```
    <body>
      <h1>Discovering Local Cafés: A Coffee Lover's Guide</h1>
      <p>By <strong>Alex Morgan</strong> | Published June 6, 2024</p>
      <img src="coffee-cup.jpg" alt="A steaming cup of coffee on a
café table">
      <p>If you love coffee as much as I do, exploring local cafés is
the perfect weekend adventure. Here's my guide to finding the best
brews in town.</p>
      <h2>Finding the Perfect Spot</h2>
      <p>Look for cafés with cozy atmospheres, friendly staff, and a
focus on quality beans. Don't be afraid to ask for recommendations!</p>
      <p>Ready to start your coffee adventure? Share your favorite
café in the comments below!</p>
    </body>
  </html>
```

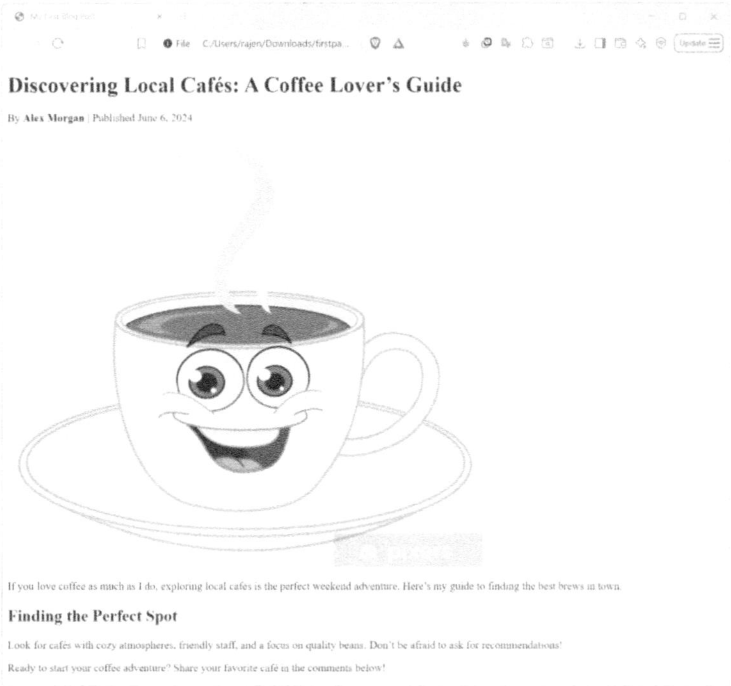

Real-World Example: Popular Blog Platforms

Websites like Medium and WordPress use similar HTML structures to organize articles. Their posts often include headings, subheadings, author details, and images, just like your example. Practicing with this template gives you a solid foundation for building your own blog or contributing to established platforms.

3. WORKING WITH LISTS AND LINKS

Imagine visiting your favorite website and easily navigating through its pages, skimming through product categories, or jumping to just the information you need. What makes this seamless experience possible? It's all about effective use of lists and links, two of the most essential building blocks in web development. In this chapter, you'll discover how these simple yet powerful tools help structure content, create user-friendly navigation, and connect your site to the wider web.

You'll start by learning to create ordered and unordered lists—perfect for organizing information, menus, or features in a clear, readable way. These lists aren't just for displaying bullet points. When combined with the right styles, they transform into elegant navigation menus, a key feature in any modern website, from sleek portfolios to thriving e-commerce platforms.

Next, you'll master the art of linking with the versatile **`<a>` tag**. You'll see how to guide users to other pages, jump to specific sections within a page (using anchors), and even open email or phone links directly from your site. These skills are not only foundational for building functional websites but are also highly valued in professional web development.

By the end of this chapter, you'll bring these concepts together in a hands-on project: building a navigation bar for your own portfolio site. This project-focused approach ensures you don't just read about theory, but gain the confidence to put your new skills to practical use. Get ready to unlock the tools that make websites interactive, connected, and truly user-focused.

3.1. Building Ordered and Unordered Lists

Lists are everywhere online, from blog post summaries to product features on e-commerce sites. Learning to create ordered and unordered lists in HTML gives you a powerful way to organize information and make your web pages clearer and easier to read. Let's dive in and practice building both types of lists, using real-world examples and hands-on coding exercises.

Why Use Lists in Web Development?

Lists help structure content logically. For example, a personal portfolio site might use a list to display programming skills, while an online recipe blog could use an ordered list to number the steps in a cooking process. HTML offers two main types of lists:

- **Unordered lists** (bulleted)
- **Ordered lists** (numbered)

Let's explore both and see how you can use them in your own projects.

Creating an Unordered List (`<ul>`)

An unordered list is perfect for items that don't have a specific order. For example, a portfolio website might list a developer's skills using bullets. Here's how to make one:

Use the `**<ul>**` tag to start the unordered list.
Wrap each item in the list with the `**<li>**` (list item) tag.

Let's see this in action:

```
<h2>My Web Development Skills</h2>
<ul>
  <li>HTML</li>
  <li>CSS</li>
  <li>JavaScript</li>
  <li>Responsive Design</li>
</ul>
```

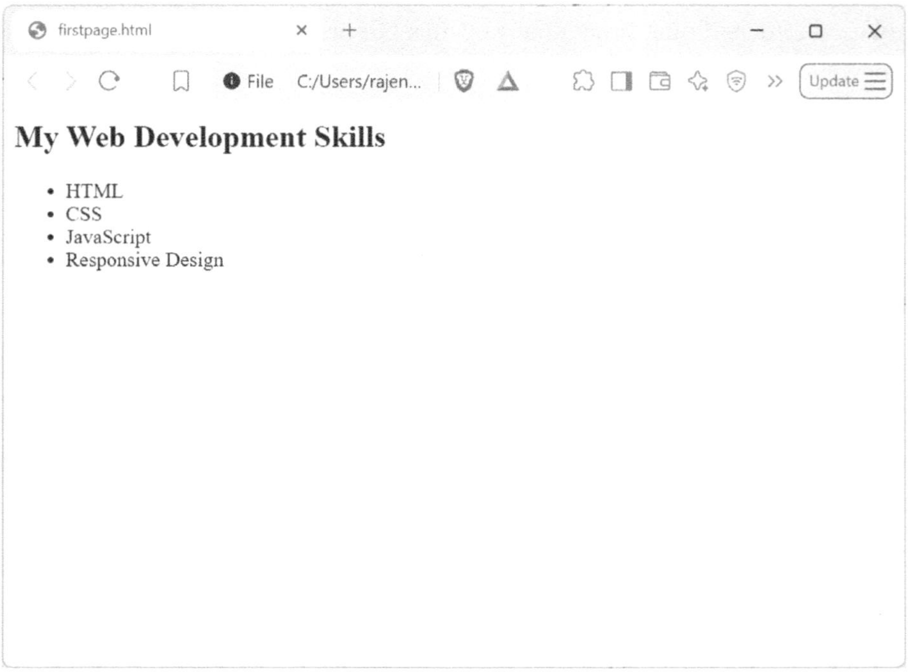

When viewed in a browser, this code displays a bulleted list of skills. The `**<ul>**` element tells the browser to use bullets, and each `**<li>**` marks a new item.

Creating an Ordered List (`<ol>`)

Ordered lists are great when the order of items matters, such as instructions or ranking. Imagine you're writing a blog post with step-by-step instructions for setting up a personal website. The `**<ol>**` tag creates a numbered list. Here's how:

Begin with the `**<ol>**` tag for "ordered list."
Add each step using `**<li>**` tags.

Example:

```
<h2>How to Set Up Your Portfolio Site</h2>
<ol>
  <li>Choose a domain name</li>
  <li>Select a web hosting provider</li>
  <li>Design your website</li>
  <li>Publish your site</li>
</ol>
```

This displays a numbered sequence, making it easy for users to follow each step in order.

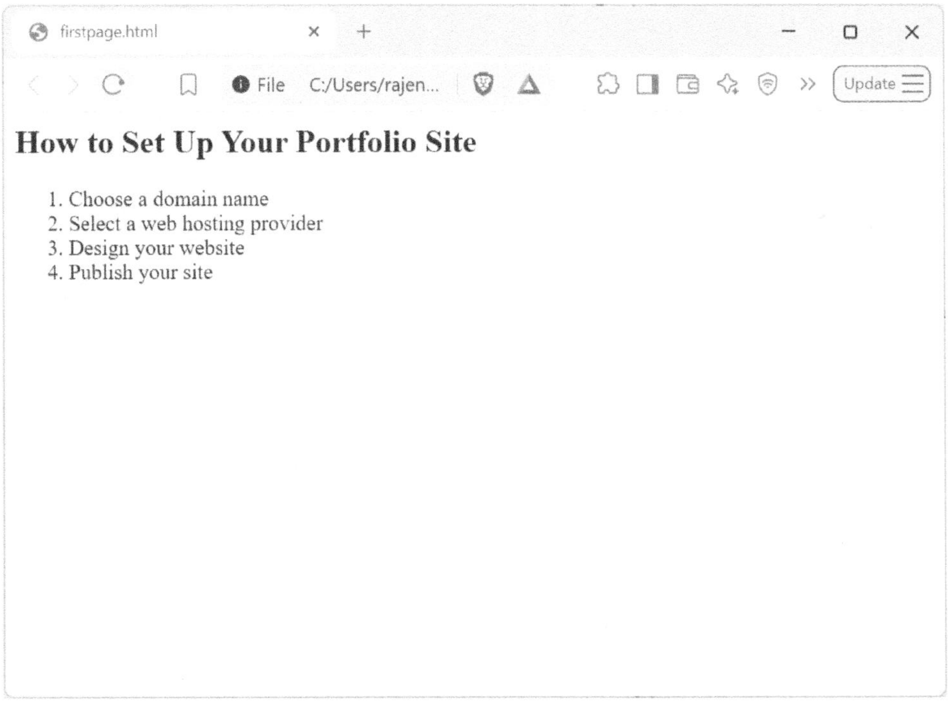

Combining Lists in Real-World Scenarios

Building lists is not just about writing code; it's about making content user-friendly. Let's look at a scenario from an e-commerce site. Imagine you're listing product features (unordered) and steps to complete a purchase (ordered):

```
<h2>Product Features</h2>
<ul>
  <li>High-resolution display</li>
  <li>Long battery life</li>
  <li>Fast charging</li>
  <li>Lightweight design</li>
</ul>

<h2>How to Buy</h2>
<ol>
  <li>Add the product to your cart</li>
  <li>Enter shipping information</li>
  <li>Choose a payment method</li>
  <li>Confirm your order</li>
</ol>
```

This approach helps shoppers quickly see product benefits and understand the steps to make a purchase.

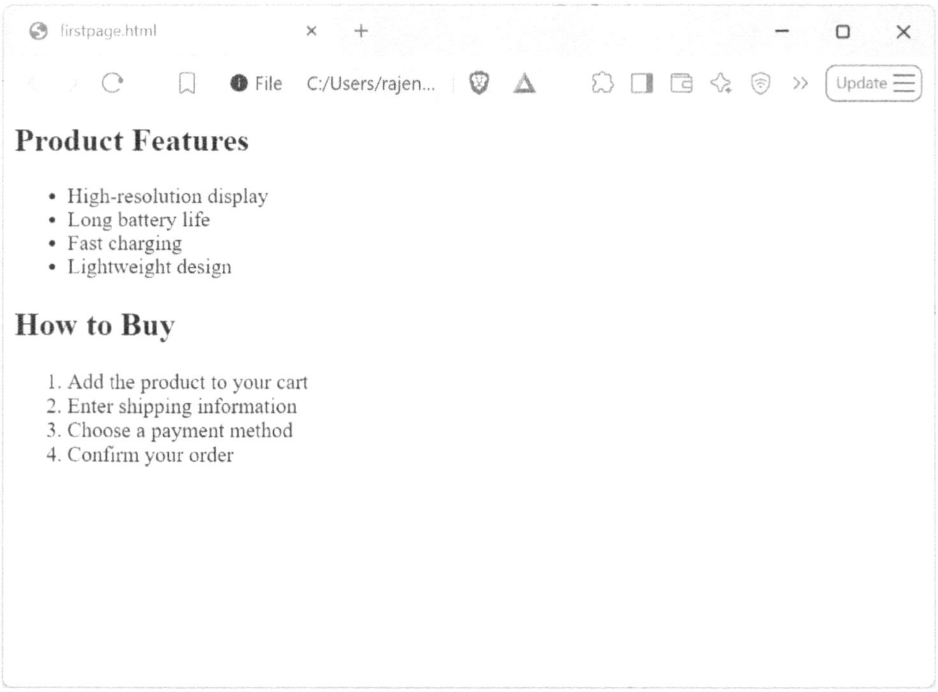

Styling Lists with CSS

HTML lists look plain by default. With CSS, you can customize their appearance to match your site's design. Here's how you can style a list for a modern look, such as for a blog sidebar showing categories:

```
<ul class= "categories">
  <li>Web Development</li>
  <li>Design Tips</li>
  <li>JavaScript Projects</li>
  <li>CSS Tricks</li>
</ul>
```

Add this CSS to style the list:

```
.categories {
  list-style-type: square; /* Changes bullet shape */  padding-left:
20px; /* Adds space before bullets */  background-color: #f9f9f9;
  border-radius: 6px;
  max-width: 220px;
}

.categories li {
  padding: 8px 0;
  color: #333;
  font-family: Arial, sans-serif;
}
```

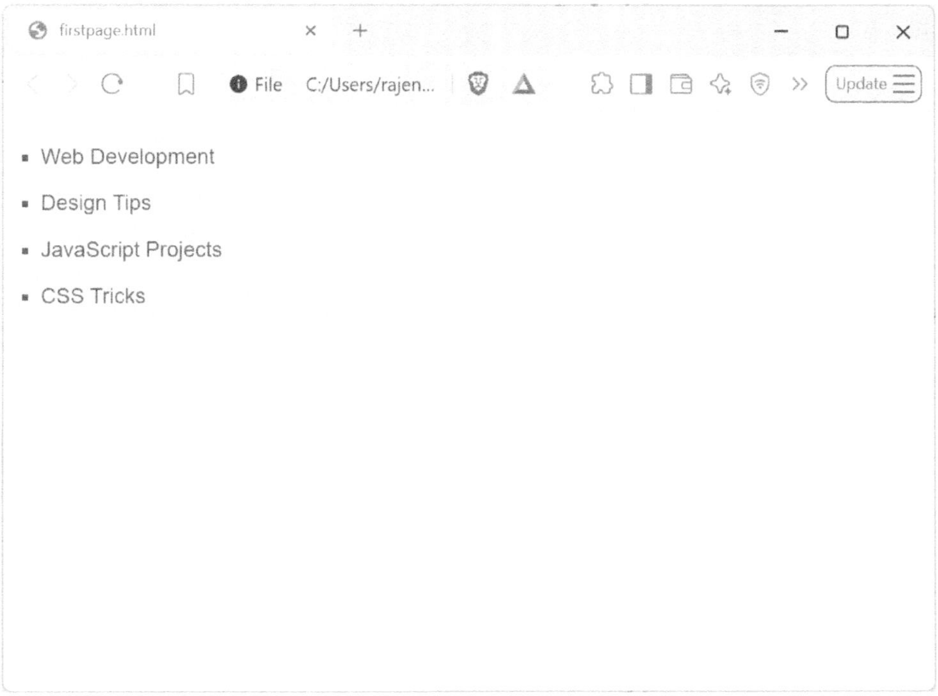

This CSS changes the default bullet style to squares, adds spacing, and applies background color and rounded corners for a clean, modern effect.

Hands-On Exercise: Build Your Own List

Practice by creating a section for your own portfolio or resume website. List your favorite programming languages (unordered), and then outline the steps you took to learn your first language (ordered).

- Open your HTML file.
- Add a heading for each list.
- Write your unordered and ordered lists using the examples above.
- Style your lists with CSS to make them stand out.

In this section, you learned how to build both **ordered** and **unordered lists** using HTML, and you saw how to style them with CSS for a more appealing look. Lists are essential for organizing information on any website, from portfolios and blogs to online shops. Practicing these concepts will help you present content clearly and professionally.

Next, you'll discover how lists can do even more: they're the foundation for creating navigation menus, one of the most important and interactive features of any website. Get ready to turn your lists into clickable menus that guide visitors through your site!

3.2. Creating Navigation Menus with Lists

Navigation menus are essential for helping visitors explore your website easily. Most professional websites, from personal portfolios to global e-commerce stores, use lists to organize their navigation links. In this section, you'll discover how to use HTML lists to create a clean, organized menu, then add CSS to make it attractive and user-friendly. Let's build your first navigation menu step by step.

Why Use Lists for Navigation?

Navigation menus typically appear as a set of links, such as "Home," "About," "Contact," or "Shop." Using a list structure not only keeps your HTML organized but also helps search engines and assistive technologies understand your site's layout.

A navigation menu is usually built with an **unordered list** (`<ul>`), where each **list item** (`<li>`) contains a navigation **link**. This structure is both readable and flexible for styling.

Step-by-Step: Building a Basic Navigation Menu

Let's start by creating a simple horizontal menu for a personal portfolio website.

Open your HTML file and add a new section for navigation.
Use a `<nav>` element to semantically mark the navigation area.
Add an unordered list with each site section as a list item.

Here's the basic HTML structure:

```
<nav>
  <ul>
    <li>Home</li>
    <li>Projects</li>
    <li>About</li>
    <li>Contact</li>
  </ul>
</nav>
```

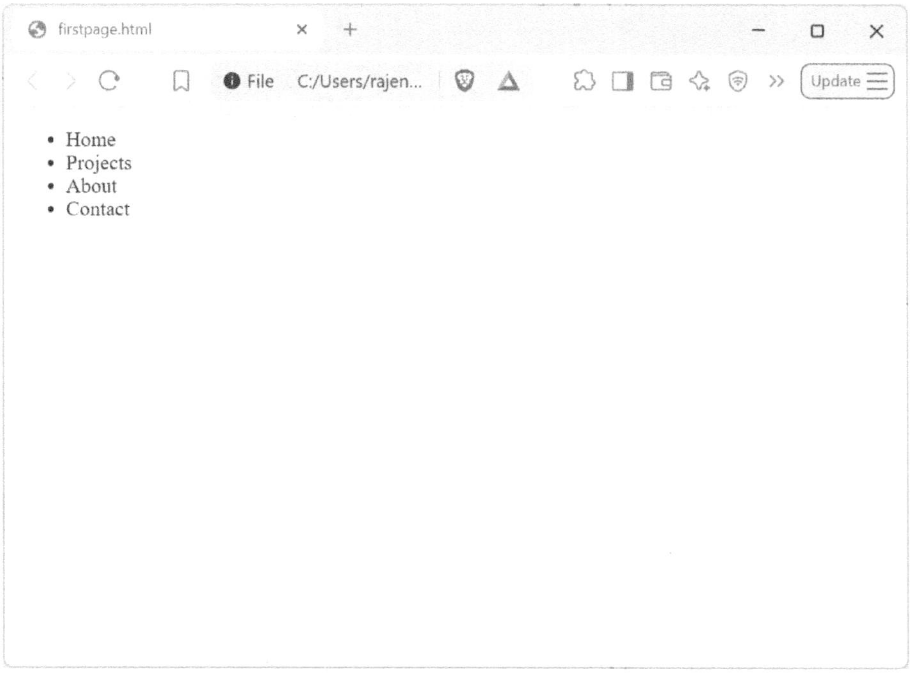

This creates a vertical list of your menu items. Next, you'll learn how to make these items look like a professional navigation bar.

Transforming Your List into a Horizontal Menu with CSS

To make your navigation more visually appealing, you'll use CSS to style the list horizontally, remove bullet points, and add spacing.

Add the following CSS inside a `<style>` tag in the `<head>` of your HTML, or in a separate CSS file:

```
nav ul {
  list-style: none; /* Removes bullet points */  padding: 0; /* Removes
default padding */  margin: 0; /* Removes default margin */  display:
flex; /* Aligns list items in a row */  background-color: #222; /* Sets
a background for the menu */}

nav li {
  padding: 16px 24px; /* Adds space around each menu item */  color:
#fff; /* Sets text color to white */  font-size: 18px; /* Increases
text size */  cursor: pointer; /* Shows pointer cursor on hover */}

/* Optional: Add a hover effect */nav li:hover {
  background-color: #444;
}
```

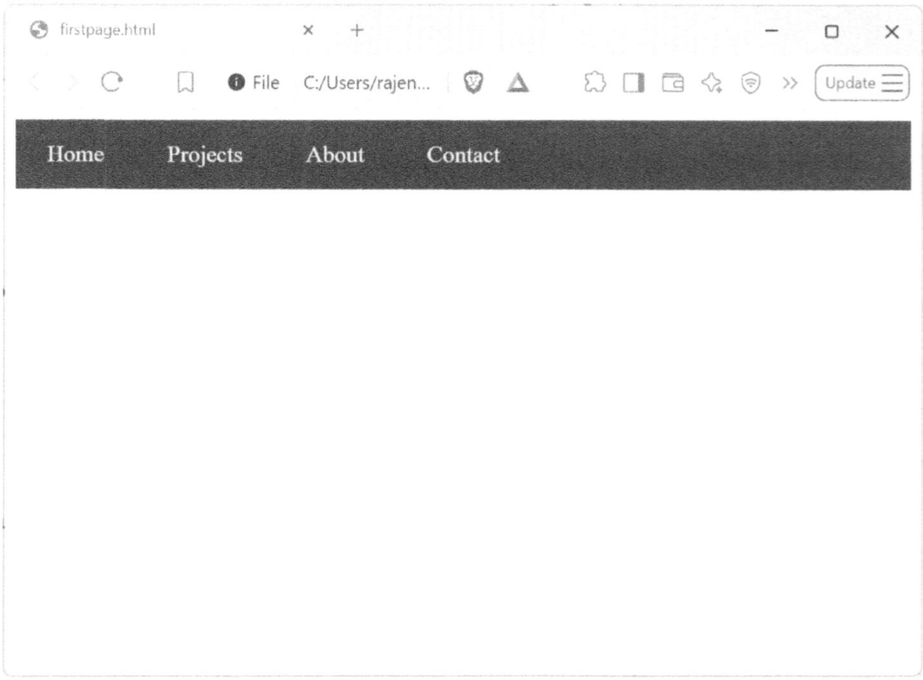

Now, your menu items will appear side by side, with a solid background and interactive hover effect. This layout is common on modern portfolio and company websites.

Real-World Example: Blog Navigation Menu

Imagine you're building a blog. Your readers need to move easily between "Home," "Articles," "Categories," and "About." Try creating a navigation bar for a blog site:

```
<nav>
  <ul>
    <li>Home</li>
    <li>Articles</li>
    <li>Categories</li>
    <li>About</li>
  </ul>
</nav>
```

Apply the same CSS as above, or change the background color and text size to match your blog's personality. Experiment with different color schemes to see how style can change the feel of your navigation.

Coding Practice: Build Your Own Navigation Menu

Hands-on practice is the best way to master these techniques. Try the following exercise to reinforce your learning:

- Choose a website idea, such as a travel blog, a small shop, or a photography portfolio.
- List the main sections your site will have (for example: "Gallery," "Shop," "Blog," "Contact").
- Use HTML to create a navigation menu with those sections as list items.
- Apply the CSS provided above, and experiment with colors, fonts, and spacing to match your site's theme.
- Test your menu on both wide and narrow screens to see your responsive design in action.

You've just transformed a plain HTML list into a stylish, responsive navigation menu—a skill found on nearly every modern website. Whether you're starting your own portfolio or planning a future e-commerce store, you'll use these techniques again and again.

Coming up next, you'll learn how to turn your navigation items into clickable links using the powerful `<a>` tag. Get ready to connect your menu to other pages and make your website truly interactive!

3.3. Linking to Other Pages: The <a> Tag

Links are what make the web truly interconnected. With just a simple click, you can jump from one page to another, visit your favorite online shop, or open a new article in a blog. In HTML, the magic behind this is the **`<a>`** tag, short for "anchor." This tag lets you create clickable links that can point to other pages, websites, or even files like PDFs and images. In this section, you'll learn how to use the **`<a>`** tag to build your own navigation menus and connect your pages, setting the stage for interactive, user-friendly websites.

What is the `<a>` Tag?

The **`<a>`** tag is one of the most important elements in HTML. It tells the browser that you want to create a hyperlink. This tag uses the **`href`** attribute (short for "hypertext reference") to specify the destination of the link.

Here's what a basic link looks like:

```
<a href= "https://www.example.com">Visit Example</a>
```

`<a>`: The anchor tag, which creates the link.
`href`: The attribute that holds the address you want to link to.
The text between the opening `<a>` and closing `</a>` is what users will click.

Creating Navigation Links for Your Website

Let's bring this to life by building a navigation menu for a portfolio site. Imagine you're creating a personal website to showcase your work. You have three pages: Home, Projects, and Contact.

Open your HTML file (for example, `index.html`).
Create a navigation section using a list and `<a>` tags:

```
<nav>
  <ul>
    <li><a href= "index.html">Home</a></li>
    <li><a href= "projects.html">Projects</a></li>
    <li><a href= "contact.html">Contact</a></li>
  </ul>
</nav>
```

Each **`<li>`** (list item) contains an **`<a>`** tag.
The **`href`** attribute links to the filename of each page in your project.
When a user clicks "Projects," for example, they'll be taken to `projects.html`.

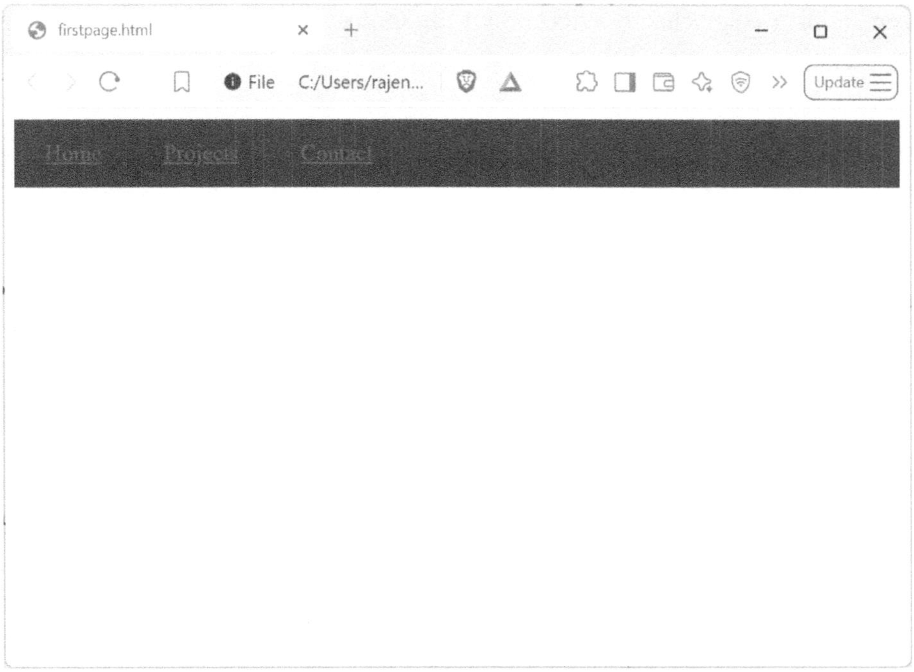

Styling Links with CSS

By default, links are blue and underlined. You can use CSS to change the color, remove underlining, or make links stand out.

Let's add some style to your navigation menu:

```
<style>
nav ul {
  list-style-type: none;
  padding: 0;
  display: flex;
  gap: 20px;
}

nav a {
  text-decoration: none; /* Removes underline */
  color: #3498db; /* Changes link color */
  font-weight: bold;
}

nav a:hover {
  color: #1abc9c; /* Changes color on mouse hover */
}
</style>
```

`**text-decoration: none;**` removes the default underline.
`**color**` changes the color of the link.
`**font-weight: bold;**` makes the link text bold.

`:hover` adds a color change effect when the mouse is over the link, improving user experience.

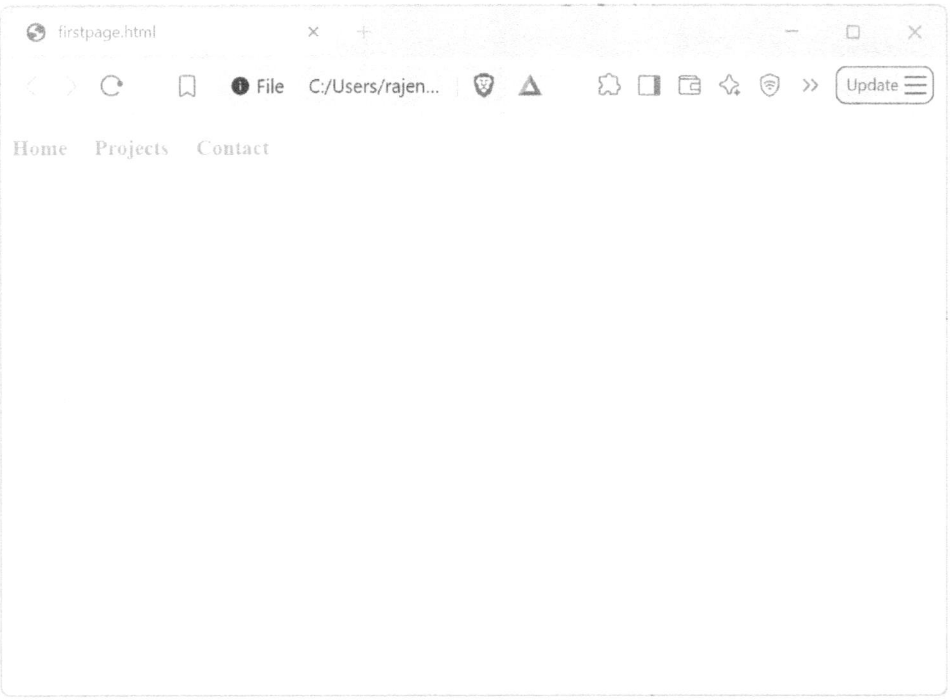

Linking to External Websites

You can also use the **`<a>`** tag to send visitors to other websites. For example, if you want to link to your GitHub profile from your portfolio:

```
<a href= "https://github.com/yourusername"target= "_blank">My
GitHub</a>
```

`target="_blank"` opens the link in a new browser tab or window, so users don't lose their place on your site.

Real-World Example: Blog Navigation

Suppose you're creating a blog. Your main page lists recent posts, and you want to provide links to individual articles:

```
<main>
  <h2>Recent Posts</h2>
  <ul>
    <li><a href= "post1.html">How I Learned HTML</a></li>
    <li><a href= "post2.html">Styling with CSS: My First Steps</a></li>
    <li><a href= "post3.html">Building My Portfolio Site</a></li>
  </ul>
</main>
```

This structure makes it easy for readers to navigate between posts, improving the usability of your blog.

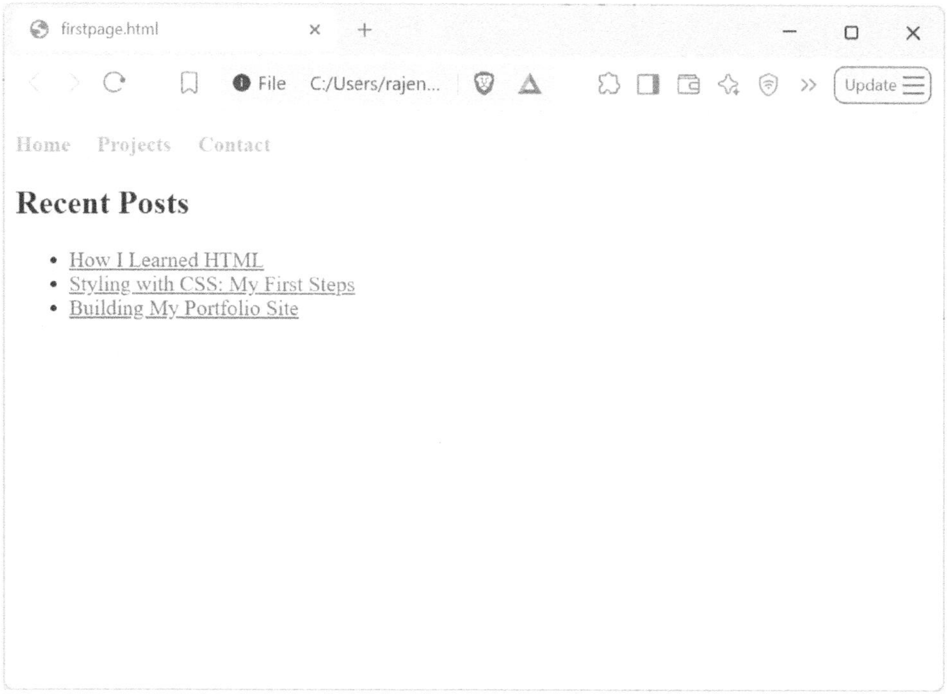

Hands-On Exercise: Build Your Own Navigation Bar

Now it's your turn! Practice by creating a simple navigation bar for a fictional online store with links to "Shop," "About Us," and "Contact." Style it so that the links are horizontal, not underlined, and change color when hovered over.

Set up your HTML:

```
<nav>
  <ul>
    <li><a href= "shop.html">Shop</a></li>
    <li><a href= "about.html">About Us</a></li>
    <li><a href= "contact.html">Contact</a></li>
  </ul>
</nav>
```

Add CSS styling:

```
<style>
nav ul {
  display: flex;
  gap: 15px;
  list-style: none;
  padding: 0;
```

```
}

nav a {
  text-decoration: none;
  color: #2d3436;
  padding: 8px 16px;
  transition: background 0.3s;
}

nav a:hover {
  background: #dfe6e9;
  color: #0984e3;
}
</style>
```

This coding practice will help you understand both the structure and style of navigation links.

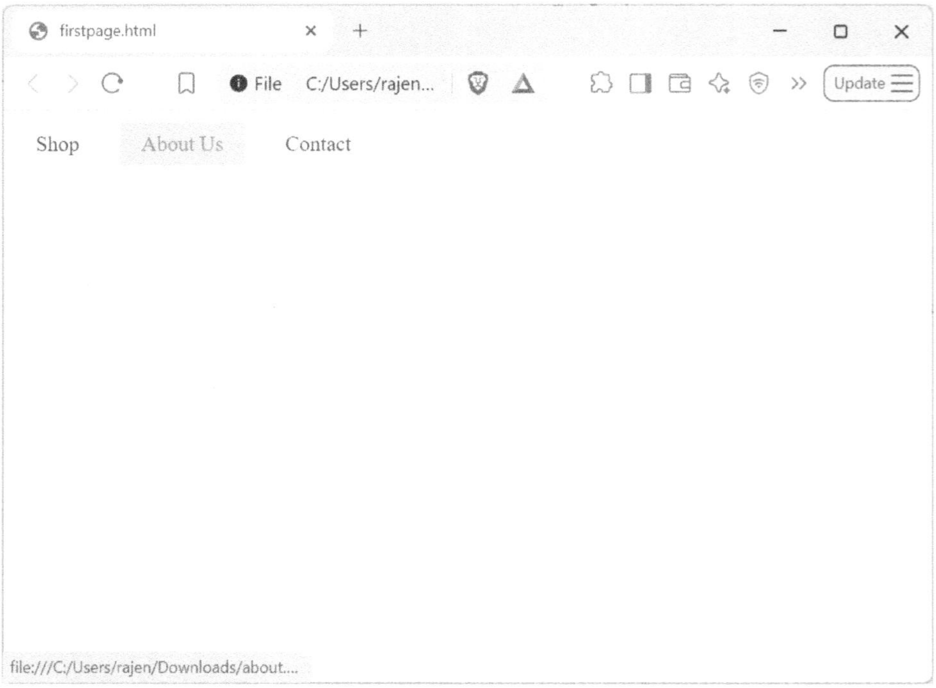

You've now learned how to use the **`<a>`** tag to create links between your web pages, add navigation menus, and link to external sites. You've also experimented with CSS to style these links, making your site more visually appealing and interactive. Take some time to build your own navigation bars and adjust the styles to your liking. The more you practice, the more confident you'll feel!

Up next, you'll unlock the power of linking within the same page using anchor links. You'll discover how to help users jump straight to specific sections, creating smooth and user-friendly navigation for longer web pages. Get ready to add another essential tool to your web development toolkit!

3.4. Linking to Sections Within a Page (Anchors)

Imagine you've built a web page with several sections: an introduction, a list of your favorite projects, and your contact information. As your pages grow longer, you'll want to help visitors find what they need quickly, without endless scrolling. This is where **anchor links** come in. Anchor links let users jump instantly to specific parts of your page, creating a smooth, professional browsing experience.

Let's dive into how you can use anchors to create in-page navigation, a skill that's essential for both user experience and modern web design.

Understanding Anchor Links

An **anchor link** is a special kind of hyperlink that points to a specific section within the same web page. To create an anchor link, you'll use two key elements:

The **id** attribute, which marks the target section.
The **href** attribute with a hash symbol (``#``) in your link, which tells the browser to scroll to the corresponding section.

Step-by-Step: Creating In-Page Navigation

Let's build a simple personal portfolio page with anchor links to each section. You can follow along and modify the example as you go.

Set Up the Structure

Start with the basic HTML structure and add three sections: About Me, Projects, and Contact.

```
<!DOCTYPE html>
<html lang= "en">
<head>
  <meta charset= "UTF-8">
  <title>My Portfolio</title>
  <style>
    body { font-family: Arial, sans-serif; }
    nav { background: #e3e3e3; padding: 10px; }
    nav a { margin: 0 15px; color: #333; text-decoration: none; }
    section { margin: 40px 0; padding: 20px; background: #f9f9f9; }
    h2 { color: #1976d2; }
  </style>
</head>
<body>

  <nav>
```

```
    <a href= "#about">About Me</a>
    <a href= "#projects">Projects</a>
    <a href= "#contact">Contact</a>
  </nav>

  <section id= "about">
    <h2>About Me</h2>
    <p>Hello! I'm learning web development and building awesome
websites.</p>
  </section>

  <section id= "projects">
    <h2>Projects</h2>
    <ul>
      <li>Personal Blog</li>
      <li>Recipe Finder App</li>
      <li>Photo Gallery</li>
    </ul>
  </section>

  <section id= "contact">
    <h2>Contact</h2>
    <p>You can reach me via email or follow me on social media.</p>
  </section>

</body>
</html>
```

Assign IDs to Sections

Notice the ``id`` attributes (`about`, `projects`, and `contact`) in each ``section`` element. These IDs act as unique markers that anchor links can target.

Create Navigation Links

In the navigation bar, each ``<a>`` tag's ``href`` uses a hash (``#``) followed by the corresponding section ID. For example, ``<a href="#projects">Projects</a>`` tells the browser to scroll to the element with ``id="projects"``.

Try It Out

Open your HTML file in a browser. Click on the “Projects” link in the navigation. The page should jump directly to your Projects section! Try the other links to see how the page moves smoothly to each part.

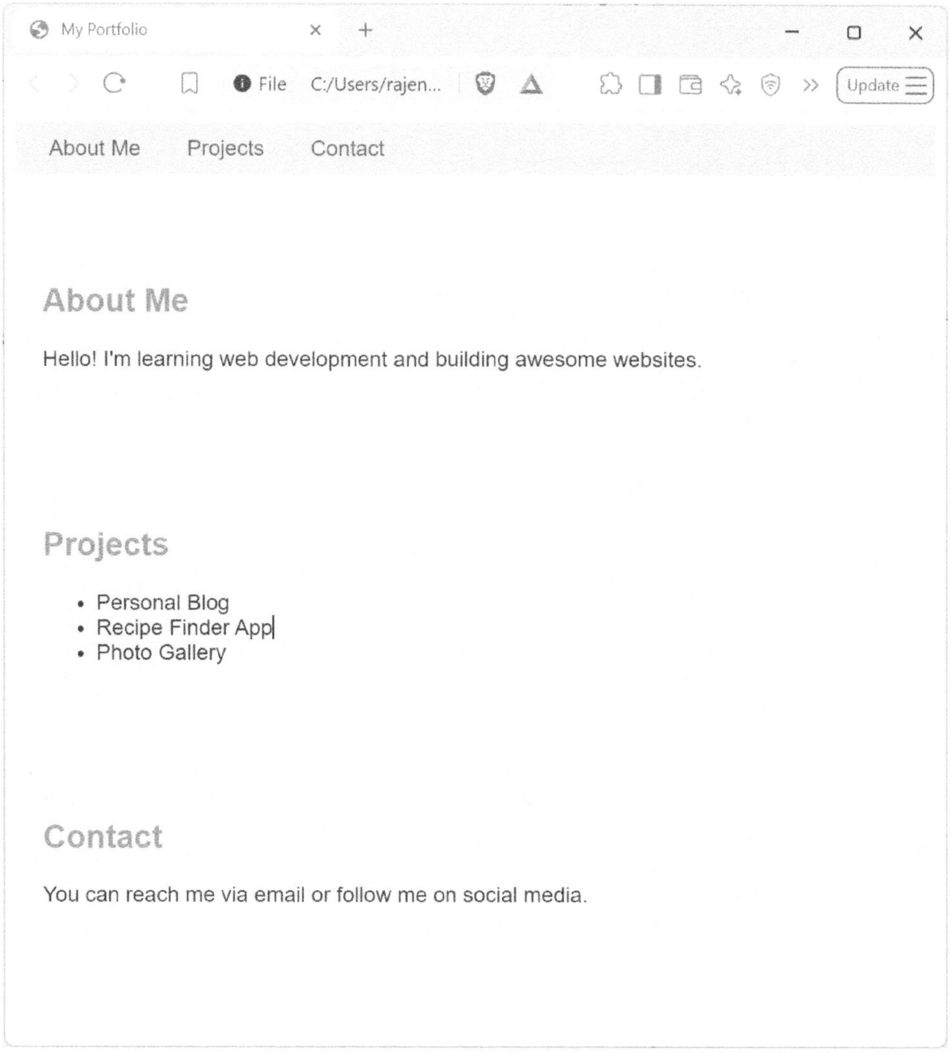

Real-World Example: Table of Contents for a Blog Post

Many blogs and documentation sites use anchor links to help readers jump between sections. Let's create a simple table of contents for a blog article.

```
<nav>
  <strong>Table of Contents:</strong>
  <a href= "#introduction">Introduction</a>
  <a href= "#tips">Top Tips</a>
  <a href= "#conclusion">Conclusion</a>
</nav>

<article>
  <h2 id= "introduction">Introduction</h2>
  <p>Welcome to my guide on smart web development...</p>

  <h2 id= "tips">Top Tips</h2>
  <ul>
```

```
    <li>Practice coding daily</li>
    <li>Use semantic HTML</li>
    <li>Test your website on different devices</li>
  </ul>

  <h2 id= "conclusion">Conclusion</h2>
  <p>Keep learning and experimenting to improve your skills!</p>
</article>
```

In this example, each heading uses a unique ``id`` so users can jump straight from the table of contents to any section.

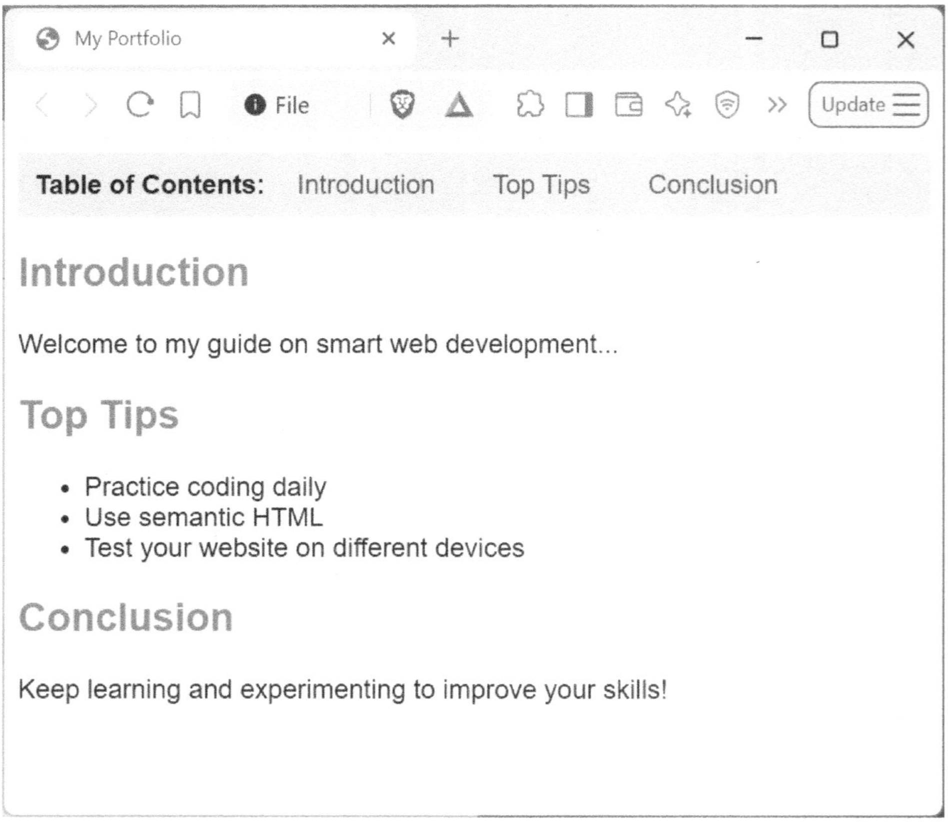

Styling Anchor Targets for Better User Experience

Sometimes, when you jump to a section, the heading might be hidden under a fixed navigation bar. You can use CSS to add some space above each anchor target for better visibility.

```
section {
  scroll-margin-top: 60px;
}
```

Add this CSS rule to your stylesheet if you have a fixed header, ensuring that the section titles are always visible after clicking an anchor link.

You've just learned how to:

- Use the ``id`` attribute to mark sections of your page.
- Create anchor links that jump to those sections for smooth in-page navigation.
- Practice this technique by building a portfolio or blog page with clear, clickable navigation.

Keep experimenting by adding more sections or creating a table of contents for any long web page you build. These skills will make your websites easier to navigate and more enjoyable for users.

Up next, you'll learn how to create links that let visitors contact you directly—by opening their email app or dialing your phone number. Get ready to make your pages even more interactive in the next section: "Linking to Email and Phone Numbers"!

3.5. Linking to Email and Phone Numbers

Have you ever visited a website and clicked a "Contact Me" link, only to see your email app pop up, ready to compose a message? Or maybe you've seen a phone number that you can tap on your mobile device to start a call instantly. These helpful, interactive features are powered by special types of HTML links. In this section, you'll learn how to easily add email and phone number links to your own web pages, making your sites more user-friendly and professional.

Let's dive in and create links that connect your visitors directly to you or your team—whether by email or phone. This is an essential step for anyone building personal portfolios, business sites, or even online stores.

How Email Links Work in HTML

To create a clickable email link, you'll use the **<a>** (anchor) tag, just like with regular web links. The difference is in the **href** attribute. Instead of starting with "http://", you'll use "mailto:", followed by the email address.

Basic Email Link Example:

```
<a href= "mailto:hello@example.com">Email Us</a>
```

When someone clicks this link, their default email client opens with a new message addressed to "hello@example.com".

Step-by-Step: Creating an Email Link

Start with the anchor tag: `<a>`
Use the **href** attribute, beginning with `mailto:`
Add the email address immediately after the colon
Insert the link text between the opening and closing anchor tags

Complete Example:

```
<a href= "mailto:support@mywebsite.com">Contact Support</a>
```

Try It Yourself:

Replace the email address with your own or a test address. Open the HTML file in your browser and click the link to see your email client launch.

Customizing Email Links with Subject and Body

You can pre-fill the subject line and even the email body for your visitors. This is a great way to guide users, especially for business sites or online portfolios.

How to Add a Subject:

Append `?subject=` to the email address in the **href**.

```
<a href= "mailto:designer@portfolio.com?subject=Project Inquiry">Email
the Designer</a>
```

How to Add a Body Message:

You can include a message body using `&body=` (note the ampersand if you already have a subject line).

```
<a href= "mailto:designer@portfolio.com?subject=Project
Inquiry&body=Hi, I am interested in your work!">Email the Designer</a>
```

Quick Tips:

Spaces in the subject or body are automatically encoded by browsers, but using `%20` for spaces ensures compatibility.
You can add more details, like cc or bcc, with `&cc=` or `&bcc=`.

Linking to Phone Numbers: The `tel:` Protocol

Making it easy for users to dial a number from your webpage is especially important for mobile visitors. Using the `tel:` protocol, you can turn any phone number into a clickable link.

Basic Phone Link Example:

```
<a href= "tel:+1234567890">Call Us Now</a>
```

When clicked (especially on a smartphone), this opens the phone app and fills in the number.

Step-by-Step: Creating a Phone Number Link

Use the anchor tag `<a>`
Set the **href** attribute to start with `tel:`
Add the full phone number (including country code for best results)
Place the visible phone number or call-to-action text inside the tags

Example with International Format:

```
<a href= "tel:+441234567890">Call Our London Office</a>
```

Best Practices:

Use the international format (`+` and country code) so users can call from anywhere.
For local audiences, you can use just the local number, but international is more universal.
Avoid extra spaces or dashes in the **href** value, though you can style the visible text for readability.

Real-World Example: Contact Section for a Portfolio Site

Let's put these skills into action by creating a simple "Contact Me" section for a personal portfolio, a common feature on web developer and designer sites.

HTML Markup:

```
<section id= "contact">
  <h2>Contact Me</h2>
  <p>
    <a href= "mailto:webdev.jane@example.com?subject=Portfolio
Inquiry">Email Me</a> |
    <a href= "tel:+15551234567">Call: (555) 123-4567</a>
  </p>
```

```
</section>
```

Basic CSS Styling:

```
#contact a {
  color: #1e90ff;
  text-decoration: none;
  font-weight: bold;
  padding: 0 8px;
  transition: color 0.2s;
}

#contact a:hover {
  color: #005fa3;
  text-decoration: underline;
}
```

How This Works:

Users can click “Email Me” to start an email with your address and a pre-filled subject.

Clicking the phone number on a mobile device will dial instantly.

The CSS styles the links to match modern web design trends, with a hover effect for interactivity.

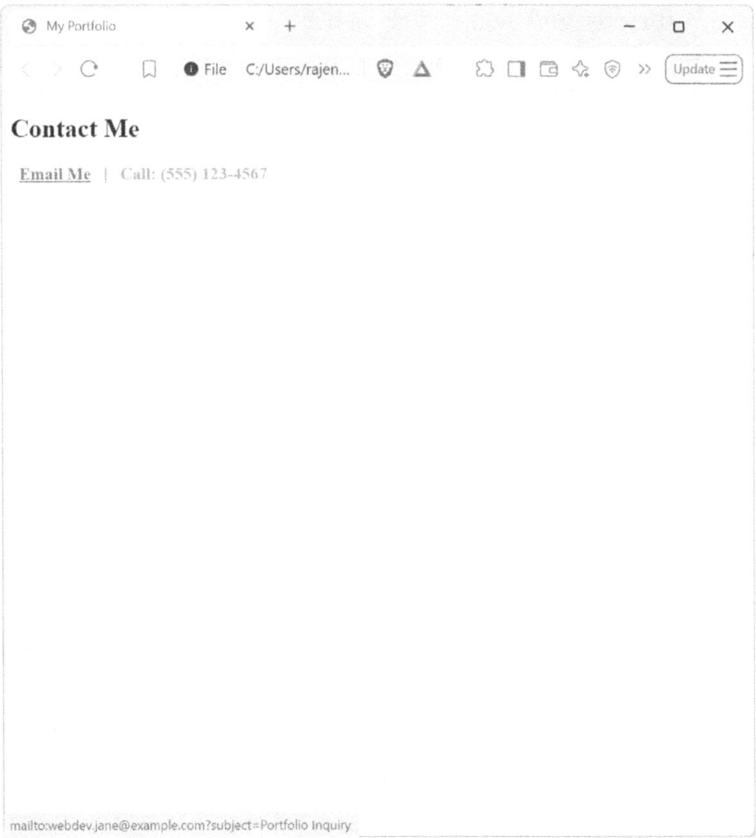

Real-World Scenario: E-commerce Store Support

Imagine you're building a small online shop. You want customers to reach you for order questions by email or phone.

HTML Example:

```
<footer>
  <p>
    Need help?
    <a href= "mailto:support@shoponline.com?subject=Order
Question">Email Support</a>
    or call
    <a href= "tel:+18005550123">1-800-555-0123</a>
  </p>
</footer>
```

This approach ensures customers can contact you easily—an essential feature for modern, user-focused websites.

You've now learned how to add interactive email and phone links to your web pages using the `mailto:` and `tel:` protocols in HTML. These features help visitors connect with you directly and make your website more functional and professional. Practice by adding contact links to your own pages and experimenting with different subjects and body messages for emails.

Next up, you'll take your site's usability to a new level by building a navigation bar for a portfolio site. Get ready to learn how to structure and style menus for seamless browsing—a key skill for every aspiring web developer!

3.6. Hands-On Exercise: Building a Navigation Bar for a Portfolio Site

A well-designed navigation bar is vital for any website, allowing visitors to move easily between different pages or sections. In this hands-on exercise, you'll create a navigation bar for a personal portfolio site using HTML lists and links. This practical example reflects real-world web development and gives you an immediate, useful project to showcase your new skills.

Understanding Navigation Bars in Modern Websites

Before writing code, let's look at how navigation bars are used on popular sites. For example, a designer's portfolio usually features a menu with links like "Home," "About," "Work," and "Contact." Similarly, blogs and e-commerce sites use navigation menus to help users quickly find articles or shop categories. By learning to build a navigation bar now, you're preparing to add professional, user-friendly features to your own projects.

Step 1: Structuring the Navigation Bar with HTML Lists

Navigation menus are typically built using HTML lists because they group related links in a structured way. We'll use an unordered list (`<ul>`) to create our menu items, with each link inside a list item (`<li>`).

Open your code editor and create a new HTML file named `portfolio.html`.
Inside the `<body>`, add a `<nav>` element to mark the navigation section.
Place a `<ul>` (unordered list) inside the `<nav>`.
Add several `<li>` elements, one for each menu item, and wrap each menu name in an `<a>` (anchor) tag for the clickable link.

Here's what your starter HTML should look like:

```
<!DOCTYPE html>
<html lang= "en">
<head>
  <meta charset= "UTF-8">
  <title>My Portfolio</title>
</head>
<body>
```

```
  <nav>
    <ul>
      <li><a href= "#home">Home</a></li>
      <li><a href= "#about">About</a></li>
      <li><a href= "#projects">Projects</a></li>
      <li><a href= "#contact">Contact</a></li>
    </ul>
  </nav>

</body>
</html>
```

The `<nav>` tag signals that this section is for navigation.

The `<ul>` groups the menu links.

Each `<li>` holds a single navigation link using the `<a>` tag.

The `href` attribute in each `<a>` points to a section of the page (using `#` plus the section name).

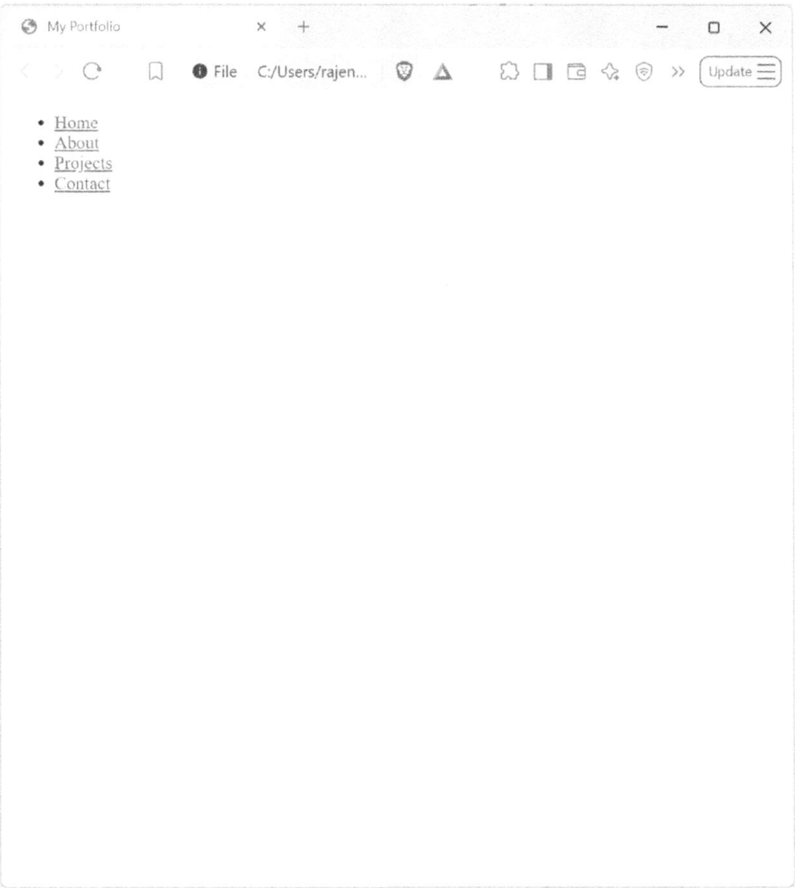

Step 2: Styling the Navigation Bar with CSS

Right now, the navigation bar looks like a plain, vertical list. Let's add some CSS to make it look sleek and horizontal, just like you see on modern portfolio websites.

Inside the `<head>`, add a `<style>` block to include your CSS.
Remove the default list styling.
Arrange the menu items side by side.
Style the links for better readability and interactivity.

Here's a basic CSS example to get you started:

```
<head>
  <meta charset= "UTF-8">
  <title>My Portfolio</title>
  <style>
    nav {
      background-color: #333;
      padding: 10px 0;
    }
    nav ul {
      list-style: none;
      margin: 0;
      padding: 0;
      display: flex;
      justify-content: center;
    }
    nav ul li {
      margin: 0 20px;
    }
    nav ul li a {
      color: #fff;
      text-decoration: none;
      font-size: 18px;
      padding: 8px 12px;
      transition: background 0.3s;
    }
    nav ul li a:hover {
      background-color: #555;
      border-radius: 4px;
    }
  </style>
</head>
```

`background-color: #333;` gives the navigation bar a dark background.
`display: flex;` and `justify-content: center;` place the menu items in a row and center them.
`list-style: none;` removes bullet points from the list.
`color: #fff;` makes the link text white, standing out against the dark background.
`:hover` adds a background effect when you mouse over a link, making the menu interactive.

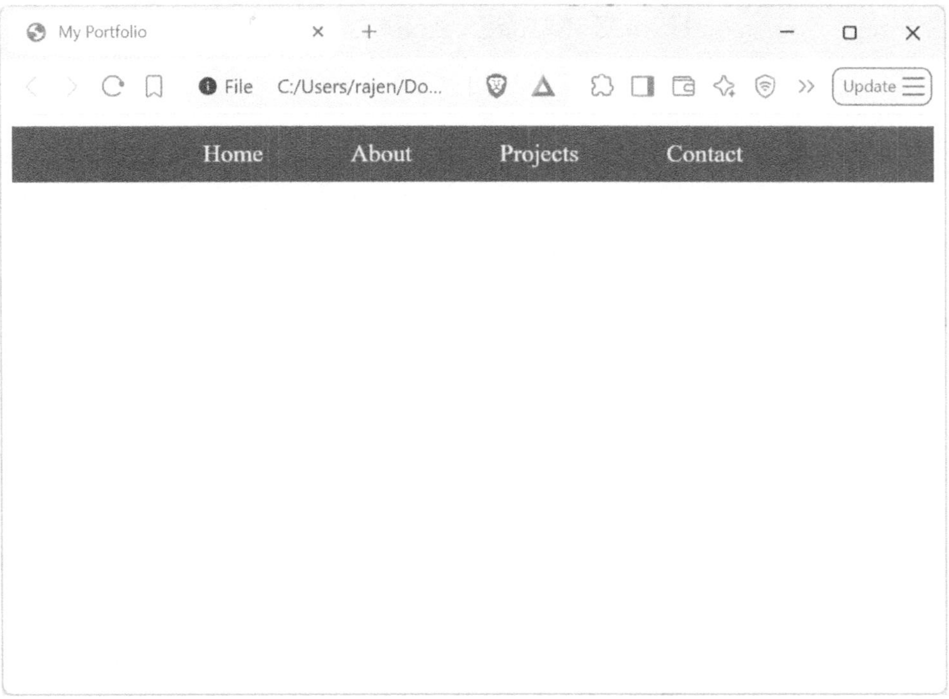

Step 3: Testing Your Navigation Bar

Open your `portfolio.html` file in a web browser. You should see a horizontal menu bar at the top, with all your navigation links spaced evenly and styled for a modern look. Hover your mouse over the links to see the interactive color change.

Step 4: Expanding the Navigation Bar

Many professional sites, like tech blogs or online stores, expand their navigation bars with extra features. You can add dropdown menus, icons, or responsive design for mobile devices as your skills grow. For now, try these beginner-friendly enhancements:

Change the colors to match your personal style.
Add more sections to the menu, such as "Blog" or "Gallery."
Link to real pages by changing the `href` values.

You've just built a navigation bar using HTML lists and links, then styled it with CSS for a polished, modern look. This exercise taught you how to:

- Structure a navigation menu with `<ul>`, `<li>`, and `<a>` tags.
- Apply CSS for layout, color, and interactive effects.
- Preview and refine your work in a browser, reinforcing the cycle of code-test-improve that's core to web development.

4. USING IMAGES AND MULTIMEDIA

A great website doesn't just deliver information—it creates an engaging experience that captures attention and brings content to life. One of the most effective ways to achieve this is by adding images and multimedia elements. In this chapter, you'll unlock the powerful tools that allow you to transform plain web pages into visually rich presentations, just like the modern sites you visit every day.

You'll learn how to display images using the versatile **img** tag and discover how to add meaningful captions with **figure** and **figcaption** for a polished, accessible look. Moving beyond static visuals, you'll explore embedding videos and audio clips, giving your pages interactivity and energy. As you progress, you'll also see how to include scalable vector graphics (SVGs) and crisp icons, which are essential for designing sharp, professional interfaces that look great on any device.

These skills are central to creating everything from eye-catching portfolios and vibrant blogs to dynamic e-commerce product galleries. By the end of this chapter, you'll not only understand the technical details but also gain hands-on practice by building your own visual gallery web page—a project designed to anchor what you've learned and boost your creative confidence.

Whether you're aiming to showcase your work, highlight products, or simply make your website more inviting, mastering multimedia is a must. Get ready to expand your toolkit and take your website designs to the next level, as you discover how images and media turn good sites into great ones.

4.1. Displaying Images with the <img> Tag

Images bring web pages to life, transforming plain text into engaging, visually rich experiences. Whether you want to showcase a product, add your profile photo, or illustrate a blog post, learning how to display images using HTML is a crucial skill for all aspiring web developers. In this section, you'll master the **`<img>` tag**, the essential building block for adding images to your web pages. Let's dive in and make your websites more vibrant and interactive.

Understanding the `<img>` Tag

The **`<img>` tag** is an inline HTML element that embeds images into your web page. Unlike some tags, `<img>` doesn't wrap around other content and doesn't have a closing tag. Instead, it uses attributes to define what image to display and how it appears.

The two most important attributes for the **`<img>` tag** are:

`src` (source): Specifies the path to the image file.
`alt` (alternative text): Provides a text description of the image for screen readers and displays if the image can't load.

Here's a basic example:

```
<img src= "images/sunset.jpg"alt= "A beautiful sunset over the mountains">
```

Adding Your First Image: Step-by-Step

Let's walk through the process of adding an image to your webpage. For this exercise, imagine you're building a personal portfolio and want to display your profile photo.

Choose or Download an Image

Save an image you'd like to use in your project folder. For this example, let's use a file called `profile.jpg` stored inside an `images` folder.

Write the HTML Code

Open your HTML file and decide where you want the image to appear. Add the `<img>` tag as shown below:

```
<img src= "images/profile.jpg"alt= "Photo of Jane Doe, Web Developer">
```

Save and View

Save your file and open it in your web browser. You should see your photo displayed at the chosen spot.

Key points to remember:

The `src` path must be correct; otherwise, the image will not show.
The `alt` text helps with accessibility and SEO.

Practical Scenario: Adding a Product Image to a Simple Online Store

Images are essential in e-commerce sites. Let's practice by adding a product photo to a simple product card.

HTML Example:

```
<div class= "product-card">
  <h2>Handmade Ceramic Mug</h2>
  <img src= "images/ceramic-mug.jpg"alt= "A handmade ceramic coffee mug
with a blue glaze">
  <p>Enjoy your favorite drink in this artisan-crafted mug.</p>
  <p><strong>Price:</strong> $18.00</p>
</div>
```

CSS Example:

```
.product-card {
  border: 1px solid #ddd;
  padding: 16px;
  max-width: 250px;
  text-align: center;
  background-color: #fafafa;
}
.product-card img {
  width: 180px;
  height: auto;
  border-radius: 8px;
  margin-bottom: 12px;
}
```

Try creating a folder named `images`, add a picture called `ceramic-mug.jpg`, and use the code above to see how images enhance your product display.

Using Images from the Web

Sometimes you might want to use images hosted online, not just those saved on your computer. You can use an absolute URL in the `src` attribute. For example:

```
<img src= "https://www.example.com/assets/logo.png"alt= "Company Logo">
```

Be sure you have permission to use online images and that they are from reliable sources.

Sizing and Styling Images with CSS

While HTML displays the image, CSS lets you control its size, shape, and appearance. Let's practice resizing and styling an image for a personal blog.

HTML Example:

```
<img src= "images/blog-header.jpg"alt= "Colorful city skyline at
night">
```

CSS Example:

```
img {
  width: 100%;
  max-width: 600px;
  height: auto;
  border: 4px solid #333;
  border-radius: 10px;
  display: block;
  margin: 24px auto;
}
```

This code ensures your image is responsive, never wider than 600 pixels, and centered with a stylish border. Responsive images look great across all devices, from phones to desktops.

Real-World Example: Portfolio Gallery

Many web developers showcase their work with image galleries. Here's how to display a small gallery of project screenshots.

HTML Example:

```
<div class= "gallery">
  <img src= "images/project1.png"alt= "Screenshot of Project 1">
  <img src= "images/project2.png"alt= "Screenshot of Project 2">
  <img src= "images/project3.png"alt= "Screenshot of Project 3">
</div>
```

CSS Example:

```
.gallery {
  display: flex;
  gap: 16px;
  justify-content: center;
}
.gallery img {
  width: 120px;
  height: 80px;
  object-fit: cover;
  border: 2px solid #ccc;
  border-radius: 6px;
```

```
  transition: border-color 0.2s;
}
.gallery img:hover {
  border-color: #0077cc;
}
```

This creates a row of equally sized images that highlight your projects and add interactivity with a hover effect.

You've just learned how to display images on your web pages using the `<img>` tag, set alternative text for accessibility, use both local and online images, and style images with CSS for a professional look. Try adding different images to your own practice pages—experiment with sizes, borders, and layouts to see what works best for your style and content.

Next, you'll take your image skills further by learning how to add captions using the `<figure>` and `<figcaption>` tags, making your visuals even more informative and accessible. Get ready to make your web pages stand out with beautiful, well-described images!

4.2. Adding Captions with <figure> and <figcaption>

Images are essential for making your web pages visually engaging, but they're even more powerful when paired with meaningful captions. Captions help users understand the context of an image, making your site more accessible and user-friendly. HTML provides two special tags, `**<figure>**` and `**<figcaption>**`, that allow you to group images with their captions in a semantic and visually organized way. In this section, you'll learn to use these tags to enhance your web pages, with hands-on coding you can practice right now.

Why Use `<figure>` and `<figcaption>`?

Before diving into code, let's understand why these tags are important:

- `**<figure>**` groups your image (or other media) and its caption together. This helps search engines and assistive technologies understand that the image and caption belong together.
- `**<figcaption>**` provides a clear, semantic way to add a caption directly associated with the image.

This approach is more accessible than simply placing a `**<p>**` tag below your image, and it's easier to style consistently with CSS.

Basic Structure: Pairing Images with Captions

Let's start by creating a simple example where you display an image with a custom caption. You'll follow these steps:

Open your HTML editor and create a new file called `figure-example.html`.
Add the following code to practice using `<figure>` and `<figcaption>`.

```
<!DOCTYPE html>
<html lang="en">
<head>
  <meta charset="UTF-8">
  <title>Figure and Figcaption Example</title>
</head>
<body>
  <figure>
    <img src="images/sunset.jpg" alt="A vibrant sunset over a calm
lake">
    <figcaption>A beautiful sunset view, perfect for
relaxation.</figcaption>
  </figure>
</body>
</html>
```

The `**<figure>**` tag wraps both the image and its caption.

The `**<figcaption>**` tag sits directly inside the `<figure>`, describing the image above it.

You can replace the `src` with any image path you have, and adjust the caption for your own photos or web projects.

Styling Figures and Captions with CSS

Now, let's make your figure and caption look great. Here's how to enhance the layout and typography with CSS:

Add a `<style>` block inside your `<head>` section:

```
<style>
  figure {
    max-width: 400px;
    margin: 2rem auto;
    padding: 1rem;
    background: #f9f9f9;
    border: 1px solid #eee;
    border-radius: 8px;
    text-align: center;
  }
  figure img {
    max-width: 100%;
    border-radius: 4px;
  }
```

```
  figcaption {
    margin-top: 0.75rem;
    color: #555;
    font-size: 1rem;
    font-style: italic;
  }
</style>
```

The `**figure**` selector centers your image and gives it a nice background and border.

The `**figure img**` selector ensures your image fits nicely within the border.

The `**figcaption**` selector styles your caption, making it easy to read and visually distinct.

Try editing the caption text or swapping the image to see how your styles stay consistent.

Real-World Example: Portfolio Image Gallery

Imagine you're building a **portfolio website** to showcase your photography or design work. You can use `<figure>` and `<figcaption>` to create a professional gallery section, where each image has a descriptive caption.

Here's a hands-on example:

```
<section>
  <h2>My Portfolio</h2>
  <figure>
    <img src="images/forest-path.jpg" alt="A winding path through a
dense green forest">
    <figcaption>Morning walk through the forest, capturing the essence
of nature.</figcaption>
  </figure>
  <figure>
    <img src="images/city-lights.jpg" alt="City skyline at night with
bright lights">
    <figcaption>City skyline illuminated by sparkling night
lights.</figcaption>
  </figure>
  <figure>
    <img src="images/beach-day.jpg" alt="Sunny day at the beach with
waves and blue sky">
    <figcaption>Relaxing day at the coast, enjoying the sun and
surf.</figcaption>
  </figure>
</section>
```

Each image is grouped with its own caption, making your gallery both attractive and accessible.

This structure is easy for visitors (and search engines) to navigate and understand.

You can further style your gallery with CSS to create columns or grids, but for now, focus on getting comfortable with the basic structure.

Real-World Coding Scenario: Blog Post with Infographic

Many **blogs** use infographics or charts to explain data or ideas. Let's add a chart with a caption to a blog post so readers know exactly what they're seeing.

```
<article>
  <h2>The Growth of Web Development</h2>
```

```
  <p>Over the past decade, web development has seen remarkable growth. The following chart illustrates this trend:</p>
  <figure>
    <img src="images/web-growth-chart.png" alt="Line chart showing growth in web development jobs from 2010 to 2024">
    <figcaption>Figure 1: Steady increase in web development job postings between 2010 and 2024.</figcaption>
  </figure>
  <p>This trend highlights the importance of learning HTML and CSS today!</p>
</article>
```

Notice how the **`<figcaption>`** gives context to the chart. This is especially helpful for readers quickly scanning your post or for those using screen readers.

Hands-On Exercise

Practice adding figures and captions to your own project:

Choose two or three images you like (they can be photos, graphics, or charts).
For each image, write a descriptive caption.
Create a new HTML section using the `<figure>` and `<figcaption>` tags for each image.

Apply the CSS styles provided above to make your figures stand out.

Experiment by changing the layout, caption text, or images. This will help you build confidence in using these semantic HTML elements.

You've now learned to use `**<figure>**` and `**<figcaption>**` to group images with meaningful captions, making your pages more attractive and accessible. Styling these elements with CSS helps your images and captions look professional and consistent across your site. Practicing with real-world examples like portfolios and blog articles prepares you to use these tags in any project.

Next, you'll take your multimedia skills even further by learning how to **embed videos and audio** directly into your web pages. Get ready to make your websites even more dynamic and interactive with rich media elements!

4.3. Embedding Videos and Audio

Adding video and audio to your website not only captivates visitors, but also provides richer ways to share information, tell stories, or showcase creative work. Whether you want to feature a product demo, share a podcast, or simply add a background music track, HTML makes it possible with just a few simple elements. In this section, you'll learn how to embed videos and audio files step by step, using real-world scenarios that you can instantly apply to your own projects.

Understanding Multimedia Elements in HTML

HTML5 introduced the **video** and **audio** elements, making it much easier to add media without plugins like Flash. These elements are now standard on modern websites and work across all major browsers and devices.

The `**<video>**` tag is used to embed video files.
The `**<audio>**` tag is used to embed audio files.
Both elements support multiple file formats and offer built-in controls (like play, pause, and volume).

Let's walk through how to use these elements in your code.

Embedding a Video: Step-by-Step

Imagine you're building a personal portfolio site and want to showcase a short introduction video about yourself. Here's how you can embed a video using HTML:

First, save your video file (for example, `intro.mp4`) in your project folder.
Insert the `**<video>**` element in your HTML, providing the file path.

Add the **controls** attribute, so visitors can play and pause the video.

```
<video src= "images/intro.mp4"controls width= "600">
  Sorry, your browser doesn't support embedded videos.
</video>
```

src: Path to your video file.
controls: Displays play, pause, and volume buttons.
width: Sets the width of the video on the page.

You can also offer multiple video formats for better browser compatibility:

```
<video controls width= "600">
  <source src= "images/intro.mp4"type= "video/mp4">
  <source src= "images/intro.webm"type= "video/webm">
  Sorry, your browser doesn't support embedded videos.
</video>
```

This way, the browser will play the first supported format it finds.

Adding a Poster Image

A poster image is a thumbnail that displays before the video plays. This is especially helpful for portfolios or e-commerce sites where you want the page to look polished.

```
<video controls width= "600"poster= "thumbnail.jpg">
  <source src= " images/intro.mp4"type= "video/mp4">
  Sorry, your browser doesn't support embedded videos.
</video>
```

poster: Sets the preview image before the video is played.

Customizing Video with CSS

Let's add a simple border and some spacing around the video, making it stand out on your page.

```
video {
  border: 3px solid #3498db;
  border-radius: 8px;
  margin-bottom: 20px;
  box-shadow: 0 4px 12px rgba(0,0,0,0.15);
}
```

Add this CSS to your stylesheet or within a `<style>` block in your HTML.

Real-World Example: Embedding a Tutorial Video in a Blog

Suppose you run a tech blog and want to include a coding tutorial video. You would add the video inside your blog post like this:

```
<article>
  <h2>How to Build Your First Web Page</h2>
  <video controls width= "500"poster= "tutorial-thumbnail.png">
    <source src= " images/webpage-tutorial.mp4"type= "video/mp4">
    <source src= " images/webpage-tutorial.ogg"type= "video/ogg">
    Your browser does not support the video tag.
  </video>
  <p>Watch this video to learn the basics of HTML and CSS.</p>
</article>
```

This approach engages your readers and makes learning more interactive.

Embedding Audio: Step-by-Step

Podcasts, music samples, or sound effects can add personality and depth to your site. The `**<audio>**` element works similarly to the video element.

Place your audio file (for example, `background-music.mp3`) in your project folder. Insert the `**<audio>**` element in your HTML, with the **controls** attribute.

```
<audio src= "background-music.mp3"controls>
  Sorry, your browser doesn't support embedded audio.
</audio>
```

You can support different audio formats for compatibility:

```
<audio controls>
  <source src= "background-music.mp3"type= "audio/mp3">
  <source src= "background-music.ogg"type= "audio/ogg">
  Your browser does not support the audio element.
</audio>
```

Customizing Audio Player Appearance

While the default audio player is provided by the browser, you can still adjust its spacing or style it with CSS for a better fit in your design.

```
audio {
  width: 300px;
  margin-top: 15px;
  display: block;
}
```

Real-World Example: Adding a Podcast Player to Your Portfolio

If you're creating a portfolio site and want to showcase your podcast episode, add an audio player like this:

```
<section>
  <h3>Listen to My Latest Podcast Episode</h3>
  <audio controls>
    <source src= "episode1.mp3"type= "audio/mp3">
    <source src= "episode1.ogg"type= "audio/ogg">
    Sorry, your browser doesn't support audio playback.
  </audio>
</section>
```

This lets visitors listen directly from your site, encouraging more engagement.

Accessibility Considerations

Including **text alternatives** ensures your site is accessible:

For videos, provide captions or a transcript.
For audio, offer a transcript below the player.

Example:

```
<p><strong>Transcript:</strong> Welcome to my podcast! In this episode,
we'll explore...</p>
```

Quick Coding Exercise: Embed Both Video and Audio

Try adding both a video and an audio player to a sample web page, using your own media files or free online samples. Apply the CSS styles above to see how they change the appearance of your media elements.

Summary: Practice Makes Perfect

You've now learned how to embed videos and audio files in your web pages using simple HTML elements and enhanced their appearance with CSS. Practice by adding multimedia to your personal or demo site, experiment with different file types, and adjust styling to match your vision.

Next, you'll discover how to add visual flair to your site with icons and SVG graphics. Get ready to bring even more creativity and modern style to your web projects in the upcoming section!

4.4. Inserting Icons and SVG Graphics

Icons and SVG graphics are powerful tools for modern web development, adding visual interest and intuitive navigation to any website. As you move beyond basic images, learning how to insert scalable icons and SVG graphics will help your projects look professional and polished on any device. In this section, you'll master the essentials of adding icons and SVGs to your web pages, using real-world examples and hands-on practice.

Understanding Icons and SVG Graphics

Before we start coding, let's clarify what icons and SVG graphics are:

Icons are small, simple images or symbols used to represent actions, features, or content. Think of the magnifying glass for "search" or the envelope for "email." They help users quickly understand your site.

SVG (Scalable Vector Graphics) is an image format that uses XML to describe shapes, lines, and colors. SVGs are sharp at any size and are perfect for logos, icons, and illustrations.

Many modern websites use SVGs for their icons because they scale beautifully and are easy to style with CSS.

Why Use Icons and SVGs?

They improve user experience by making navigation clearer.
SVGs load quickly and look crisp on all screen sizes and resolutions.
You can style and animate SVGs with CSS for interactive effects.

Let's see how to use both built-in web icons and custom SVG graphics.

Adding Icons with Font Libraries: Font Awesome

One of the easiest ways to add icons is by using a popular icon library like **Font Awesome**. This library provides hundreds of ready-to-use icons you can insert with a single line of HTML.

Let's add a Font Awesome icon to a navigation bar, just like you'd see on a portfolio or personal website.

Include Font Awesome in Your HTML

Add the following line inside the `<head>` of your HTML file to load Font Awesome from a Content Delivery Network (CDN):

```
<link rel= "stylesheet"href=
"https://cdnjs.cloudflare.com/ajax/libs/font-
awesome/6.4.0/css/all.min.css">
```

Insert an Icon

To display a simple home icon (often used for the homepage link), use the `<i>` element with appropriate classes:

```
<nav>
  <ul>
    <li><a href= "#"><i class= "fas fa-home"></i> Home</a></li>
    <li><a href= "#"><i class= "fas fa-user"></i> About</a></li>
    <li><a href= "#"><i class= "fas fa-envelope"></i> Contact</a></li>
  </ul>
</nav>
```

The classes `fas fa-home`, `fas fa-user`, and `fas fa-envelope` tell Font Awesome which icons to display.

Style the Navigation Bar with CSS

Let's make the navigation look a bit nicer:

```
nav ul {
  list-style: none;
  padding: 0;
  display: flex;
  gap: 1.5rem;
}

nav a {
  text-decoration: none;
  color: #333;
  font-size: 1.2rem;
  display: flex;
  align-items: center;
}

nav i {
  margin-right: 0.5rem;
  color: #0077cc;
}
```

Now, your navigation menu has stylish, easy-to-understand icons!

Embedding Custom SVG Graphics Directly in HTML

While icon libraries are convenient, sometimes you want complete control over your graphics. SVG files are perfect for custom icons, logos, or illustrations.

Here's how to embed an SVG graphic directly in your HTML:

Copy the SVG Code

Suppose you want to add a simple heart icon to a “Like” button, as seen on e-commerce or blog comment sections. You can find free SVG icons on sites like [Heroicons](https://heroicons.com/) or [SVG Repo](https://www.svgrepo.com/). Here’s a basic heart SVG:

```
<button class= "like-btn">
  <svg width= "24"height= "24"viewBox= "0 0 24 24"fill= "none"
       xmlns= "http://www.w3.org/2000/svg">
    <path d= "M12 21s-6.5-5.5-9-9.5C2.5 7 5 4 8.5 4c1.7 0 3.4 1 4.5
2.6C14.1 5 15.8 4 17.5 4 21 4 23.5 7 21 11.5 18.5 15.5 12 21 12 21z"
         stroke= "#e63946"stroke-width= "2"fill= "#fff"/>
  </svg>
  Like
</button>
```

Style the SVG with CSS

Now, let’s add some style to make the like button interactive:

```
.like-btn {
  background: none;
  border: 1px solid #e63946;
  padding: 0.5rem 1rem;
  border-radius: 24px;
  color: #e63946;
  font-size: 1rem;
  display: flex;
  align-items: center;
  cursor: pointer;
  transition: background 0.2s;
}

.like-btn:hover {
  background: #ffe5e9;
}

.like-btn svg {
  margin-right: 0.5rem;
  transition: fill 0.2s;
}

.like-btn:hover svg path {
  fill: #e63946;
  stroke: #fff;
}
```

When the user hovers over the button, the heart fills with color, creating engaging feedback.

Real-World Example: Blog Social Share Icons

Let's put your new skills into practice with a scenario. Imagine you're building a blog post page and want to add social media icons for sharing. You can use a combination of icon libraries and SVG graphics for a sleek look.

Here's how you might add Twitter and Facebook icons using Font Awesome in your blog's footer:

```
<footer>
  <p>Share this post:</p>
  <a href= "https://twitter.com/"target= "_blank">
    <i class= "fab fa-twitter"></i>
  </a>
  <a href= "https://facebook.com/"target= "_blank">
    <i class= "fab fa-facebook"></i>
  </a>
</footer>
```

And a quick CSS snippet to style the icons:

```
footer a {
  margin-right: 1rem;
  color: #555;
  font-size: 1.5rem;
  transition: color 0.2s;
}

footer a:hover {
  color: #1da1f2; /* Twitter blue */}
```

This technique is used on thousands of blogs and online magazines to encourage sharing and boost engagement.

Tips for Working with SVGs and Icons

SVGs can be edited directly in your HTML, giving you full control over shapes and colors.

Use CSS to animate SVGs for interactive effects, such as changing fill color on hover.

Keep your SVG code tidy; only include the shapes you need for your icon.

Always provide alternative text or descriptive labels for icons, especially if they convey important meaning.

Quick Practice: Replace a PNG Logo with an SVG

Try updating your site's logo from a static image (like PNG or JPEG) to an SVG. Find or create an SVG logo, then place the SVG code directly inside your `<header>`. Use CSS to adjust its size and color, ensuring it looks sharp on all screens.

You've just learned how to insert both icon fonts and SVG graphics, making your web pages stylish, modern, and responsive. By practicing these techniques, you'll gain confidence in designing navigation menus, buttons, and engaging UI elements found on professional websites.

Up next, you'll put these new skills to the test by creating a visual gallery web page. Get ready to combine images, icons, and creative layouts to showcase your coding abilities in a real-world project!

4.5. Hands-On Exercise: Creating a Visual Gallery Web Page

Before we dive in, let's imagine you're building a beautiful online portfolio or a travel blog. Both rely on eye-catching images to tell a story and engage visitors. In this hands-on exercise, you'll learn how to create a visual gallery web page. You'll gain practical experience adding images, styling them with CSS, and arranging them in a gallery layout. By the end, you'll have a project you can personalize and expand, just like the image galleries you see on popular portfolio and travel websites.

Preparing Your Image Files

To start, you'll need a few image files. You can use your own photos or download free images from trusted sites like Unsplash or Pexels. Save your images in a dedicated folder named `images` within your project directory for easy access.

For this exercise, let's assume you have three images:

`sunset.jpg`
`mountain.jpg`
`forest.jpg`

Building the HTML Structure for the Gallery

Let's set up the basic HTML for your gallery page. You'll use the **`<img>`** tag to display images and a container **`<div>`** to organize your gallery.

Open your code editor and create a new file named `gallery.html`.
Add the following HTML structure:

```
<!DOCTYPE html>
<html lang= "en">
<head>
  <meta charset= "UTF-8">
  <title>My Visual Gallery</title>
  <link rel= "stylesheet"href= "gallery.css">
</head>
<body>
  <h1>My Visual Gallery</h1>
```

```
  <div class= "gallery">
    <img src= "images/sunset.jpeg"alt= "Beautiful sunset over the
ocean">
    <img src= "images/mountain.jpeg"alt= "Snow-capped mountain peak">
    <img src= "images/forest.jpeg"alt= "Lush green forest">
  </div>
</body>
</html>
```

The `**<div class="gallery">**` groups your images, making them easier to style.

The `**alt**` attribute provides a text description, which is important for accessibility and for search engines.

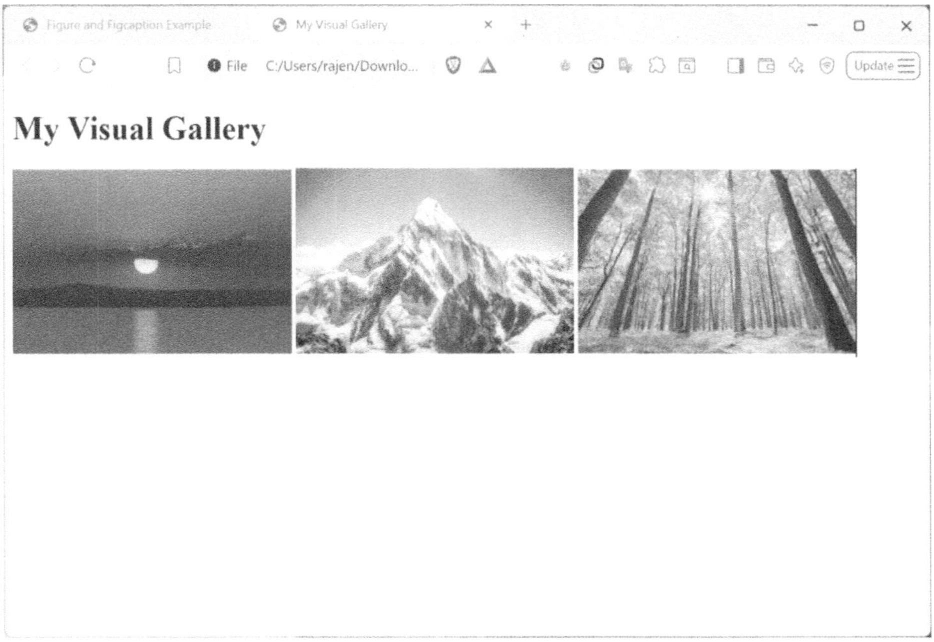

Styling Your Gallery with CSS

Now, let's make your gallery look polished and professional using **CSS**. You'll create a separate CSS file to keep your styles organized.

In the same project folder, create a new file named `gallery.css`.

Add the following CSS to style your gallery:

```
body {
  font-family: Arial, sans-serif;
  background-color: #f9f9f9;
  margin: 0;
  padding: 20px;
}

h1 {
  text-align: center;
```

```
  color: #333;
}

.gallery {
  display: flex;
  justify-content: center;
  gap: 20px;
  margin-top: 30px;
}

.gallery img {
  width: 250px;
  height: 170px;
  object-fit: cover;
  border-radius: 10px;
  box-shadow: 0 4px 12px rgba(0,0,0,0.1);
  transition: transform 0.2s, box-shadow 0.2s;
  cursor: pointer;
}

.gallery img:hover {
  transform: scale(1.05);
  box-shadow: 0 6px 24px rgba(0,0,0,0.2);
}
```

The **`.gallery`** class uses **Flexbox** to arrange images in a row with spacing between them.

`object-fit: cover;` ensures your images fill their boxes neatly without distortion.

The **`:hover`** effect adds interactivity, making your gallery more engaging.

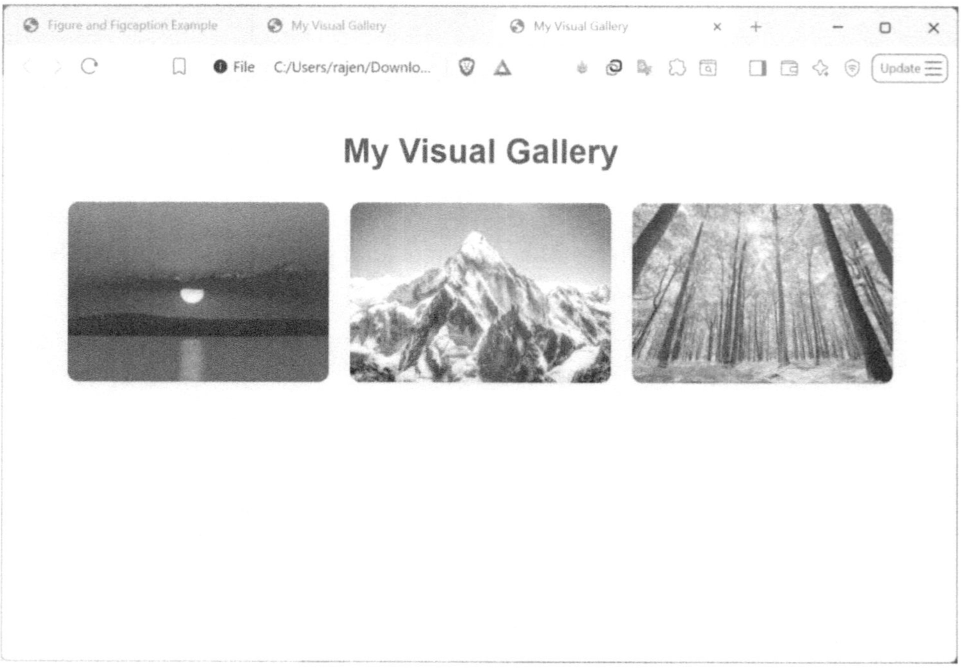

Making Your Gallery Responsive

A modern web gallery should look great on any device, from large screens to smartphones. Let's add a responsive touch using a CSS media query:

```
@media (max-width: 700px) {
  .gallery {
    flex-direction: column;
    align-items: center;
  }
  .gallery img {
    width: 90vw;
    max-width: 350px;
    height: auto;
    margin-bottom: 20px;
  }
}
```

This code changes the gallery layout to a column on small screens, making images stack vertically.

`width: 90vw;` makes each image fill most of the viewport width on mobile devices.

Customizing Your Gallery

You can expand your gallery by adding more images or captions. For example, to add a caption under each image, nest each image and caption inside a **`<figure>`** and use a **`<figcaption>`**:

```
<div class= "gallery">
  <figure>
    <img src= "images/sunset.jpeg"alt= "Beautiful sunset over the
ocean">
    <figcaption>Sunset at the beach</figcaption>
  </figure>
  <figure>
    <img src= "images/mountain.jpeg"alt= "Snow-capped mountain peak">
    <figcaption>Majestic Mountain</figcaption>
  </figure>
  <figure>
    <img src= "images/forest.jpeg"alt= "Lush green forest">
    <figcaption>Green Forest</figcaption>
  </figure>
</div>
```

Then update your CSS for the new structure:

```
.gallery {
  display: flex;
  justify-content: center;
  gap: 20px;
  margin-top: 30px;
}
```

```
.gallery figure {
  display: flex;
  flex-direction: column;
  align-items: center;
  margin: 0;
}

.gallery figcaption {
  margin-top: 8px;
  font-size: 1em;
  color: #555;
}
```

This approach is used on many modern photo portfolio websites, giving each image context and a professional look.

Real-World Inspiration: Travel Blog and Portfolio Sites

Many travel blogs, such as "Nomadic Matt," use image galleries to showcase destinations. Similarly, creative portfolios like "Behance" feature neat, responsive galleries to highlight photography or design work. By following these steps, you are learning the same techniques used in real-world websites.

Try It Yourself: Expand Your Gallery

To reinforce your skills, try these challenges:

Add more images to your gallery.
Experiment with different border styles or hover effects in your CSS.
Replace the images and captions with your own photos and descriptions.

Hands-On Summary

By building a visual gallery, you've learned to use the **`<img>`** tag for images, organize content with containers, and apply CSS for layout, style, and responsiveness. You've also seen how adding hover effects and captions can make your gallery more engaging and accessible. Practicing these techniques will help you create attractive, user-friendly web pages.

Next, you'll wrap up this chapter by reviewing the key concepts you've learned and taking on a creative coding challenge to test your new skills. Get ready to bring everything together and see how far you've come as a web developer!

5. ORGANIZING CONTENT WITH HTML ELEMENTS

Every great website relies on more than just content—it depends on how that content is organized and presented to visitors. As you continue your journey toward becoming a skilled web developer, this chapter focuses on one of the most essential skills you'll need: structuring your web pages for clarity, usability, and professionalism. Here, you'll discover how to group information using versatile elements like **<div>** and **<span>**, and how to take your skills further with modern, meaningful **semantic HTML5 elements** such as **<header>**, **<nav>**, and **<article>**.

Mastering these building blocks will empower you to create web pages that are not only visually appealing but also easy for users—and search engines—to navigate. You'll see firsthand how semantic tags make your code cleaner and more readable, laying the groundwork for websites that stand out in today's competitive digital landscape. We'll also introduce you to the fundamentals of accessible web design, ensuring your creations are usable for everyone, including visitors with disabilities.

Whether your goal is to design a personal portfolio, launch a blog, or build a news site, the techniques you learn here are used by professional web developers every day. By the end of this chapter, you'll put your new skills into practice with a hands-on project: structuring a news or magazine article using the concepts and tags you've learned. Get ready to take the next step in transforming your ideas into organized, professional-quality web pages.

5.1. Dividing Content with <div> and <span>

When you start building web pages, you'll notice that you need ways to group and organize content. Whether it's wrapping a bunch of text to apply a background color, creating columns for a portfolio gallery, or highlighting a small piece of text within a paragraph, HTML offers two incredibly useful elements: **<div>** and **<span>**.

These elements are called "generic containers." They don't change how your content looks on their own, but they give you the power to organize, group, and style different parts of your webpage with CSS. Let's explore how to use them effectively, step by step.

What Are <div> and <span>?

Before you jump into code, it's important to understand what makes **<div>** and **<span>** special:

<div> (short for "division") is a block-level element. It starts on a new line and stretches the full width of its parent container. Use **<div>** to group larger chunks of content.

<span> is an inline element. It stays within the flow of surrounding text, only taking up as much space as its content. Use **<span>** to style or target small pieces of content within a line.

Think of **<div>** as a box to organize sections, while **<span>** is a highlighter for words or phrases.

Using <div> to Group Webpage Sections

Let's try a practical example that mimics a simple personal portfolio layout you might find on a real website. Imagine you want to separate your introduction, portfolio projects, and contact information.

Open your code editor and follow these steps:

Start with a basic HTML structure:

```
<!DOCTYPE html>
<html>
<head>
  <title>My Portfolio</title>
  <link rel="stylesheet" href="styles.css">
</head>
<body>
  <!-- Content will go here -->
</body>
</html>
```

Inside the `` `<body>` ``, add three sections using `` `<div>` ``:

```
<body>
  <div>
    <h1>Jane Doe</h1>
    <p>Welcome to my web portfolio!</p>
  </div>
  <div>
    <h2>Projects</h2>
    <ul>
      <li>Responsive Website</li>
      <li>Photo Gallery</li>
    </ul>
  </div>
  <div>
    <h2>Contact Me</h2>
    <p>Email: jane@example.com</p>
  </div>
</body>
```

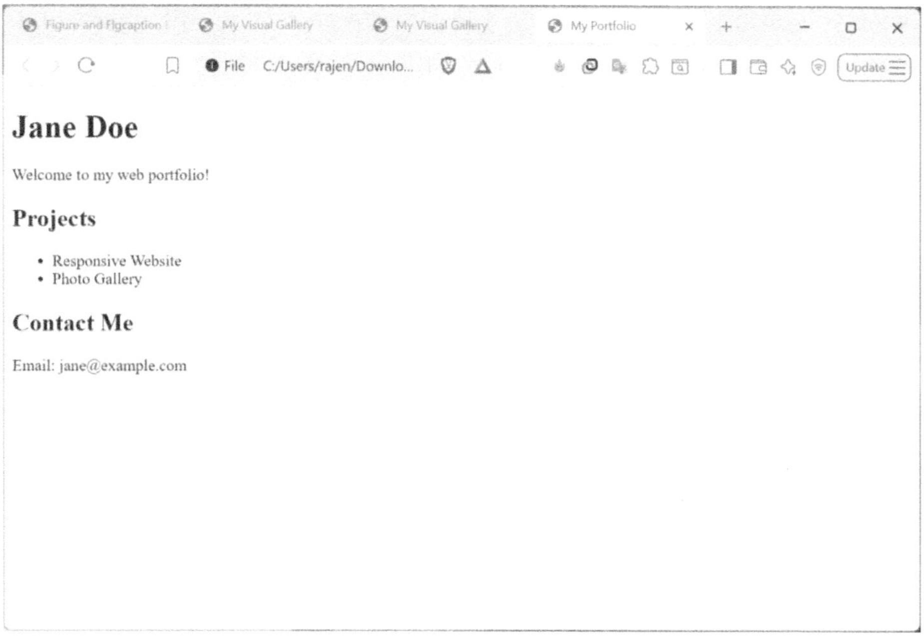

Now, let's style these divisions to make them stand out. Create a file named `styles.css` and add:

```
div {
  border: 2px solid #4CAF50;
  padding: 16px;
  margin-bottom: 16px;
  border-radius: 8px;
  background-color: #f9f9f9;
}
```

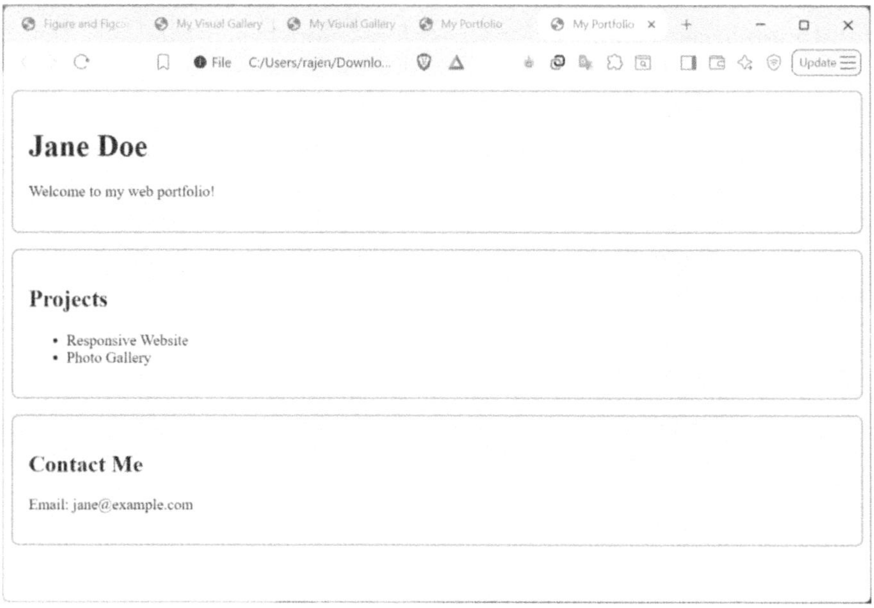

When you open your HTML file in a browser, each section will appear in its own box, separated by some space. This is how many real-world websites start organizing content, such as portfolios or e-commerce product listings.

Highlighting Text with <span>

Sometimes you need to style or emphasize just a word or phrase within a sentence. That's where **<span>** shines. For example, you might want to highlight your name or a special offer.

Let's update the contact section from the previous example to highlight the email address.

Wrap the email address in a `<span>`:

```
<div>
  <h2>Contact Me</h2>
  <p>Email: <span>jane@example.com</span></p>
</div>
```

Add a new CSS rule to make the email address stand out:

```
span {
  color: #e91e63;
  font-weight: bold;
  background-color: #fffbe7;
  padding: 2px 6px;
  border-radius: 4px;
}
```

Now, when you check your page, the email address will be highlighted with a pink color and a soft background, making it easy for visitors to spot.

Combining <div> and <span> for Real-World Layouts

Let's look at another scenario—a simple blog post layout. Many blogs use **<div>** to group the post content, sidebar, and footer, while **<span>** might highlight the date or author name.

Try building a simple blog post section:

Add the following inside your `<body>`:

```
<div>
  <h2>How to Learn HTML & CSS</h2>
  <p>
    <span>Posted on June 10, 2024</span> by <span>Jane Doe</span>
  </p>
  <p>
    Learning HTML and CSS is the first step to becoming a web
```

```
developer. Practice daily, build projects, and don't be afraid to
experiment!
    </p>
  </div>
```

In your CSS file, style the blog post and the metadata:

```
  div {
    max-width: 600px;
    margin: 32px auto;
    padding: 24px;
    background: #e3f2fd;
    border: 1px solid #2196f3;
  }

  span {
    font-size: 0.95em;
    color: #1565c0;
    font-style: italic;
  }
```

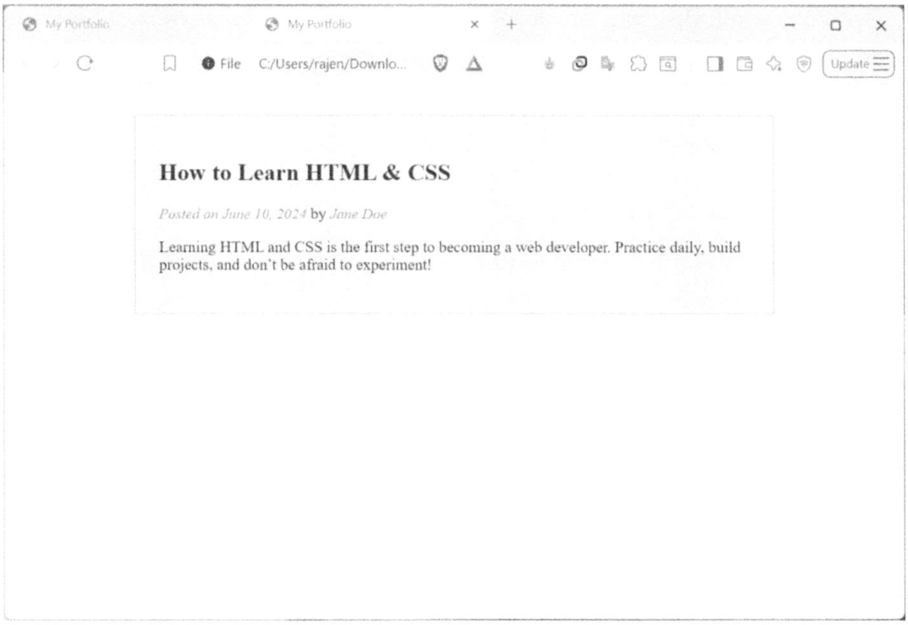

You'll see the post title, the metadata (date and author) styled differently, and the content all grouped neatly.

Exercise: Organize a Simple Webpage

Put your skills to the test by creating a simple "About Me" page using **<div>** and **<span>**:

Create three **<div>** sections for "About Me," "Hobbies," and "Favorite Quote."

Use **<span>** to highlight your name in the introduction, a hobby you love most, and the author of the quote.
Style the **<div>** and **<span>** elements with CSS to make each section visually distinct.

Experiment with different background colors, borders, and text styles. Don't hesitate to refresh your browser and see how each change impacts the look and feel of your page!

Hands-On Summary

By practicing with **<div>** and **<span>**, you've learned how to structure and highlight content, providing a flexible foundation for any webpage. These elements are essential for grouping sections and pinpointing specific pieces of text for styling, making your pages clearer and more attractive. Keep experimenting with grouping and inline highlighting to see how small changes can create professional, organized layouts.

Up next, you'll discover an even smarter way to organize your HTML: **semantic HTML5 elements**. These elements go beyond generic containers, helping browsers and users understand the true meaning and structure of your content. Get ready to build web pages that are not only beautiful but also meaningful!

5.2. Using Semantic HTML5 Elements

When you build a web page, organizing your content is crucial—not just for visual appeal, but for clarity and accessibility. Semantic HTML5 elements help you give meaning to your web page sections, telling browsers, search engines, and assistive technologies what each part of your site represents. Let's dive in and practice using these powerful tags to make your sites smarter and easier to maintain.

What Are Semantic HTML5 Elements?

Semantic elements are HTML tags that describe their meaning in a clear way, both to humans and machines. Unlike generic containers like \`<div>\`, semantic elements such as **\`<header>\`**, **\`<nav>\`**, **\`<main>\`**, **\`<section>\`**, **\`<article>\`**, **\`<aside>\`**, and **\`<footer>\`** communicate the role of the content inside them.

Why does this matter? Semantic HTML:

Improves readability for developers, making your code easier to understand.
Helps search engines (like Google) better index your site's content.
Enhances accessibility, assisting screen readers in navigating your site.

Exploring Key Semantic HTML5 Elements

Let's look at the most common semantic elements you'll use:

`<header>`: Represents the introductory content or a group of navigation links.
`<nav>`: Contains navigation links for your website.
`<main>`: Holds the main content unique to the page.
`<section>`: Defines a standalone section within the page.
`<article>`: For independent, self-contained content (like a blog post or news article).
`<aside>`: Contains content related to the main content, like a sidebar or pull quote.
`<footer>`: Contains footer information, like copyright or contact details.

Hands-On Coding: Structuring a Simple Portfolio Page

Let's practice by building the skeleton of a personal portfolio site. This will help you see how semantic elements create a meaningful structure.

Open your favorite code editor and create a new file called `portfolio.html`.
Type the following code, paying attention to each semantic element and its purpose:

```
<!DOCTYPE html>
<html lang= "en">
<head>
  <meta charset= "UTF-8">
  <title>My Portfolio</title>
</head>
<body>
  <header>
    <h1>Jane Doe</h1>
    <p>Web Developer & Designer</p>
  </header>

  <nav>
    <ul>
      <li><a href= "#about">About Me</a></li>
      <li><a href= "#projects">Projects</a></li>
      <li><a href= "#contact">Contact</a></li>
    </ul>
  </nav>

  <main>
    <section id= "about">
      <h2>About Me</h2>
      <p>Hello! I build modern, responsive websites. Welcome to my
portfolio.</p>
    </section>

    <section id= "projects">
      <h2>Projects</h2>
      <article>
        <h3>Personal Blog</h3>
        <p>A clean, readable blog site built with HTML5 and CSS3.</p>
      </article>
      <article>
        <h3>Online Store</h3>
        <p>A simple e-commerce homepage with a modern layout.</p>
      </article>
```

```
    </section>

    <aside>
      <h2>Quick Facts</h2>
      <ul>
        <li>Based in Tokyo</li>
        <li>Loves CSS art</li>
        <li>Speaks Japanese and English</li>
      </ul>
    </aside>
  </main>

  <footer>
    <p>&copy; 2024 Jane Doe</p>
  </footer>
</body>
</html>
```

This structure organizes your content so anyone (and any browser or search engine) can understand what each part of your page represents.

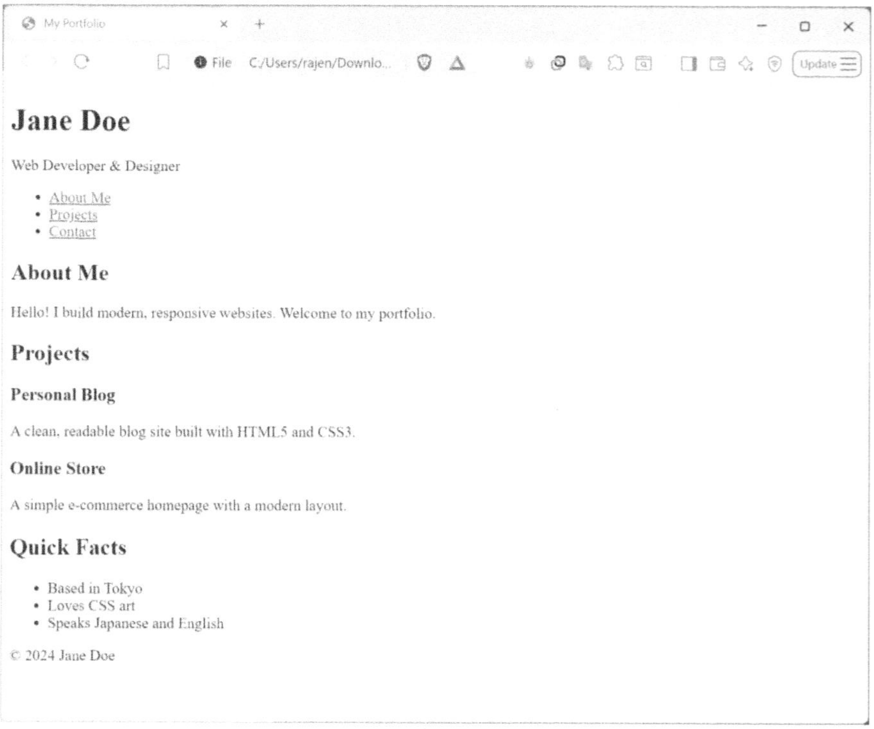

Adding Simple CSS for Clarity

Let's add a touch of CSS to visually separate these sections and reinforce their roles. Create a new file called `style.css` and link it in your HTML `<head>`:

```
<link rel= "stylesheet"href= "style.css">
```

Now, add the following CSS:

```
header, nav, main, section, article, aside, footer {
  border: 1px solid #ccc;
  margin: 10px;
  padding: 10px;
}

header {
  background: #f2f2f2;
}

nav {
  background: #e0e7ff;
}

main {
  background: #fff;
}

aside {
  background: #f1faee;
  float: right;
  width: 200px;
}

footer {
  background: #eeeeee;
  text-align: center;
  font-size: 0.9em;
}
```

Open your `portfolio.html` in a web browser to see how each semantic element creates a clear, organized structure. Notice how the **`<aside>`** stands out as a sidebar, and each **`<section>`** and **`<article>`** is visually distinct.

Real-World Scenario: Simple Blog Page

Let's apply semantic HTML to a basic blog post structure, a common scenario for many web projects:

```
<article>
  <header>
    <h2>Mastering Semantic HTML5</h2>
    <p>Posted by Jane on June 10, 2024</p>
  </header>
  <section>
    <p>Semantic HTML5 elements make your pages easier to read and
maintain...</p>
  </section>
  <footer>
    <p>Tags: HTML5, Web Development</p>
  </footer>
</article>
```

Here, the `**<article>**` wraps an entire blog post, while its `**<header>**`, `**<section>**`, and `**<footer>**` provide structure and clarity.

Hands-On Exercise: Mark Up Your Own Page

Try organizing a simple webpage about your favorite hobby using at least four different semantic HTML5 elements. For example, create a `**<header>**` for the page title, a `**<nav>**` for links, a `**<main>**` with `**<section>**`s, and a `**<footer>**`. Experiment with the CSS above to highlight the structure.

Key Takeaways and Practice

Semantic HTML5 elements help make your web pages meaningful, accessible, and easier to manage. By practicing with tags like `**<header>**`, `**<nav>**`, `**<main>**`, `**<section>**`, `**<article>**`, `**<aside>**`, and `**<footer>**`, you're building the foundation for well-structured, professional websites. Try modifying the portfolio or blog layout, and notice how your code becomes more readable and organized.

Next, you'll learn how these semantic tags play a key role in creating attractive layouts. In the following section, we'll use your new skills to build simple, visually appealing web page layouts with semantic tags and CSS—get ready to see your code come to life!

5.3. Creating Simple Layouts with Semantic Tags

When you visit your favorite websites, have you noticed how the content is neatly organized into headers, navigation menus, articles, and footers? This structure is no accident—it's crafted using **semantic HTML tags**. These tags don't just make your code more readable for you; they help browsers, search engines, and assistive technologies understand your page's structure. In this section, you'll learn how to create simple, meaningful layouts using semantic tags, combining them with basic CSS for a touch of style.

Let's bring these concepts to life by building a basic layout for a personal portfolio homepage. As you follow along, you'll see how each semantic tag serves a clear purpose, and you'll gain practical experience you can apply to many types of websites, from blogs to e-commerce stores.

What Are Semantic HTML Tags?

Semantic tags clearly describe their meaning in a human- and machine-readable way. Unlike generic `<div>` or `<span>` tags, semantic tags such as **`<header>`**, **`<nav>`**, **`<main>`**, **`<section>`**, **`<article>`**, and **`<footer>`** define the different parts of a web page.

`<header>`: Contains introductory content or navigation links.
`<nav>`: Encloses the site's main navigation.
`<main>`: Holds the primary content unique to that page.
`<section>`: Groups related content together.
`<article>`: Contains independent, self-contained content.
`<footer>`: Placed at the bottom, usually for copyright, contact, or site info.

Using these tags makes your code easier to maintain and improves accessibility for users with screen readers.

Building a Simple Portfolio Layout

Let's build a simple, professional-looking portfolio homepage. This example will help you understand how semantic tags create a logical page structure.

Open your text editor and create a new file named `portfolio.html`.
Start with the basic HTML5 structure:

```
<!DOCTYPE html>
<html lang="en">
<head>
  <meta charset="UTF-8">
  <meta name="viewport" content="width=device-width, initial-
scale=1.0">
  <title>My Portfolio</title>
  <link rel="stylesheet" href="styles.css">
</head>
<body>

</body>
</html>
```

Add semantic elements inside the `<body>`:

```
<body>
  <header>
    <h1>Jane Doe</h1>
    <p>Web Developer & Designer</p>
  </header>
  <nav>
    <ul>
      <li><a href="#about">About</a></li>
      <li><a href="#projects">Projects</a></li>
      <li><a href="#contact">Contact</a></li>
    </ul>
  </nav>
  <main>
    <section id="about">
      <h2>About Me</h2>
      <p>I build creative and accessible websites for all users.</p>
    </section>
    <section id="projects">
      <h2>Featured Projects</h2>
      <article>
        <h3>Responsive Blog</h3>
        <p>A clean, mobile-friendly blog built with HTML & CSS.</p>
      </article>
      <article>
        <h3>Online Store</h3>
        <p>An e-commerce demo site showcasing modern design
trends.</p>
      </article>
    </section>
  </main>
  <footer>
    <p>&copy; 2024 Jane Doe. All rights reserved.</p>
  </footer>
</body>
```

Notice how each area of the page is clearly defined and easy to identify.

Adding CSS for a Clean Layout

Next, let's add some simple CSS to `styles.css` to visually separate these sections and reinforce the layout.

Create a new file called `styles.css` in the same folder as your HTML file. Add the following CSS:

```
body {
  font-family: Arial, sans-serif;
  margin: 0;
  background: #f9f9f9;
  color: #333;
}

header {
  background: #1e90ff;
  color: white;
  padding: 2rem 1rem;
  text-align: center;
}

nav {
  background: #333;
}

nav ul {
  list-style: none;
  margin: 0;
  padding: 0;
  display: flex;
  justify-content: center;
}

nav li {
  margin: 0 1rem;
}

nav a {
  color: #fff;
  text-decoration: none;
  padding: 0.5rem 1rem;
  display: block;
}

nav a:hover {
  background: #1e90ff;
  border-radius: 4px;
}

main {
  max-width: 800px;
```

```
  margin: 2rem auto;
  padding: 1rem;
  background: white;
  border-radius: 8px;
  box-shadow: 0 2px 8px rgba(0,0,0,0.05);
}

section {
  margin-bottom: 2rem;
}

article {
  background: #f1f1f1;
  padding: 1rem;
  border-radius: 6px;
  margin-bottom: 1rem;
}

footer {
  background: #222;
  color: #ccc;
  text-align: center;
  padding: 1rem;
  position: sticky;
  bottom: 0;
  width: 100%;
}
```

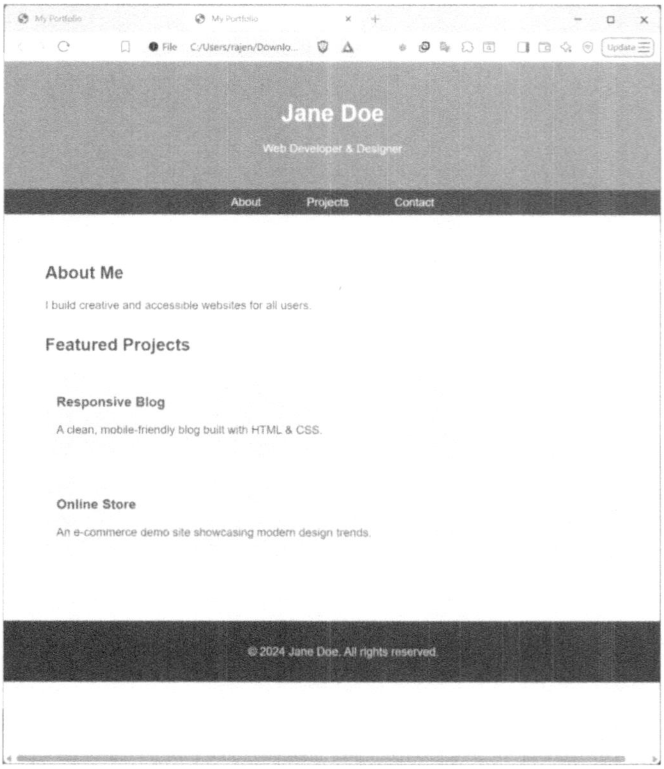

This style sheet gives each semantic section its own visual identity. The header is bold and prominent, the navigation stands out, the main content area is clean and easy to read, and the footer is understated but clear. When you open the `portfolio.html` file in your browser, you'll see a page that's both structured and visually appealing—just like real-world websites.

Real-World Example: Simple Blog Home Layout

Need another practical scenario? Imagine you're setting up a basic blog homepage. The structure will look familiar:

```
<body>
  <header>
    <h1>My Travel Blog</h1>
  </header>
  <nav>
    <ul>
      <li><a href="#destinations">Destinations</a></li>
      <li><a href="#tips">Travel Tips</a></li>
    </ul>
  </nav>
  <main>
    <section id="destinations">
      <h2>Featured Destinations</h2>
      <article>
        <h3>Paris, France</h3>
        <p>Exploring the city of lights and its cuisine.</p>
      </article>
      <article>
        <h3>Kyoto, Japan</h3>
        <p>Cherry blossoms and ancient temples await.</p>
      </article>
    </section>
    <section id="tips">
      <h2>Travel Tips</h2>
      <p>Pack light and always carry a power bank!</p>
    </section>
  </main>
  <footer>
    <p>Follow me on Instagram @mytravelblog</p>
  </footer>
</body>
```

Try combining this HTML with the CSS above and watch your blog homepage take shape.

You've just learned how to use semantic HTML tags to organize your web pages into meaningful sections, making your code cleaner and your websites easier to maintain. By combining these tags with simple CSS, you can create attractive layouts that mirror real-world portfolio and blog pages. Practice building your own homepage or blog using these techniques, and experiment with the CSS to customize your layout.

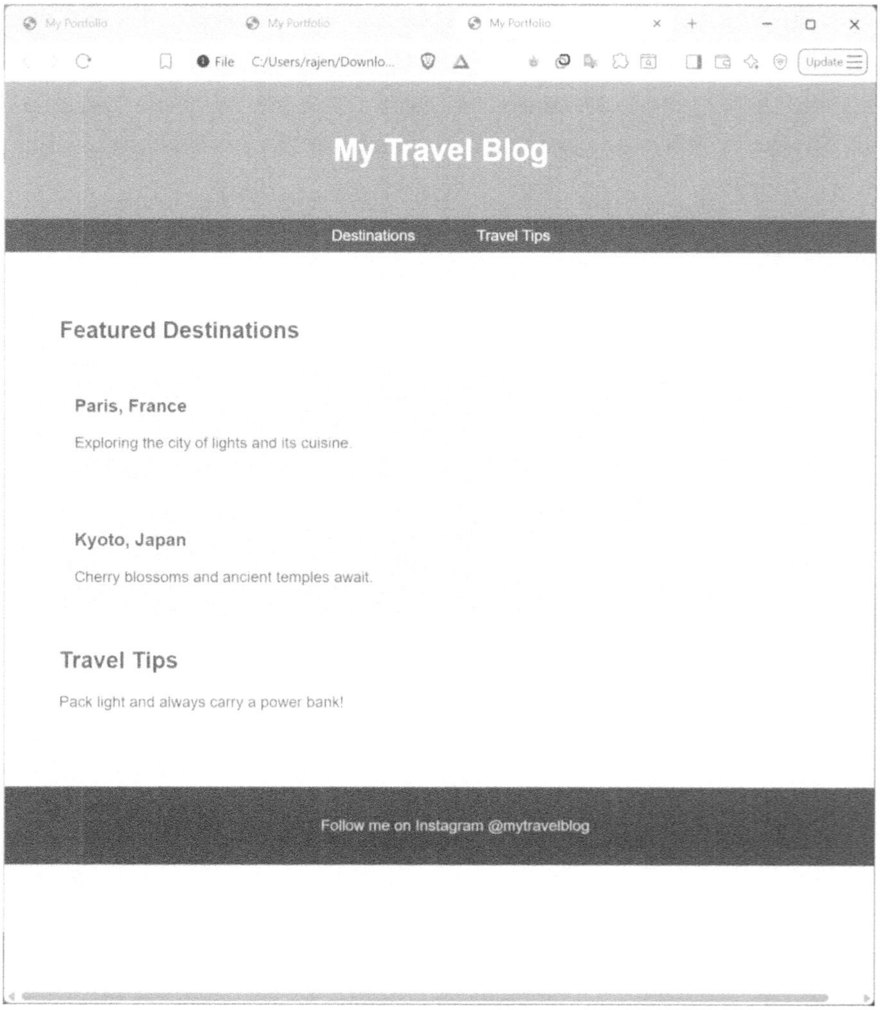

Next, we'll dive into **Accessible Web Design Fundamentals**, where you'll discover how to make your layouts welcoming and usable for all visitors, including those with disabilities. Get ready to unlock the secrets of accessible, inclusive web development!

5.4. Accessible Web Design Fundamentals

When you build a website, you're not just arranging text and images on a screen. You're creating experiences for people from all walks of life, including those who browse with assistive technologies like screen readers or use keyboard navigation instead of a mouse. This is where **accessible web design** becomes essential, ensuring your content is easy for everyone to find, read, and interact with. In this section, you'll learn how to use HTML elements in ways that make your sites both organized and accessible.

You'll get hands-on practice as you create a simple but accessible feature section for a portfolio site, and see how small steps can make a big impact.

Why Accessibility Matters in Web Development

Accessibility in web design means making your website usable for as many people as possible, regardless of their abilities or the devices they use. Properly structured HTML helps:

People who use screen readers understand your content's flow and meaning.
Users who rely on keyboard navigation move efficiently through your site.
Search engines better index your pages, which improves discoverability.

By learning accessible coding habits now, you'll create professional, future-ready sites.

The Importance of Semantic HTML Elements

Semantic HTML elements are tags that clearly describe their meaning in the code. For example, using a `**<nav>**` tag for navigation or a `**<main>**` tag for main content. These elements help both browsers and assistive technologies identify what each part of your page does.

Here are some essential semantic elements for accessibility:

`**<header>**`: Defines introductory content or a set of navigation links.
`**<nav>**`: Contains the site's navigation links.
`**<main>**`: Holds the main content unique to the page.
`**<section>**`: Groups related content together.
`**<article>**`: Represents a self-contained piece of content.
`**<aside>**`: Contains content related to the main content, like sidebars.
`**<footer>**`: Defines footer content for a page or section.

Let's see how these look in practice.

Building an Accessible Feature Section for a Portfolio

Suppose you're creating a simple portfolio site to showcase your projects. Let's make a "Featured Projects" section that's accessible and easy to navigate.

Start by opening your HTML file and adding a `**<section>**` element with a heading. Use `**<article>**` elements for each project, providing structure and clarity.

```
<section aria-labelledby= "featured-projects">
  <h2 id= "featured-projects">Featured Projects</h2>
```

```
  <article>
    <h3>Personal Blog Redesign</h3>
    <p>A clean, mobile-friendly blog built using HTML and CSS. Improved
accessibility by adding keyboard navigation.</p>
    <a href= "https://yourblog.com"aria-label= "Visit Personal Blog
Redesign project">View Project</a>
  </article>

  <article>
    <h3>Online Art Portfolio</h3>
    <p>Showcases digital illustrations with descriptive alt text for
each artwork to support screen readers.</p>
    <a href= "https://yourartsite.com"aria-label= "Visit Online Art
Portfolio project">View Project</a>
  </article>
</section>
```

Notice how:

Each project is wrapped in an `**<article>**` element, making it clear and distinct.
The `**aria-labelledby**` attribute on the section helps assistive technologies link the section to its heading.
Each link uses `**aria-label**` to give more context to screen readers.

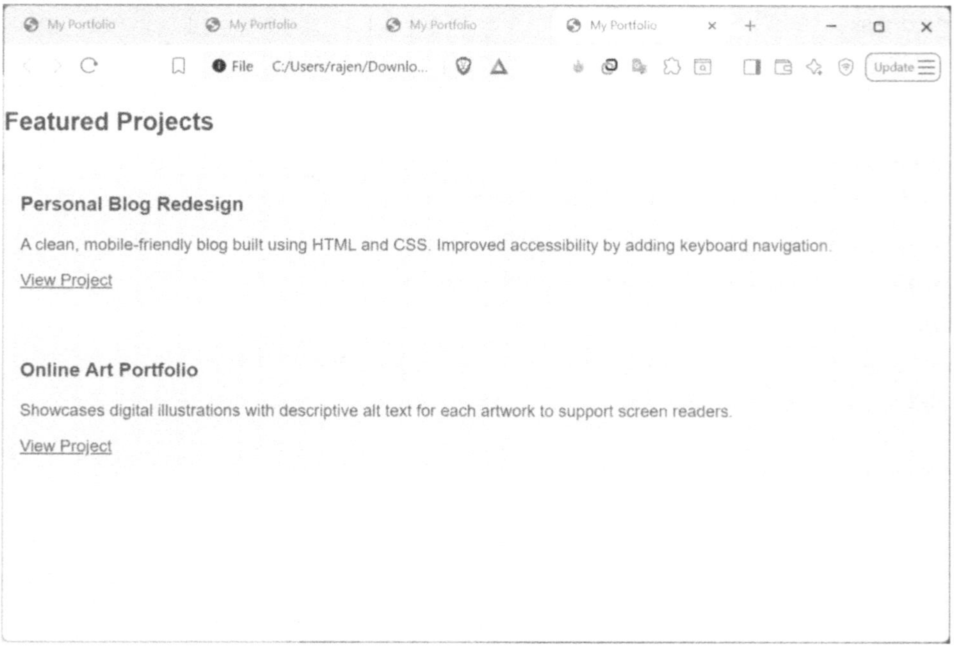

Adding Accessible Images with Alt Text

If you include images, always use the `**alt**` attribute to describe their content. This way, users who can't see the images will still know what they represent.

Here's how you might add an accessible image to one of your articles:

```
<article>
  <h3>Online Art Portfolio</h3>
  <img src= "digital-illustration.jpg"alt= "Vivid digital painting of a
sunset landscape"/>
  <p>Showcases digital illustrations with descriptive alt text for each
artwork to support screen readers.</p>
  <a href= "https://yourartsite.com"aria-label= "Visit Online Art
Portfolio project">View Project</a>
</article>
```

Enhancing Navigation for Keyboard Users

Many users rely on the keyboard (using the Tab key) to move through your site. To support this, make sure your navigation is structured with the `**<nav>**` element and uses clear, descriptive text.

Let's add an accessible navigation bar to your portfolio example:

```
<nav aria-label= "Main site navigation">
  <ul>
    <li><a href= "#about">About Me</a></li>
    <li><a href= "#projects">Projects</a></li>
    <li><a href= "#contact">Contact</a></li>
  </ul>
</nav>
```

Tips for accessible navigation:

Use a `**<nav>**` element with `**aria-label**` to describe its purpose.
Group links in an unordered list (`**<ul>**`) for clarity.
Ensure all links can be reached and used with just the keyboard.

Styling for Readability and Accessibility

Let's add some simple CSS to improve readability and usability for all visitors.

```
body {
  font-family: Arial, sans-serif;
  line-height: 1.6;
  color: #222;
  background-color: #f5f5f5;
}

section {
  margin: 2rem auto;
  max-width: 700px;
  background: white;
  padding: 1.5rem;
  border-radius: 8px;
}

h2, h3 {
```

```
  color: #005a9c;
}

a {
  color: #0066cc;
  text-decoration: underline;
}
a:focus, a:hover {
  outline: 2px solid #ff9800;
  background-color: #fff3e0;
}
```

This CSS improves accessibility by:

Using high-contrast colors for text and backgrounds.
Providing clear focus indicators for links, making keyboard navigation visible.
Choosing readable fonts and appropriate line spacing.

Real-World Example: Accessible E-Commerce Storefront

Imagine you're building a simple e-commerce homepage. Structuring content with semantic HTML and adding accessible features could look like this:

```
<header>
  <h1>Artisan Coffee Shop</h1>
  <nav aria-label= "Store navigation">
    <ul>
      <li><a href= "#menu">Menu</a></li>
      <li><a href= "#about">Our Story</a></li>
      <li><a href= "#contact">Contact</a></li>
    </ul>
  </nav>
</header>

<main>
  <section id= "menu"aria-labelledby= "menu-heading">
    <h2 id= "menu-heading">Our Coffee Menu</h2>
    <article>
      <h3>Espresso</h3>
      <img src= "espresso.jpg"alt= "Cup of espresso with crema"/>
      <p>Rich, bold espresso made from freshly roasted beans.</p>
    </article>
    <article>
      <h3>Latte</h3>
      <img src= "latte.jpg"alt= "Latte with heart-shaped foam art"/>
      <p>Smooth, creamy latte topped with artful foam.</p>
    </article>
  </section>
</main>
```

This example uses semantic HTML to organize content, descriptive alt text for images, and accessible navigation — all key components of inclusive design.

You've just learned how semantic HTML elements, descriptive attributes, and thoughtful structure can make your websites more accessible to everyone. By practicing these accessible web design fundamentals, you're not only following best practices but also making the web a more inclusive place.

Next, get ready to apply everything you've learned as you build and code a real-world news or magazine article layout. In the upcoming section, you'll put your skills to the test by structuring headlines, images, and bylines using accessible HTML, reinforcing your understanding while tackling a fun, practical challenge!

5.5. Hands-On Exercise: Structuring a News or Magazine Article

Websites like The New York Times or BBC News rely on well-organized HTML to present articles in a clear, accessible way. As a budding web developer, mastering the art of structuring content using HTML elements is crucial for creating readable, professional webpages. In this hands-on exercise, you'll build the foundation of a news or magazine article, using essential HTML tags to give structure and meaning to your content.

Understanding the Building Blocks of an Article

Before you dive into code, let's explore the common elements that make up a news or magazine article online:

Headline: The article's title, grabbing the reader's attention.
Byline: The author's name and publication date.
Featured Image: An eye-catching visual related to the article.
Article Body: The main content, broken into paragraphs and sometimes sections.
Subheadings: Break up the text for easier reading.
Related Links or Quotes: Enhance the article's depth and context.

Each of these parts uses specific HTML elements to ensure both people and browsers can understand the page's structure.

Setting Up Your HTML Document

Let's start by creating a simple HTML file for your article. Open your code editor and follow along.

Create a new file called `news-article.html`.
Begin with the basic HTML structure:

```
<!DOCTYPE html>
<html lang="en">
<head>
```

```
  <meta charset="UTF-8">
  <title>Sample News Article</title>
  <link rel="stylesheet" href="styles.css">
</head>
<body>

</body>
</html>
```

This template gives you a clean slate for your article content.

Structuring the Article with Semantic Elements

Professional news sites use semantic elements to organize their articles. Semantic HTML tags describe the meaning of the content, making your code easier to understand and more accessible to assistive technologies.

Let's break down the article into its core sections.

Inside the `` `<body>` ``, add the following structure:

```
<article>
  <header>
    <h1>Climate Change: The World Responds</h1>
    <p class="byline">By Jane Doe | June 15, 2024</p>
    <img src="images/climate-change.jpg" alt="Melting glacier with
polar bears">
  </header>

  <section>
    <h2>Global Efforts to Reduce Emissions</h2>
    <p>
      Leaders from around the globe have gathered to discuss actionable
steps in combating climate change. Innovative policies and renewable
energy initiatives are at the forefront of these discussions.
    </p>
  </section>

  <section>
    <h2>Local Communities Take Action</h2>
    <p>
      Grassroots organizations are making significant strides, from
tree planting campaigns to educational workshops. These efforts
highlight the importance of local involvement in addressing global
issues.
    </p>
    <blockquote>
      "Every community has a role to play," says Dr. Lin, an
environmental scientist.
    </blockquote>
  </section>

  <footer>
    <p>For more stories like this, visit <a
href="https://www.bbc.com/news/science_and_environment">BBC Science &
Environment</a>.</p>
```

```
  </footer>
</article>
```

Let's review the structure:

`<article>` wraps the entire news story.
`<header>` contains the headline, byline, and a featured image.
`<section>` tags break the body into logical parts, each with a subheading using `<h2>`.
`<blockquote>` highlights a notable quote.
`<footer>` offers additional links or context.

This semantic approach mirrors real-world news sites, making your content more professional and accessible.

Adding Style with CSS

Now, give your article some visual polish. Create a file named `styles.css` in the same folder as your HTML file.

Add the following CSS to style your article:

```
body {
  font-family: Arial, sans-serif;
  background: #f8f9fa;
  margin: 0;
  padding: 24px;
}

article {
  max-width: 700px;
  margin: 0 auto;
  background: #fff;
  border-radius: 8px;
  box-shadow: 0 2px 8px rgba(0,0,0,0.07);
  padding: 32px;
}

header {
  border-bottom: 1px solid #e0e0e0;
  margin-bottom: 24px;
  padding-bottom: 16px;
}

h1 {
  font-size: 2.2em;
  margin: 0 0 8px 0;
}

.byline {
  color: #888;
  font-size: 1em;
```

```
  margin-bottom: 16px;
}

header img {
  width: 100%;
  max-height: 300px;
  object-fit: cover;
  border-radius: 6px;
  margin-top: 16px;
}

section {
  margin-bottom: 24px;
}

h2 {
  font-size: 1.3em;
  margin-bottom: 10px;
  color: #205081;
}

blockquote {
  margin: 16px 0;
  padding: 12px 24px;
  background: #eef6fa;
  border-left: 4px solid #47a3da;
  font-style: italic;
  color: #333;
}

footer {
  border-top: 1px solid #e0e0e0;
  padding-top: 16px;
  font-size: 0.95em;
  color: #555;
}
```

With these styles, your article will look clean and modern, much like those found on leading news platforms. You can experiment with colors, fonts, and spacing to further customize your design.

Real-World Practice: Building a Magazine Article Layout

Imagine you're creating an online magazine feature about travel destinations. Try tweaking the article's structure and content to fit this scenario:

Change the headline and byline.
Replace the image with a scenic travel photo.
Add a new section highlighting top destinations, using a list (`<ul>` and `<li>`).
Include a quote from a travel expert.

For example, your new section might look like this:

```
<section>
  <h2>Top Destinations of 2024</h2>
  <ul>
    <li>Kyoto, Japan</li>
    <li>Cape Town, South Africa</li>
    <li>Reykjavik, Iceland</li>
  </ul>
</section>
```

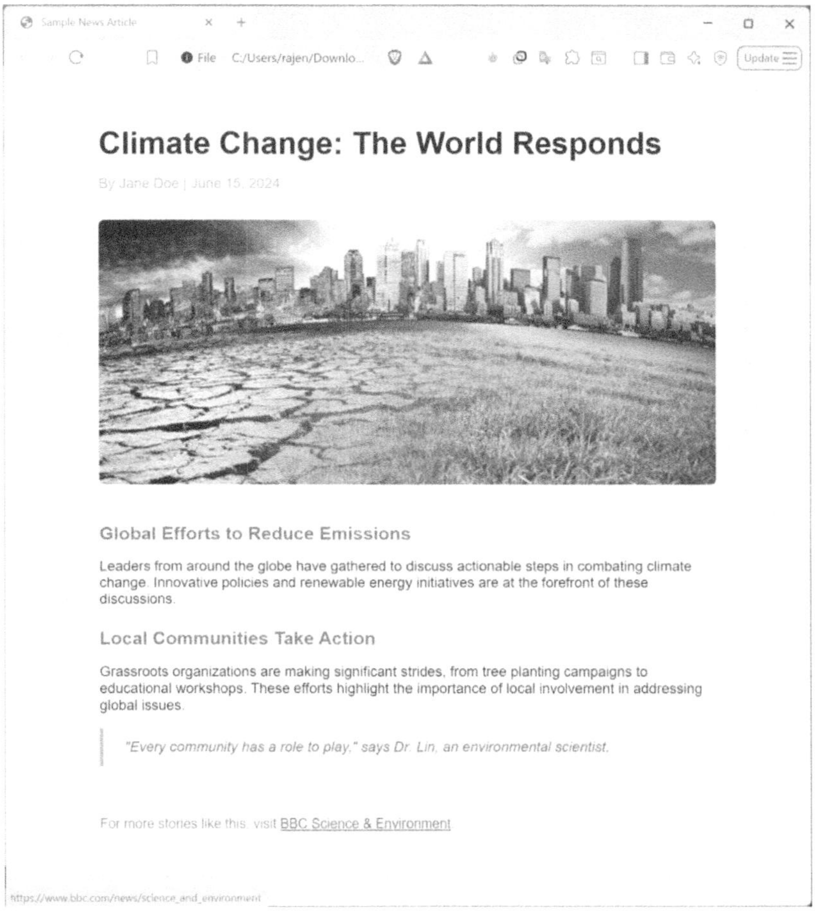

This exercise helps you see how flexible HTML is for structuring all kinds of web articles, from breaking news to feature stories.

In this exercise, you've learned how to use semantic HTML elements like `<article>`, `<header>`, `<section>`, and `<footer>` to organize a news or magazine article. You practiced structuring headlines, bylines, images, and quotes, then added CSS to make the article visually appealing. This foundation will help you build accessible, well-structured content for any website project.

6. INTRODUCTION TO CSS STYLING

Imagine visiting your favorite website and instantly noticing the vibrant colors, polished typography, and inviting layouts that make the content come alive. What sets these sites apart isn't just their structure—it's the creative styling that turns plain HTML into something beautiful and engaging. In this chapter, you'll discover how to transform basic web pages into eye-catching, modern sites using CSS, or Cascading Style Sheets. CSS is the language that lets you express your unique design vision and create professional-looking websites that stand out in today's digital landscape.

You're about to unlock the secrets of CSS selectors, learning how to target specific HTML elements for styling with ease and precision. From choosing the perfect color palette and adding dynamic backgrounds, to refining the look of your text, you'll start to see how small changes with CSS can make a big difference in a site's appearance. As you master borders, margins, and padding, you'll gain control over spacing and layout, essential for creating visually balanced and user-friendly pages.

To put your new skills into practice, you'll work through a hands-on project: styling a simple blog homepage. This real-world example will help you apply what you learn step by step, building your confidence and reinforcing core web design techniques. By the end of this chapter, you'll have the practical tools and creative inspiration to elevate your web projects—whether you're crafting a personal portfolio, starting a blog, or taking the first steps toward a career in web development. Let's dive in and discover how CSS brings your websites to life.

6.1. Understanding CSS Selectors

Cascading Style Sheets (CSS) is what transforms a plain HTML page into an engaging, beautifully designed website. As you start styling your web pages, you'll use **CSS selectors** to target specific HTML elements and control how they look. Mastering selectors is the first step to writing efficient, powerful CSS that brings your designs to life.

What Are CSS Selectors?

A **CSS selector** tells the browser which HTML elements you want to style. Think of selectors as a way to "point" at parts of your webpage so you can change their appearance. Selectors can target single elements, groups of elements, or elements with specific characteristics like classes or IDs.

Let's look at how selectors work in a simple scenario: styling a personal portfolio homepage.

Basic Selectors: Targeting Elements by Name

The most fundamental CSS selector is the **element selector**. This selector matches every instance of a specific HTML element on your page. For example, you might want to style all your headings or paragraphs the same way.

Example:

Suppose you want to make every `<h2>` heading on your portfolio bold and blue.

HTML:

```
<h2>About Me</h2>
<h2>Projects</h2>
```

CSS:

```
h2 {
  color: blue;
  font-weight: bold;
}
```

This CSS rule will apply to every `<h2>` element, making them all blue and bold. Try creating a new HTML file, add a couple of headings, and apply this CSS in a `<style>` tag between the `<head>` tags.

Class Selectors: Styling Groups of Elements

Sometimes, you want to style only certain elements—even if they are the same type. This is where the **class selector** shines. Classes are reusable and let you target multiple elements.

HTML:

```
<p class= "highlight">I am a passionate web developer.</p>
<p>This is my blog section.</p>
<p class= "highlight">Contact me for collaboration!</p>
```

CSS:

```
.highlight {
  background-color: #ffeeba;
  border-left: 4px solid #ff8800;
  padding: 10px;
```

```
}
```

Here, only the paragraphs with `class="highlight"` get the special background and border. Classes are perfect for highlighting testimonials, important notes, or special announcements on your website.

ID Selectors: Styling Unique Elements

Each **ID selector** targets one specific element with a unique identifier. Use IDs sparingly, mainly for elements that appear only once per page, like a site header or a main footer.

HTML:

```
<h1 id= "main-title">Welcome to My Portfolio</h1>
```

CSS:

```
#main-title {
  font-size: 2.5em;
  text-align: center;
  color: #333366;
}
```

This rule only affects the `<h1>` with `id="main-title"`. IDs are great for branding elements, main banners, or call-to-action buttons.

Combining Selectors for Precision

You can combine selectors to be more specific. For instance, you might want to style links inside a navigation menu differently from other links.

HTML:

```
<nav>
  <ul>
    <li><a href= "#about"class= "nav-link">About</a></li>
    <li><a href= "#projects"class= "nav-link">Projects</a></li>
    <li><a href= "#contact">Contact</a></li>
  </ul>
</nav>
```

CSS:

```
.nav-link {
  text-transform: uppercase;
  color: #008080;
  text-decoration: none;
```

```
  font-weight: bold;
}
```

Here, only links with the `nav-link` class get the uppercase, bold look. This helps you keep your navigation links visually distinct.

Descendant Selectors: Targeting Nested Elements

A **descendant selector** lets you target elements that are nested inside others. This is especially useful for complex layouts, like an e-commerce product grid or a blog with multiple sections.

Example: Styling All Paragraphs Inside a Section

HTML:

```
<section>
  <h2>Featured Project</h2>
  <p>This project showcases my best work.</p>
  <p>It uses HTML, CSS, and JavaScript.</p>
</section>
```

CSS:

```
section p {
  color: #555555;
  font-style: italic;
}
```

This rule styles only the paragraphs inside a `<section>`, letting you control context-specific styling.

Attribute Selectors: Targeting by HTML Attributes

CSS can target elements based on any attribute, not just class or id. This is powerful for styling forms, product listings, or interactive components.

Example: Styling External Links

HTML:

```
<a href= "https://external.com"target= "_blank">External Site</a>
<a href= "#about">About Me</a>
```

CSS:

```
a[target= "_blank"] {
  color: #d9534f;
```

```
  font-weight: bold;
  text-decoration: underline;
}
```

Here, any link that opens in a new tab (`target="_blank"`) gets a special style, helping users spot external links easily.

Practice Scenario: Styling a Simple Blog Post

Let's try a mini-project. Imagine you're building a basic blog post page. Use different selectors to style the page.

HTML:

```
<article>
  <h2 class= "post-title">My First Blog Post</h2>
  <p class= "author">By Jane Doe</p>
  <p>This is my first post as a web developer. Stay tuned for more!</p>
  <a href= "https://twitter.com/janedoe"class= "social-link"target=
"_blank">Follow me on Twitter</a>
</article>
```

CSS:

```
.post-title {
  color: #2c3e50;
  margin-bottom: 0.5em;
}

.author {
  font-size: 0.9em;
  color: #888888;
}

article p {
  line-height: 1.6;
}

.social-link[target= "_blank"] {
  background-color: #1da1f2;
  color: #fff;
  padding: 0.4em 0.8em;
  border-radius: 4px;
  text-decoration: none;
  display: inline-block;
}
```

Experiment with these selectors by editing the HTML and CSS, observing how each rule affects the page. Try adding another paragraph or a new class to see the results.

You've learned how to use CSS selectors to precisely target, group, and style elements on your web pages. From basic element selectors to classes, IDs, and attribute selectors, these tools let you bring structure and personality to your designs. Practice by creating a simple web page and experimenting with different selectors to see how they interact.

Next, you'll discover how to add color and backgrounds to your site, bringing your layouts alive with vibrant palettes and dynamic visual effects. Get ready to make your web pages truly stand out!

6.2. Adding Color and Backgrounds

Color is one of the most powerful tools in web design. With CSS, you can transform a plain web page into something vibrant, welcoming, and memorable simply by controlling the color of text, backgrounds, and even entire sections. In this section, you'll get hands-on with adding color and backgrounds to your HTML elements. These skills are fundamental for building visually engaging sites, whether you're designing a personal portfolio or helping an online shop stand out.

Understanding CSS Color Syntax

Before diving in, it's important to understand how CSS handles color. You can specify colors in several different ways:

Color names: Common names like `red`, `blue`, or `green`.
Hex codes: Six-digit codes starting with `#`, such as `#3498db` (a shade of blue).
RGB values: Using the `rgb()` function like `rgb(52, 152, 219)`.
RGBA values: Like RGB, but with an alpha channel for transparency, e.g., `rgba(52, 152, 219, 0.5)`.

Let's see how to apply each of these to your HTML elements.

Setting Text and Background Colors

Color is often the first thing users notice on a website. With CSS, you can set both the **color** (for text) and the **background-color** (for backgrounds) properties.

Suppose you're creating a simple promotional banner for an e-commerce homepage. Here's how you might add color:

Create a basic HTML structure:

```
<div class= "promo-banner">
  <h2>Summer Sale!</h2>
  <p>Get up to 50% off on all items. Shop now!</p>
```

```
</div>
```

Add CSS to style the banner:

```
.promo-banner {
  color: white; /* Sets the text color */       background-color:
#e67e22; /* A warm orange background */       padding: 20px;
  border-radius: 10px;
  text-align: center;
}
```

Here, `color` changes the text color to white.
`background-color` uses a hex code to set a lively orange background.
`border-radius` and `padding` help the banner look modern and inviting.

Try changing the color values to see how different combinations affect the mood of your banner.

Using RGB and RGBA for More Control

Sometimes, you want more flexibility or transparency. The **rgba()** color function lets you choose a color and set its transparency, which is useful for overlays or highlighting content without hiding what's behind it.

Let's say you want to overlay a semi-transparent color on a portfolio image. Here's how:

HTML markup for your portfolio card:

```
<div class= "portfolio-card">
  <img src= "your-photo.jpg"alt= "Project Screenshot">
  <div class= "overlay">
    <h3>My Project</h3>
  </div>
</div>
```

CSS for the overlay effect:

```
.portfolio-card {
  position: relative;
  width: 300px;
  height: 200px;
}
.portfolio-card img {
  width: 100%;
  height: 100%;
  display: block;
}
.overlay {
```

```
    position: absolute;
    top: 0; left: 0;
    width: 100%; height: 100%;
    background-color: rgba(44, 62, 80, 0.7); /* Semi-transparent dark
blue */       color: #fff;
    display: flex;
    align-items: center;
    justify-content: center;
    font-size: 1.5em;
  }
```

The `.overlay` uses `background-color: rgba(44, 62, 80, 0.7)` to create a dark blue overlay that's 70% opaque.

This technique is popular in modern global web design trends, especially for portfolios, as it spotlights project names without obscuring the background image.

Adding Background Images

CSS also allows you to use images as backgrounds, giving your web pages extra depth and personality. This is especially helpful for blogs or landing pages looking to create a memorable first impression.

Let's create a stylish header for a personal blog:

HTML for the blog header:

```
<header class= "blog-header">
  <h1>Welcome to My Travel Blog</h1>
  <p>Adventures from around the world</p>
</header>
```

CSS to add a background image:

```
  .blog-header {
    background-image: url( 'images/beach.jpg');
    background-size: cover; /* Scales the image to cover the area */
background-position: center; /* Centers the image */       color: #fff;
    text-align: center;
    padding: 60px 20px;
    box-shadow: 0 4px 10px rgba(0,0,0,0.2);
  }
```

`background-image` sets the image.

`background-size: cover` ensures the image covers the entire header area, which is great for responsive layouts.

`background-position: center` keeps the main part of the image visible.

`box-shadow` adds subtle depth, making the header stand out.

Experiment with different images and see how they affect the atmosphere of your site.

Combining Multiple Backgrounds

You can even layer multiple backgrounds using a comma-separated list. This is useful for creating visual effects, such as adding a gradient over an image.

For example, let's enhance the blog header with a gradient overlay:

```
.blog-header {
  background-image:
    linear-gradient(rgba(0,0,0,0.4), rgba(0,0,0,0.4)), /* semi-
transparent black overlay */     url( 'images/beach.jpg');
  background-size: cover;
  background-position: center;
  color: #fff;
  text-align: center;
  padding: 60px 20px;
}
```

The `linear-gradient` overlays a semi-transparent black, improving text readability against bright images.

In this section, you learned to use CSS to add color and backgrounds to your web pages, making them visually engaging and unique. You explored how to use color values, background images, and layering techniques for real-world scenarios like promotional banners, portfolio overlays, and blog headers. Try customizing the examples above with your favorite colors and images to reinforce your skills and develop your own web style.

Up next, you'll learn how to use CSS to change the appearance of your text, including font styles, sizes, and spacing, giving you even more creative control over your web designs. Get ready to make your site's content truly shine!

6.3. Changing Text Appearance with CSS

When you visit a modern website, you'll notice how clear, engaging text makes all the difference. Whether it's a bold headline on a portfolio homepage or elegantly styled product descriptions on an online store, CSS (Cascading Style Sheets) is the secret behind these visually appealing experiences. In this section, you'll learn how to transform plain HTML text into eye-catching, readable content using CSS. This hands-on guide will walk you through essential techniques to change fonts, colors, sizes, weights, and line spacing, laying the groundwork for truly professional web pages.

Why Text Styling Matters in Web Design

Text is the backbone of communication on the web. Good styling improves readability, highlights important information, and reflects your site's personality. Imagine a blog with tiny, hard-to-read text or an e-commerce store with clashing colors—users won't stay long. With CSS, you control how every word looks and feels, ensuring your site is both beautiful and accessible.

Linking CSS to Your HTML

Before you can style text, ensure your HTML file is connected to a CSS stylesheet. This separation allows you to keep your content and design organized.

Here's how to connect a CSS file to your HTML:

```
<!DOCTYPE html>
<html lang= "en">
<head>
  <meta charset= "UTF-8">
  <title>My First Styled Page</title>
  <link rel= "stylesheet"href= "styles.css">
</head>
<body>
  <h1>Welcome to My Web Page!</h1>
  <p>This is a simple paragraph to demonstrate text styling in CSS.</p>
</body>
</html>
```

In this example, `styles.css` is the file where you'll write your CSS rules.

Changing Font Family

The **font-family** property lets you specify which typeface your text should use. This is perfect for giving your site a unique look or matching a brand's identity.

Try this step-by-step exercise:

Open your `styles.css` file.
Add the following rule to style all paragraphs:

```
p {
  font-family: 'Arial', sans-serif;
}
```

You'll immediately see all `<p>` elements use the Arial font (or the default sans-serif if Arial isn't available).

Tip: Always include a generic family (like `sans-serif` or `serif`) as a fallback in case the preferred font isn't available on the user's device.

Adjusting Font Size

With the **font-size** property, you can make text larger or smaller. This is especially useful for headlines, subheadings, or drawing attention to key information.

Let's practice:

In your CSS file, add styles for headings and paragraphs:

```
h1 {
  font-size: 2.5em;
}

p {
  font-size: 1.1em;
}
```

Here, `em` units scale with the font size of the parent element, making your design more flexible.

Setting Font Weight and Style

Bold or italicized text can emphasize key points or create visual contrast. Use **font-weight** for boldness and **font-style** for italics.

Example:

```
h1 {
  font-weight: bold;
}

strong {
  font-weight: 700; /* Numeric value for extra bold */}

em {
  font-style: italic;
}
```

Try adding `<strong>` and `<em>` tags to your HTML:

```
<p>This is a <strong>bold</strong> and <em>italic</em> example.</p>
```

Refresh your browser to see the changes.

Applying Color to Text

Color grabs attention and sets the mood of your site. Use the **color** property to style text. You can use color names, hex codes, or RGB values.

Examples:

```
h1 {
  color: #2E8B57; /* A deep green */}

p {
  color: rgb(60, 60, 60); /* A dark gray */}
```

Experiment by changing color values and observing how they affect your page's feel.

Enhancing Readability with Line Height and Letter Spacing

Sometimes, adjusting the space between lines or letters makes paragraphs easier to read, especially for blogs or content-heavy sites.

line-height controls vertical spacing between lines.
letter-spacing controls horizontal spacing between letters.

Try this in your CSS:

```
p {
  line-height: 1.6; /* More space between lines */  letter-spacing:
0.5px; /* Slightly expanded letters */}
```

Real-World Scenario: Styling a Blog Post

Imagine you're designing a personal blog. You want the title to stand out, the author's name to be subtle, and the main content to be easy on the eyes.

Here's a practical example using HTML and CSS:

HTML:

```
<article>
  <h2>Exploring the Wonders of Web Development</h2>
  <p class= "author">By Jane Doe</p>
  <p class= "content">
    Learning to style your text with CSS is the first step to creating
beautiful, inviting web pages. With just a few lines of code, you can
transform plain text into something special!
  </p>
</article>
```

CSS:

```
h2 {
  font-family: 'Georgia', serif;
  font-size: 2em;
  color: #1a202c;
  font-weight: bold;
}
```

```
.author {
  font-size: 0.9em;
  color: #7d7d7d;
  font-style: italic;
}

.content {
  font-family: 'Verdana', sans-serif;
  font-size: 1.15em;
  line-height: 1.8;
  color: #333333;
}
```

Try copying this code into your project and modify the fonts, colors, or spacing to match your personal style.

Hands-On Exercise: Style Your Own Portfolio Headline

Let's put your new skills into practice! Imagine you're building a portfolio homepage. Add the following HTML and use CSS to make your headline pop and your tagline inviting:

HTML:

```
<section>
  <h1>Sam Lee - Web Developer</h1>
  <p class= "tagline">Crafting clean, modern websites for creative
minds.</p>
</section>
```

Now, style it in your `styles.css`:

```
h1 {
  font-family: 'Trebuchet MS', sans-serif;
  color: #005f73;
  font-size: 2.8em;
  font-weight: 800;
  letter-spacing: 1px;
}

.tagline {
  font-family: 'Calibri', sans-serif;
  color: #0a9396;
  font-size: 1.25em;
  font-style: italic;
  line-height: 1.5;
}
```

Refresh your browser and see your professional-looking headline and tagline.

You've just learned how to change font families, sizes, weights, colors, and spacing using CSS. These text styling basics are essential for making your web pages attractive and readable. Practice by mixing different styles and experimenting with real website scenarios like blogs, portfolios, or even the homepage of your favorite e-commerce store.

Up next, you'll discover how to add borders, margins, and padding. These layout tools will give your text and other elements some breathing room and help you create visually organized, modern designs. Get ready to unlock even more creative control over your web pages!

6.4. Adding Borders, Margins, and Padding

Before your website can look polished and professional, you need to master the art of spacing and outlining elements. Borders, margins, and padding are essential CSS techniques that help you control how content appears and how much space it takes up on the page. These tools let you shape the flow of your web pages, making them easier to read and more visually appealing. In this section, you'll learn how to add clear boundaries, create breathing room, and control the spacing between your HTML elements.

Understanding the CSS Box Model

Every HTML element is treated as a box by your browser. The CSS box model describes how these boxes are structured and how their size and spacing are determined. Here's a quick breakdown:

Content: The innermost part, where your text or images go.
Padding: The space between the content and the border.
Border: The outline that wraps the padding and content.
Margin: The space between the border and other elements.

Visualizing this model will help you understand how borders, margins, and padding work together.

Adding Borders to Elements

Borders help define and separate sections of your web page, making content easier to scan. You can customize a border's width, style, and color.

Let's start by adding a border to a card in a portfolio site:

```
<!-- HTML --><div class= "portfolio-card">
  <h3>My Project</h3>
  <p>This is a brief description of my web development project.</p>
```

```
</div>
```

Now, add CSS to style the border:

```
/* CSS */.portfolio-card {
  border: 2px solid #3498db; /* 2 pixels wide, solid style, blue color
*/  border-radius: 8px; /* Rounded corners for a modern look */
padding: 16px; /* Add some space inside the card */}
```

border: Combines the border width, style, and color.
border-radius: Rounds the corners of the border for a softer appearance.

Borders aren't just for containers. Try adding them to images, navigation bars, or even buttons to make interactive elements stand out.

Setting Padding for Inner Spacing

Padding controls the space **inside** an element, between its content and border. This makes text less cramped and improves readability.

Let's refine our portfolio card by customizing the padding:

```
.portfolio-card {
  border: 2px solid #3498db;
  border-radius: 8px;
  padding: 24px 12px; /* 24px top/bottom, 12px left/right */}
```

You can set padding for each side individually:

```
.portfolio-card {
  padding-top: 24px;
  padding-right: 12px;
  padding-bottom: 24px;
  padding-left: 12px;
}
```

This gives you fine control over where you want more or less space.

Using Margins for Outer Spacing

Margins create space **outside** an element's border, separating it from other elements on the page. Margins help prevent sections from colliding and keep your layout organized.

Suppose you want to space out multiple cards in a project gallery:

```
<!-- HTML --><div class= "portfolio-card">
  <h3>Project One</h3>
```

```
  <p>Description of project one.</p>
</div>
<div class= "portfolio-card">
  <h3>Project Two</h3>
  <p>Description of project two.</p>
</div>
```

```
/* CSS */.portfolio-card {
  border: 2px solid #3498db;
  border-radius: 8px;
  padding: 20px;
  margin-bottom: 24px; /* Adds space below each card */}
```

You can also set margins for all sides at once:

```
.portfolio-card {
  margin: 20px; /* 20px space on all sides */}
```

Or, target specific sides for custom layouts:

```
.portfolio-card {
  margin-top: 0;
  margin-right: 0;
  margin-bottom: 24px;
  margin-left: 0;
}
```

Practical Scenario: Styling a Call-to-Action Button

Borders, margins, and padding are especially important for key interface elements like buttons. Well-spaced, visually distinct buttons encourage users to click and interact with your site.

Here's how you might style a call-to-action button for a small e-commerce product page:

```
<!-- HTML --><button class= "buy-now">Buy Now</button>
```

```
/* CSS */.buy-now {
  background-color: #e67e22;
  color: #fff;
  font-size: 18px;
  border: 3px solid #d35400;
  border-radius: 5px;
  padding: 12px 32px; /* More width for a prominent look */  margin-
top: 18px; /* Separates the button from product info */  cursor:
pointer; /* Changes cursor to indicate interactivity */}
```

Notice how the **padding** gives the button generous space, while the **margin-top** keeps it from crowding other content. The **border** and **border-radius** add definition and style.

Real-World Example: Blog Post Layout

Let's apply these concepts to a simple blog post snippet. Borders, margins, and padding work together to create a clean, readable post preview.

```
<!-- HTML --><article class= "blog-snippet">
  <h2>Welcome to My Blog</h2>
  <p>Discover tips, tutorials, and insights on modern web development
and design trends.</p>
</article>


/* CSS */.blog-snippet {
  border-left: 5px solid #2ecc71; /* A colored left border for emphasis
*/  background-color: #f9f9f9;
  padding: 20px 30px;
  margin-bottom: 32px;
}
```

The **border-left** highlights the post and draws attention.
Padding creates space inside the article for comfortable reading.
Margin-bottom ensures each post snippet stands apart from the next.

You've just taken a vital step in web design by learning how to add borders for definition, padding for inner space, and margins for outer separation. These simple yet powerful CSS properties give you control over the look and feel of every element on your site. Experiment with different border styles, play with spacing, and see how these changes improve your layouts.

In the next section, you'll put these skills into action by styling a simple blog homepage. Get ready to combine borders, margins, and padding to create a visually engaging, user-friendly site layout that looks great on any screen!

6.5. Hands-On Exercise: Styling a Simple Blog Homepage

Now that you've explored the basics of CSS and how it works with HTML, it's time to put these skills into practice. In this hands-on exercise, you'll style a simple blog homepage using CSS. This project will help you see how CSS properties like color, spacing, borders, and layout come together to create a visually appealing webpage. By the end, you'll have a homepage that looks professional and is easy to read, just like the homepages of popular sites such as Medium or a tech blog.

Setting Up Your Blog Homepage Structure

Before you dive into styling, you need a basic HTML structure for your blog homepage. Picture a simple layout: a header with the blog title, a navigation bar, a main section with blog post previews, and a footer.

Here's the starter HTML you'll use for this exercise:

```
<!DOCTYPE html>
<html lang= "en">
<head>
  <meta charset= "UTF-8">
  <title>My Simple Blog</title>
  <link rel= "stylesheet"href= "styles.css">
</head>
<body>
  <header>
    <h1>My Simple Blog</h1>
    <nav>
      <ul>
        <li><a href= "#">Home</a></li>
        <li><a href= "#">Posts</a></li>
        <li><a href= "#">About</a></li>
      </ul>
    </nav>
  </header>
  <main>
    <article>
      <h2>First Blog Post</h2>
      <p>Welcome to my first blog post! Here I'll share my journey into
web development.</p>
      <a href= "#">Read more</a>
    </article>
    <article>
      <h2>Learning HTML & CSS</h2>
      <p>Today I learned how to build web pages using HTML and style
them with CSS.</p>
      <a href= "#">Read more</a>
    </article>
  </main>
  <footer>
    <p>&copy; 2024 My Simple Blog</p>
  </footer>
</body>
</html>
```

Save this code as `index.html` in your project folder. Next, you'll create the `styles.css` file where all your CSS styling will go.

Adding Basic CSS Styling

CSS lets you control the look and feel of your blog. You'll start by making the layout more attractive and easier to read.

Create a file named `styles.css` in the same folder as your HTML. Add the following basic styles:

```
body {
  font-family: Arial, sans-serif;
  background-color: #f5f6fa;
  margin: 0;
  padding: 0;
  color: #222;
}

header {
  background-color: #353b48;
  color: #fff;
  padding: 24px 0;
  text-align: center;
}

nav ul {
  list-style: none;
  padding: 0;
  margin: 12px 0 0 0;
  display: flex;
  justify-content: center;
  gap: 24px;
}

nav a {
  color: #fbc531;
  text-decoration: none;
  font-weight: bold;
}

nav a:hover {
  text-decoration: underline;
}

main {
  max-width: 700px;
  margin: 40px auto 0 auto;
  padding: 0 16px;
}

article {
  background-color: #fff;
  border: 1px solid #dcdde1;
  border-radius: 8px;
  margin-bottom: 28px;
  padding: 22px;
  box-shadow: 0 2px 8px rgba(0,0,0,0.04);
}

article h2 {
  margin-top: 0;
  color: #273c75;
}
```

```
article a {
  color: #0097e6;
  text-decoration: none;
}

article a:hover {
  text-decoration: underline;
}

footer {
  background-color: #353b48;
  color: #fff;
  text-align: center;
  padding: 18px 0;
  margin-top: 60px;
  font-size: 0.95em;
}
```

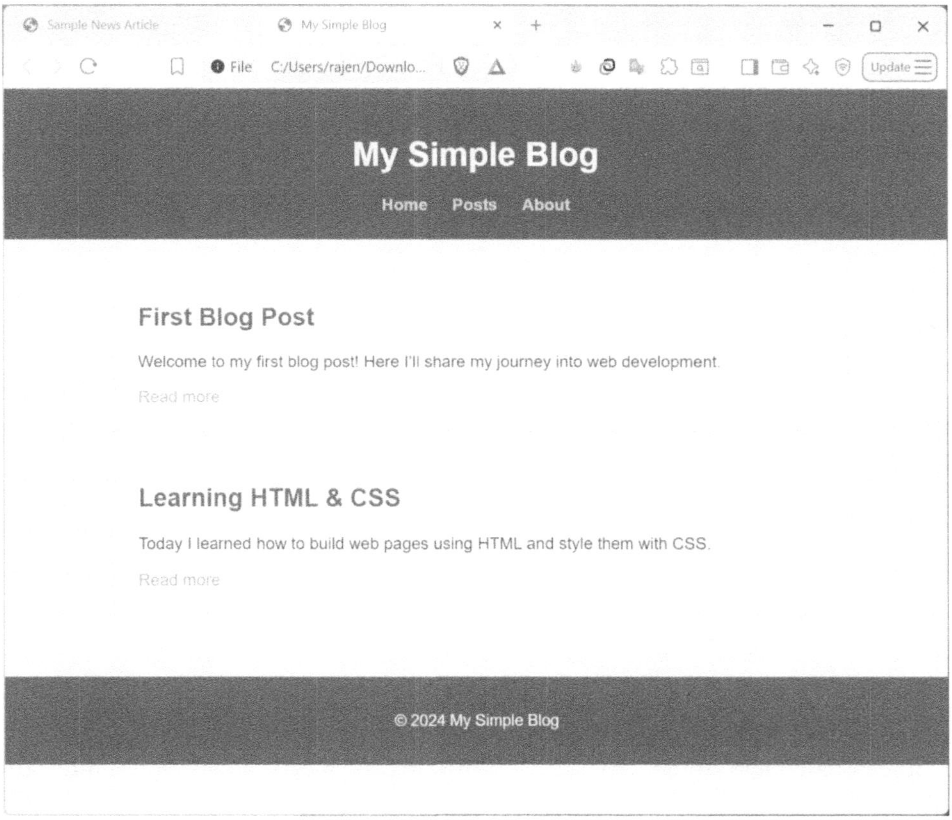

As you apply these styles, notice how each CSS property changes the appearance of the blog homepage:

background-color sets the page and section colors for contrast and readability.
font-family and **color** improve text readability.

padding and **margin** create space around content, preventing it from feeling cramped.
border and **border-radius** add subtle definition and soften article edges.
box-shadow gives a gentle depth to articles, making them stand out.

Customizing Your Blog with Layout and Spacing

Let's explore how layout and spacing make your blog feel modern and organized, a common trend on real sites like personal portfolios and minimalist blogs.

Try adjusting the spacing and layout to see how it impacts the design:

Change the main section's maximum width for wider screens:

```
main {
  max-width: 850px;
}
```

Increase the **gap** between navigation links for better spacing:

```
nav ul {
  gap: 36px;
}
```

Add more **padding** to articles for a comfortable reading area:

```
article {
  padding: 30px;
}
```

Experiment with these properties. Adjust the numbers and refresh your browser to see changes in real time. This hands-on tweaking is how experienced developers refine their site designs.

Enhancing Blog Posts with Consistent Style

A user-friendly blog is not just about layout; it's also about consistent, attractive styling. Below are some tips for refining your blog's look, commonly used by professional web designers:

Use a consistent color palette for text and backgrounds.
Style links so they are easy to identify and interact with.
Add hover effects to buttons and links for a modern, interactive feel.
Use margins and padding to create a clean separation between sections.

You can enhance your articles with a subtle hover effect:

```
article:hover {
  box-shadow: 0 4px 16px rgba(0,0,0,0.10);
  transition: box-shadow 0.2s;
}
```

This provides a visual cue when users move their mouse over each blog post, similar to the effect you see on many popular blogging platforms.

Hands-On Coding Summary

In this exercise, you created and styled a simple blog homepage using real HTML and CSS. You learned how to:

Structure a basic blog page with header, navigation, main content, and footer.
Apply CSS properties to control color, spacing, borders, and layout.
Use hover effects and box shadows to add interactivity and depth.
Experiment with spacing and alignment for a modern, readable design.

Try customizing fonts, colors, and spacing even more to match your unique style or to mimic a favorite website. This practice is key to developing your own web design skills.

Up next, in the conclusion of the chapter, you'll reflect on the concepts you've mastered and take on a coding challenge that reinforces your new CSS skills, preparing you for even more advanced web development adventures!

7. ADVANCED TEXT AND FONT STYLING

Typography is one of the most powerful tools in a web developer's toolkit. The way you style text and choose fonts can instantly transform a basic webpage into a visually engaging experience, capturing your visitors' attention and shaping their first impressions. In this chapter, you'll unlock a new level of design control by mastering advanced text and font styling techniques. Whether you're building a modern portfolio, a personal blog, or a polished e-commerce site, understanding how to creatively use fonts and text effects is essential for making your site stand out.

You'll begin by exploring how to enhance your web pages with beautiful, professional fonts using Google Fonts, along with strategies for selecting web-safe fonts that ensure your text always looks great—no matter where or how your site is viewed. Next, you'll learn to style links and craft subtle yet engaging hover effects, making your navigation both intuitive and inviting. With hands-on practice, you'll gain confidence in adjusting text spacing, alignment, and layout, giving you the tools to create clean, readable, and visually balanced pages.

To put your new skills into action, you'll work through a guided project: designing a stylish, resume-inspired webpage. This practical exercise will help you connect each technique to real-world web development scenarios, demonstrating how advanced text styling can elevate any site's appearance and usability. By the end of this chapter, you'll not only be able to impress visitors with beautiful typography but also build the confidence and creativity needed to tackle professional-quality web projects. Let's dive in and start shaping your words into something truly memorable.

7.1. Using Google Fonts and Web-Safe Fonts

Fonts play a crucial role in shaping the look and feel of your website. The right font choices can make your site appear professional, friendly, or creative, while poor font selections might hurt readability and user experience. As you progress with CSS styling, it's important to understand both **web-safe fonts** and the powerful world of **Google Fonts**. In this section, you'll learn how to use both options skillfully, ensuring your text looks great on any device and browser.

What Are Web-Safe Fonts?

Before exploring online font libraries, let's clarify what **web-safe fonts** are. Web-safe fonts are a set of fonts that are commonly installed on most computers and devices.

Because they're almost always available, you can be confident your website text will display consistently for most visitors.

Some popular web-safe fonts include:

- Arial
- Verdana
- Times New Roman
- Georgia
- Courier New

When you use these fonts, you reduce the risk of your text looking different on other people's screens.

Setting Web-Safe Fonts with CSS

Let's put web-safe fonts into practice using CSS. The main property you'll use is **font-family**. It's a good idea to specify a **font stack**: a prioritized list of fonts, so the browser can use the first one it finds.

Suppose you're creating a simple portfolio website. You want your headings to appear in **Georgia**, and your body text in **Verdana**.

```
<!DOCTYPE html>
<html lang= "en">
<head>
  <meta charset= "UTF-8">
  <title>My Web-Safe Portfolio</title>
  <style>
    h1, h2, h3 {
      font-family: Georgia, 'Times New Roman', Times, serif;
    }
    p {
      font-family: Verdana, Geneva, Tahoma, sans-serif;
    }
  </style>
</head>
<body>
  <h1>Jane Doe Portfolio</h1>
  <h2>About Me</h2>
  <p>Hello! I'm Jane, a web developer specializing in accessible, user-
friendly sites. Welcome to my online portfolio.</p>
</body>
</html>
```

Notice the use of **font-family** for both headings and paragraphs. Each list ends with either `serif` or `sans-serif` as a fallback, ensuring your text stays readable even if the other fonts are unavailable.

Why Use Google Fonts?

While web-safe fonts are reliable, they can also feel a bit generic. Today's websites often feature distinctive, eye-catching typography using online font libraries. **Google Fonts** is a free, widely-used service that offers hundreds of stylish fonts optimized for the web. With Google Fonts, you can set your site apart and better match your brand or desired mood.

Adding Google Fonts to Your Project

Let's walk through the process of using Google Fonts in your HTML and CSS.

Choose a Font from Google Fonts

Visit fonts.google.com and browse the font library. Let's say you want to use **Montserrat** for headings and **Roboto** for body text—both are modern, popular choices seen on many professional sites.

Get the Embed Link

- Search for "Montserrat" and "Roboto."

- Select the weights/styles you want (for example, Regular 400 and Bold 700).

- Click "Get embed code" or "< >" to copy the provided link.

The embed code looks like this:

```
<link href=
"https://fonts.googleapis.com/css2?family=Montserrat:wght@400;700&famil
y=Roboto:wght@400;700&display=swap"rel= "stylesheet">
```

Insert the Link in Your HTML

Place the copied `<link>` tag inside the `<head>` section of your HTML file, before your CSS.

Apply the Google Fonts in CSS

Update your **font-family** rules in CSS to use Montserrat and Roboto, with web-safe fallbacks.

Here's a complete example for a simple blog homepage:

```
<!DOCTYPE html>
<html lang= "en">
<head>
```

```
  <meta charset= "UTF-8">
  <title>My Blog with Google Fonts</title>
  <link href=
"https://fonts.googleapis.com/css2?family=Montserrat:wght@400;700&famil
y=Roboto:wght@400;700&display=swap"rel= "stylesheet">
  <style>
    body {
      font-family: 'Roboto', Arial, sans-serif;
      margin: 40px;
      background: #fafafa;
    }
    h1, h2, h3 {
      font-family: 'Montserrat', 'Arial Black', Arial, sans-serif;
      font-weight: 700;
      letter-spacing: 1px;
    }
    p {
      font-family: 'Roboto', Arial, sans-serif;
      font-size: 1.1em;
      line-height: 1.6;
      color: #333;
    }
  </style>
</head>
<body>
  <h1>Welcome to My Coding Blog</h1>
  <h2>Why I Love Learning Web Development</h2>
  <p>Web development gives you the power to turn ideas into reality.
With just a few lines of code, you can share your thoughts, showcase
your projects, and connect with the world.</p>
</body>
</html>
```

Now your headings use Montserrat and your body text uses Roboto, both loaded from Google Fonts. If, for any reason, Google Fonts can't load, the browser will fall back to Arial or other sans-serif fonts.

Combining Web-Safe and Google Fonts for Flexibility

It's best practice to always add web-safe fonts as fallbacks, even when using Google Fonts. This approach guarantees that your site remains readable if users experience network issues, or if Google Fonts is blocked in their region.

For example, a font stack like this:

```
font-family: 'Lato', 'Helvetica Neue', Helvetica, Arial, sans-serif;
```

means your website will use Lato from Google Fonts first, then look for "Helvetica Neue," then Helvetica, and so on. This increases the consistency and reliability of your design.

Hands-On Exercise: Build a Stylish Product Card

Let's apply your new skills by creating a product card for a small e-commerce site. You'll use Google Fonts for a modern look and provide web-safe fallbacks.

Select Your Fonts

Choose **Poppins** for headings and **Open Sans** for body text.

Add Google Fonts to Your HTML

Go to Google Fonts and search for Poppins and Open Sans. Choose the weights you want (e.g., 400 and 600), then copy the embed link:

```
<link href=
"https://fonts.googleapis.com/css2?family=Open+Sans:wght@400;600&family
=Poppins:wght@400;600&display=swap"rel= "stylesheet">
```

Write the HTML and CSS for the Product Card

```
<!DOCTYPE html>
<html lang= "en">
<head>
  <meta charset= "UTF-8">
  <title>Product Card Example</title>
  <link href=
"https://fonts.googleapis.com/css2?family=Open+Sans:wght@400;600&family
=Poppins:wght@400;600&display=swap"rel= "stylesheet">
  <style>
    body {
      background: #f0f4f8;
      font-family: 'Open Sans', Verdana, Geneva, sans-serif;
      display: flex;
      justify-content: center;
      align-items: center;
      height: 100vh;
    }
    .product-card {
      background: #fff;
      border-radius: 12px;
      box-shadow: 0 2px 8px rgba(0,0,0,0.1);
      max-width: 320px;
      padding: 24px;
      text-align: center;
    }
    .product-title {
      font-family: 'Poppins', Arial, 'Helvetica Neue', sans-serif;
      font-weight: 600;
      font-size: 1.5em;
      margin-bottom: 10px;
    }
    .product-description {
      font-family: 'Open Sans', Verdana, Geneva, sans-serif;
      font-size: 1em;
```

```
      color: #555;
      margin-bottom: 16px;
    }
    .price {
      font-family: 'Poppins', Arial, 'Helvetica Neue', sans-serif;
      font-weight: 600;
      color: #2c9c69;
      font-size: 1.2em;
    }
  </style>
</head>
<body>
  <div class= "product-card">
    <div class= "product-title">Wireless Headphones</div>
    <div class= "product-description">Enjoy crisp sound and all-day
comfort with our latest wireless headphones. Perfect for work, travel,
or relaxation.</div>
    <div class= "price">$89.99</div>
  </div>
</body>
</html>
```

This product card uses Poppins for the title and price, and Open Sans for the description. Notice how the font stacks always include web-safe options after the Google Fonts.

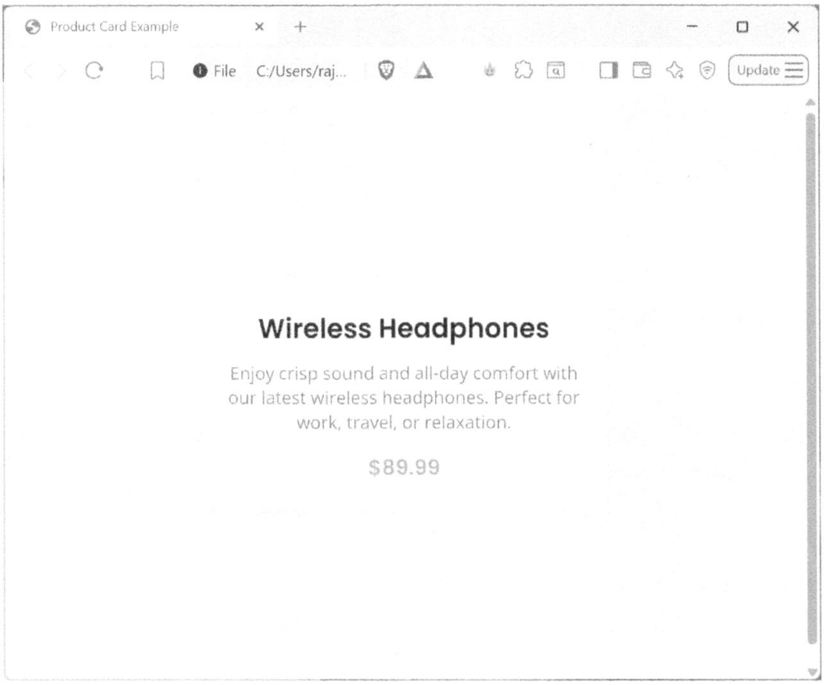

Real-World Inspiration: Global Design Trends

Many popular sites blend unique fonts with web-safe fallbacks. For example, a cultural blog might use Google Fonts to create a friendly, approachable feel with **Quicksand** for headings and **Source Sans Pro** for body text. An international e-commerce site could rely on **Noto Sans** and **Noto Serif** to support multiple languages with consistent style.

Experimenting with different Google Fonts can help your projects stand out and feel up-to-date, while always ensuring accessibility and reliability.

Coding Summary and Practice

In this section, you learned how to set web-safe fonts for basic reliability, and how to enhance your site's personality with Google Fonts. You practiced loading and applying Google Fonts, building versatile font stacks, and creating a modern product card. Try swapping in different Google Fonts on your own projects, and always provide web-safe fallbacks for best results.

Up next, you'll discover how to style links and add engaging hover effects, bringing interactivity and polish to your web pages. Get ready to make your site truly stand out for users!

7.2. Styling Links and Hover Effects

Links are the lifeblood of the web, guiding visitors from one page to another or connecting them to valuable resources. By default, browsers style links in a basic way, often using blue underlined text for unvisited links and purple for visited links. While this makes links easy to spot, it lacks personality and doesn't always match your site's unique style. As you grow as a web developer, learning to customize link appearance and add interactive hover effects will dramatically improve your site's look and user experience.

Let's dive into practical CSS techniques for transforming ordinary links into visually appealing, interactive elements that capture attention and invite clicks.

Understanding Link States

Before customizing links, it's important to know about **link states**. CSS provides pseudo-classes that let you style links differently based on how users interact with them:

:link targets a link that has not yet been visited.
:visited targets a link the user has already clicked.
:hover applies styles when a user moves their mouse over the link.
:active applies styles when a link is being clicked.

By combining these pseudo-classes, you can create clear, accessible, and attractive navigation.

Basic Link Styling with CSS

Start by creating a simple HTML snippet with several links. This will serve as our practice ground.

```
<!DOCTYPE html>
<html lang= "en">
<head>
  <meta charset= "UTF-8">
  <title>Link Styling Practice</title>
  <link rel= "stylesheet"href= "styles.css">
</head>
<body>
  <h2>Explore Our Site</h2>
  <nav>
    <a href= "portfolio.html">Portfolio</a> |
    <a href= "blog.html">Blog</a> |
    <a href= "shop.html">Shop</a>
  </nav>
</body>
</html>
```

Suppose you want your links to match your site's color scheme, have no underline, and stand out more. Here's how you can achieve this in your `styles.css` file:

```
a:link,
a:visited {
  color: #3498db;
  text-decoration: none;
  font-weight: bold;
  transition: color 0.3s;
}
```

Let's break this down:

color changes the link text color.
text-decoration: none removes the default underline.
font-weight: bold makes links stand out.
transition: color 0.3s adds a smooth color change effect.

Adding Hover Effects for Interactivity

One of the best ways to make links interactive is by styling the **:hover** state. When visitors move their cursor over a link, it should respond visually. This signals that the link is clickable and enhances user engagement.

Update your CSS to add a hover effect:

```
a:hover {
  color: #e67e22;
  text-decoration: underline;
  cursor: pointer;
}
```

Here's what happens:

The link color changes to a warm orange.
The underline reappears on hover, making it clear this is an interactive element.
The **cursor: pointer** ensures the pointer icon changes, reinforcing clickability.

Real-World Scenario: Navigation Bar for a Portfolio Site

Let's apply these techniques in a real-world scenario. Imagine you're building a navigation bar for a personal portfolio website. You want the links to look modern and respond with a subtle background highlight when hovered.

First, structure your navigation bar using HTML:

```
<nav class= "main-nav">
  <a href= "about.html">About Me</a>
  <a href= "projects.html">Projects</a>
  <a href= "contact.html">Contact</a>
</nav>
```

Now, add CSS to style the links as buttons and create a smooth hover effect:

```
.main-nav a:link,
.main-nav a:visited {
  display: inline-block;
  padding: 10px 20px;
  margin: 0 8px;
  color: #2c3e50;
  background-color: #ecf0f1;
  border-radius: 5px;
  text-decoration: none;
  font-size: 1.1rem;
  transition: background 0.3s, color 0.3s;
}

.main-nav a:hover {
  background-color: #2c3e50;
  color: #ecf0f1;
  box-shadow: 0 2px 8px rgba(44, 62, 80, 0.15);
}
```

This approach creates button-like links that change background and text color on hover, making navigation feel responsive and polished.

Advanced Hover Effects: Underline Animation

Modern websites often use animated underline effects for links, adding a touch of sophistication. Let's create a sliding underline that appears when a user hovers over a link, inspired by trending web design portfolios.

Add the following HTML for your navigation:

```
<nav class= "animated-links">
  <a href= "#">Home</a>
  <a href= "#">Portfolio</a>
  <a href= "#">Contact</a>
</nav>
```

Now, use CSS to achieve the animated underline:

```
.animated-links a {
  position: relative;
  color: #34495e;
  text-decoration: none;
  padding-bottom: 4px;
  transition: color 0.3s;
}

.animated-links a::after {
  content: "";
  display: block;
  position: absolute;
  left: 0;
  bottom: 0;
  height: 2px;
  width: 0;
  background: #1abc9c;
  transition: width 0.3s;
}

.animated-links a:hover {
  color: #1abc9c;
}

.animated-links a:hover::after {
  width: 100%;
}
```

Explanation:

position: relative allows absolute positioning of the pseudo-element.
::after creates a line beneath the link.
On hover, the underline smoothly expands from left to right, adding a dynamic touch.

Scenario: E-Commerce Product Links

Consider a global e-commerce site showcasing product categories. For accessibility and brand consistency, you want hovered links to not only change color, but also slightly increase in size.

```
<div class= "product-categories">
  <a href= "#">Electronics</a>
  <a href= "#">Fashion</a>
  <a href= "#">Home & Living</a>
</div>
```

CSS:

```
.product-categories a:link,
.product-categories a:visited {
  color: #222;
  text-decoration: none;
  padding: 8px 16px;
  font-size: 1rem;
  border-radius: 4px;
  transition: background 0.2s, transform 0.2s;
}

.product-categories a:hover {
  background: #f6e58d;
  color: #30336b;
  transform: scale(1.1);
}
```

This effect draws attention to hovered links, making them feel tactile and interactive, perfect for high-traffic shopping sites.

Hands-On Exercise: Customizing Link Styles

Try these steps to practice your new skills:

Create a set of navigation links for a fictional travel blog.
Style the links with your favorite color scheme.
Add a hover effect that changes both background and text color.
Experiment with font size and border radius for a unique look.
Bonus: Try using the **::after** pseudo-element to add an animated underline.

Here's a starter template:

```
<nav class= "travel-nav">
  <a href= "#">Destinations</a>
  <a href= "#">Tips</a>
  <a href= "#">Gallery</a>
```

```
</nav>
```

And a sample CSS to get you started:

```
.travel-nav a {
  color: #136a8a;
  text-decoration: none;
  font-size: 1.2rem;
  margin: 0 10px;
  padding: 8px 16px;
  border-radius: 4px;
  transition: background 0.4s, color 0.4s;
}

.travel-nav a:hover {
  background: #fbc531;
  color: #fff;
}
```

Feel free to customize the colors and effects to match your own blog vision!

Coding Summary

You've just learned how to transform plain links into engaging, interactive elements using CSS. By mastering pseudo-classes like **:hover**, exploring animated effects with **::after**, and applying real-world styling scenarios, you're adding a professional touch to your web pages. Practice these techniques on your projects to solidify your understanding and develop your unique web design style.

Next, you'll discover how to control text spacing and alignment with CSS, giving you even more power to organize content and create visually harmonious layouts. Get ready to make your text look polished and perfectly positioned!

7.3. Controlling Text Spacing and Alignment

When you want your web pages to look professional and readable, controlling the spacing and alignment of your text is essential. CSS provides several properties to fine-tune the way your text appears, ensuring content is both visually appealing and easy to navigate. In this section, you'll learn how to use CSS to adjust the space between lines, letters, and words, as well as how to align text within its container. These techniques are crucial for everything from blog articles to product descriptions in online stores, and mastering them will give your sites a polished, modern feel.

Understanding Line Spacing with `line-height`

Every paragraph or heading on a website consists of multiple lines of text. The space between these lines—known as **line height**—can impact readability. Increasing the line height makes text less cramped and easier for users to scan.

You control line height using the `line-height` property in CSS. This property can take unitless numbers, percentages, or specific length units like pixels.

Let's see it in action:

```
<!DOCTYPE html>
<html lang= "en">
<head>
  <meta charset= "UTF-8">
  <title>Line Height Example</title>
  <style>
    .spacious-text {
      line-height: 2;
      font-size: 1.1rem;
      color: #333;
    }
  </style>
</head>
<body>
  <p class= "spacious-text">
    Carefully chosen line spacing improves readability and gives your
text a clean, modern appearance. Try adjusting the line height and see
how it changes the feel of your content.
  </p>
</body>
</html>
```

Try changing the `line-height` value to see how the text spacing grows or shrinks. For most body text, a value between 1.4 and 2 provides comfortable reading.

Adjusting Space Between Letters and Words

Beyond line spacing, you may want to control the distance between individual letters or entire words for styling or branding purposes. CSS offers two useful properties:

letter-spacing: Controls the space between characters.
word-spacing: Controls the space between words.

These properties are especially helpful for headings, logos, or when following a particular design trend.

Here's how you can apply them:

```
<!DOCTYPE html>
<html lang= "en">
<head>
  <meta charset= "UTF-8">
  <title>Letter and Word Spacing</title>
  <style>
    .wide-letters {
      letter-spacing: 3px;
      font-size: 2rem;
      color: #1a73e8;
      font-weight: bold;
    }
    .spaced-words {
      word-spacing: 20px;
      font-size: 1.2rem;
      color: #444;
    }
  </style>
</head>
<body>
  <h1 class= "wide-letters">Portfolio</h1>
  <p class= "spaced-words">Web Developer Designer Creator</p>
</body>
</html>
```

Notice how the letters in "Portfolio" are spaced out, giving a stylish, branded look. The paragraph demonstrates increased word spacing, which can be useful for taglines or navigation menus. Experiment with different values to match the tone of your site.

Aligning Text Horizontally

Aligning text is a fundamental part of web design. Whether you want your content left-aligned, centered, or right-aligned, CSS makes it simple using the `text-align` property.

The most common values are:

left: Aligns text to the left (default for most Western languages)
center: Centers the text within its container
right: Aligns text to the right
justify: Stretches lines so that each line has equal width (except the last one)

Let's walk through a typical web scenario: centering a heading and justifying a block of text for a blog post.

```
<!DOCTYPE html>
<html lang= "en">
<head>
  <meta charset= "UTF-8">
  <title>Text Alignment Example</title>
  <style>
```

```
    h2 {
      text-align: center;
      font-size: 2rem;
      margin-bottom: 0.5em;
      color: #4b0082;
    }
    .blog-content {
      text-align: justify;
      max-width: 600px;
      margin: 0 auto;
      font-size: 1.1rem;
      color: #222;
    }
  </style>
</head>
<body>
  <h2>Welcome to My Blog</h2>
  <div class= "blog-content">
    <p>
      Justified text creates clean left and right edges, commonly used
in magazines and professional blogs. Try using different alignments for
headings, navigation, or body text to see what feels right for your
project.
    </p>
  </div>
</body>
</html>
```

This example shows a centered heading and a justified paragraph, similar to what you might see on a popular blog or news site. You can apply `text-align` to headings, paragraphs, lists, and other block-level elements.

Vertical Alignment for Inline Elements

Text not only aligns horizontally but can also be adjusted vertically, especially when dealing with images or inline elements such as icons within a line of text. The `vertical-align` property helps you control the vertical placement of these elements relative to the text baseline.

Common values for `vertical-align` include:

baseline (default): Aligns with the baseline of the surrounding text
middle: Aligns to the middle of the line
top: Aligns with the top of the tallest element on the line
bottom: Aligns with the lowest element on the line

Here's how to vertically align an icon next to some text, a technique often seen in modern navigation bars or feature lists:

```
<!DOCTYPE html>
<html lang= "en">
<head>
  <meta charset= "UTF-8">
  <title>Vertical Align Example</title>
  <style>
    .icon-text img {
      vertical-align: middle;
      width: 24px;
      height: 24px;
    }
    .icon-text {
      font-size: 1.2rem;
      color: #00796b;
    }
  </style>
</head>
<body>
  <p class= "icon-text">
    <img src= "https://cdn-icons-png.flaticon.com/512/25/25694.png"alt=
"star icon">
    Featured Project
  </p>
</body>
</html>
```

Try replacing `middle` with `top` or `bottom` and observe how the icon shifts in relation to the text.

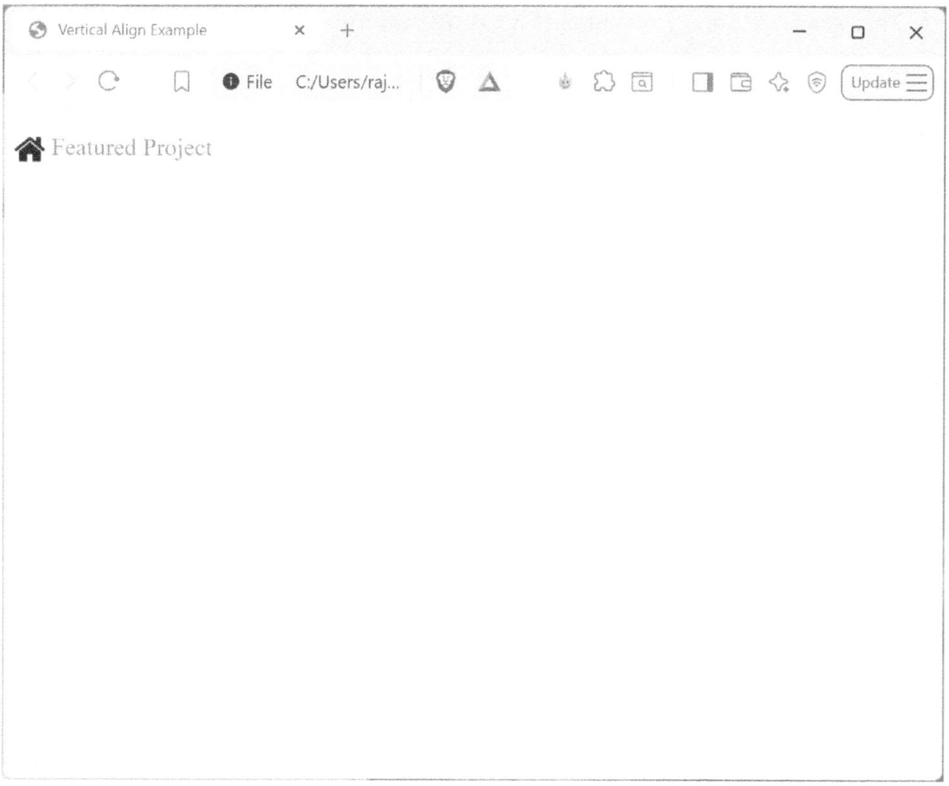

Real-World Scenario: E-Commerce Product Listings

Let's apply what you've learned to a practical example: styling product names and prices in an e-commerce product grid. Clean spacing and alignment make shopping sites easier to use.

```
<!DOCTYPE html>
<html lang= "en">
<head>
  <meta charset= "UTF-8">
  <title>Product Listing Example</title>
  <style>
    .product {
      width: 250px;
      border: 1px solid #e0e0e0;
      border-radius: 5px;
      padding: 16px;
      margin: 12px auto;
      font-family: Arial, sans-serif;
    }
    .product-name {
      font-size: 1.1rem;
      letter-spacing: 1px;
      font-weight: bold;
      text-align: center;
      margin-bottom: 0.5em;
    }
```

```
    .product-price {
      font-size: 1rem;
      color: #388e3c;
      text-align: right;
      word-spacing: 8px;
    }
    .product-desc {
      font-size: 0.95rem;
      line-height: 1.7;
      color: #555;
      text-align: left;
    }
  </style>
</head>
<body>
  <div class= "product">
    <div class= "product-name">Wireless Headphones</div>
    <div class= "product-price">$49.99</div>
    <div class= "product-desc">
      Lightweight, long-lasting, and premium sound quality.<br>
      Perfect for music lovers on the go.
    </div>
  </div>
</body>
</html>
```

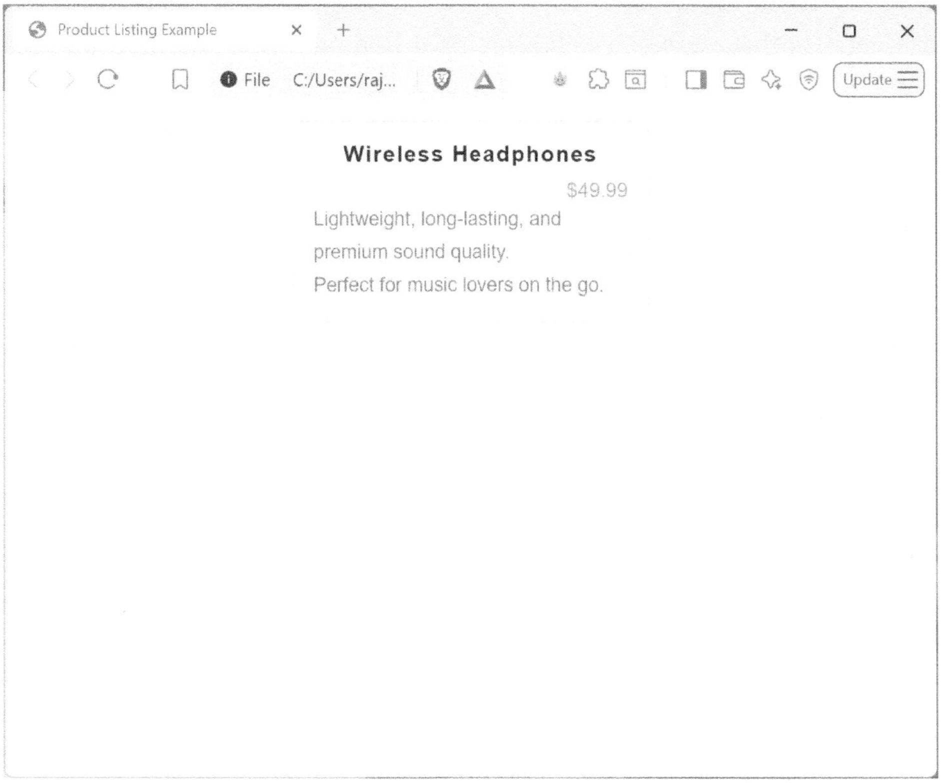

Notice how each piece of text uses different spacing and alignment for clarity and emphasis. This approach is used by leading e-commerce sites to display information in an organized and visually appealing way.

Quick Hands-On Summary

You've just explored powerful CSS properties for controlling text spacing and alignment, including `line-height`, `letter-spacing`, `word-spacing`, `text-align`, and `vertical-align`. Practicing these techniques will make your content much easier to read and more visually attractive. Try adjusting these properties on your own web pages, playing with different combinations to see how they affect your layouts.

Up next, you'll put your new skills into practice as you design a stylish resume page. Get ready to combine text styling, spacing, and alignment to create a professional, standout web resume!

7.4. Hands-On Exercise: Designing a Stylish Resume Page

A resume page is one of the most practical and rewarding projects for new web developers. Not only does it help you practice advanced text and font styling, but it also results in a real-world asset you can use to showcase your skills, education, and experience. In this exercise, you'll combine the power of CSS with semantic HTML to design a visually appealing, professional resume that stands out—much like the personal portfolio pages and "About Me" sections you see on popular web designer sites.

Let's build your stylish resume page step by step, focusing on advanced CSS properties for text and font styling, layout techniques, and modern design trends.

Planning Your Resume Structure

Before jumping into the code, it's helpful to outline what sections your resume will include. Most resumes have these key parts:

- Header with your name and job title
- Contact details
- Summary or profile section
- Work experience
- Education
- Skills

You might have seen similar layouts on global freelancer platforms or in the "Our Team" pages of creative agencies. Structuring your content this way keeps it readable and professional.

Setting Up the Basic HTML

We'll begin by writing the HTML structure for your resume. This provides the content foundation that you'll style using CSS.

Create a new HTML file called `resume.html`.

Add the basic HTML document structure and include semantic tags for the main sections.

```
<!DOCTYPE html>
<html lang= "en">
<head>
  <meta charset= "UTF-8">
  <title>My Stylish Resume</title>
  <link rel= "stylesheet"href= "resume.css">
</head>
<body>
  <header>
    <h1>Jordan Lee</h1>
    <p class= "job-title">Front-End Web Developer</p>
    <p class= "contact-info">jordan.lee@email.com | (555) 123-4567 |
Seattle, WA</p>
  </header>
  <section class= "summary">
    <h2>Profile</h2>
    <p>Creative developer with a passion for building beautiful, user-
friendly web experiences. Skilled in HTML, CSS, and modern web
technologies.</p>
  </section>
  <section class= "experience">
    <h2>Experience</h2>
    <div class= "job">
      <h3>Web Developer <span>@ PixelWorks Studio</span></h3>
      <p class= "date">March 2022 - Present</p>
      <ul>
        <li>Designed and launched responsive websites for global
clients.</li>
        <li>Collaborated with designers to implement modern UI
trends.</li>
      </ul>
    </div>
    <div class= "job">
      <h3>Junior Developer <span>@ CodeLab</span></h3>
      <p class= "date">June 2020 - Feb 2022</p>
      <ul>
        <li>Maintained and updated e-commerce platforms.</li>
        <li>Assisted with cross-browser compatibility testing.</li>
      </ul>
    </div>
  </section>
  <section class= "education">
    <h2>Education</h2>
    <p>B.S. in Computer Science, University of Washington (2020)</p>
  </section>
  <section class= "skills">
    <h2>Skills</h2>
```

```
    <ul>
      <li>HTML & CSS</li>
      <li>Responsive Design</li>
      <li>JavaScript Basics</li>
      <li>Version Control (Git)</li>
    </ul>
  </section>
</body>
</html>
```

This structure mimics what you might find on professional networking sites or the "Meet the Team" pages of global design studios. Now, let's make it visually impressive with CSS.

Applying Advanced Text and Font Styling with CSS

Next, you'll apply advanced CSS properties to enhance the typography and layout of your resume. Create a new file named `resume.css` in the same folder as your HTML file.

Start by selecting a modern web font and setting base styles for the page.

```
@import url(
'https://fonts.googleapis.com/css?family=Montserrat:400,700|Roboto:400,
700&display=swap');

body {
  font-family: 'Roboto', Arial, sans-serif;
  background: #f9f9f9;
  color: #222;
  margin: 0;
  padding: 0;
  line-height: 1.6;
}
```

Here, the `@import` rule loads fonts from Google Fonts, a common practice for modern, professional-looking sites.

The `font-family` property sets a clean, readable base font.

Style the header for impact and clarity.

```
header {
  background: #262a33;
  color: #fff;
  text-align: center;
  padding: 2.5rem 1rem 1.5rem 1rem;
  border-bottom: 4px solid #3ac569;
}

h1 {
  font-family: 'Montserrat', Arial, sans-serif;
  font-size: 2.8rem;
  font-weight: 700;
  margin: 0;
  letter-spacing: 1.5px;
  text-transform: uppercase;
}

.job-title {
  font-size: 1.25rem;
  font-weight: 400;
  margin-top: 0.5rem;
  color: #3ac569;
```

```
  letter-spacing: 1px;
}

.contact-info {
  font-size: 1rem;
  margin-top: 0.7rem;
  opacity: 0.85;
}
```

`text-transform: uppercase` and `letter-spacing` create a bold, memorable name display, similar to personal portfolio sites.

The colored border and accent text add a modern touch.

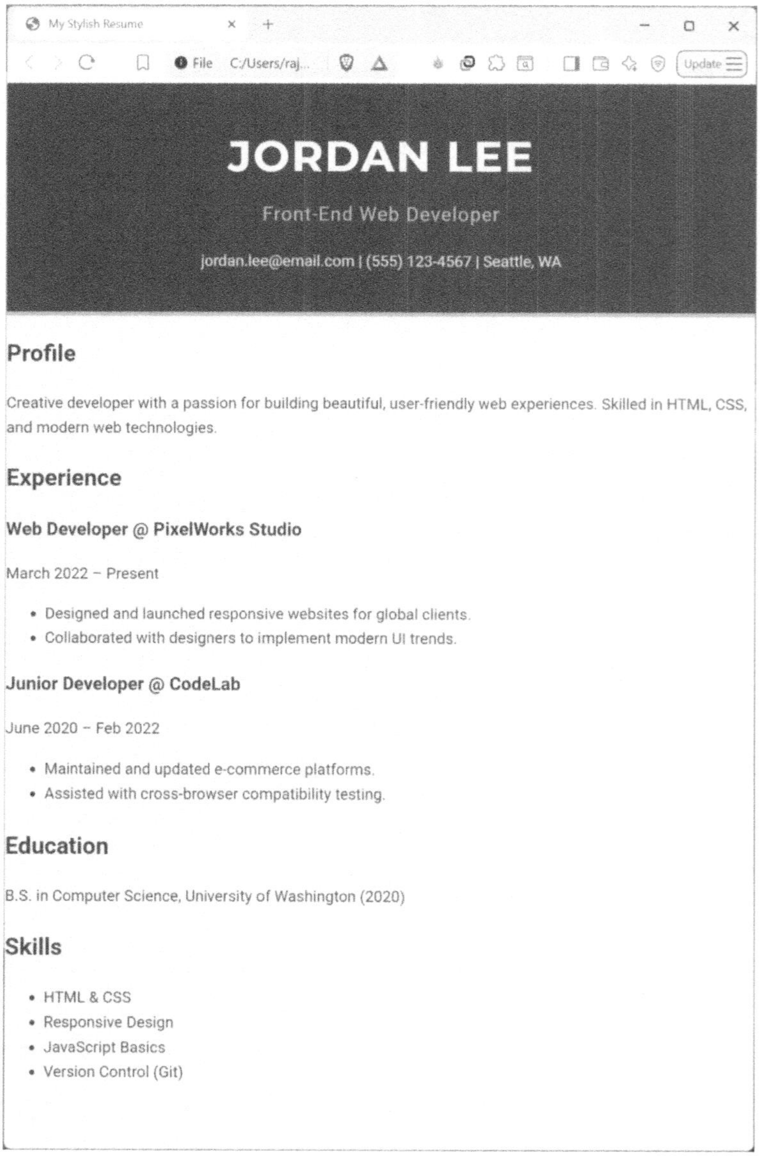

Enhance section headers and layout for readability.

```
section {
  max-width: 700px;
  margin: 2.5rem auto;
  background: #fff;
  padding: 2rem 2.5rem;
  border-radius: 15px;
  box-shadow: 0 4px 16px rgba(0,0,0,0.06);
}

h2 {
  font-family: 'Montserrat', Arial, sans-serif;
  color: #262a33;
  font-size: 1.6rem;
  font-weight: 700;
  border-left: 5px solid #3ac569;
  padding-left: 0.8rem;
  margin-bottom: 1.2rem;
  letter-spacing: 0.5px;
  background: linear-gradient(90deg, #e9ffd9 0%, #fff 100%);
}
```

Section containers with rounded corners and subtle shadows echo the clean, card-like style popular on tech company team pages.

Style job titles, dates, and lists for clarity and professional polish.

```
.job h3 {
  font-size: 1.15rem;
  margin: 0.5rem 0 0.1rem 0;
  font-weight: 700;
  display: flex;
  justify-content: space-between;
  align-items: baseline;
}

.job h3 span {
  font-size: 1rem;
  color: #3ac569;
  font-weight: 400;
  margin-left: 1rem;
}

.date {
  font-size: 0.95rem;
  color: #888;
  margin-bottom: 0.4rem;
  font-style: italic;
}

.job ul {
  margin-top: 0.1rem;
  margin-bottom: 1.2rem;
  padding-left: 1.4rem;
  color: #222;
}
```

These styles improve hierarchy and scan-ability, drawing inspiration from job board profiles and résumé builders.

Add finishing touches to the skills section and overall page.

```
.skills ul {
  display: flex;
  flex-wrap: wrap;
  gap: 0.7rem;
  list-style: none;
  padding: 0;
  margin: 0;
}

.skills li {
  background: #e9ffd9;
  color: #262a33;
  padding: 0.5rem 1.2rem;
  border-radius: 20px;
  font-size: 1rem;
  font-weight: 500;
  box-shadow: 0 2px 8px rgba(58,197,105,0.09);
  margin-bottom: 0.6rem;
```

```
}
```

Using badges for skills reflects trends seen in developer portfolios and e-commerce team pages.

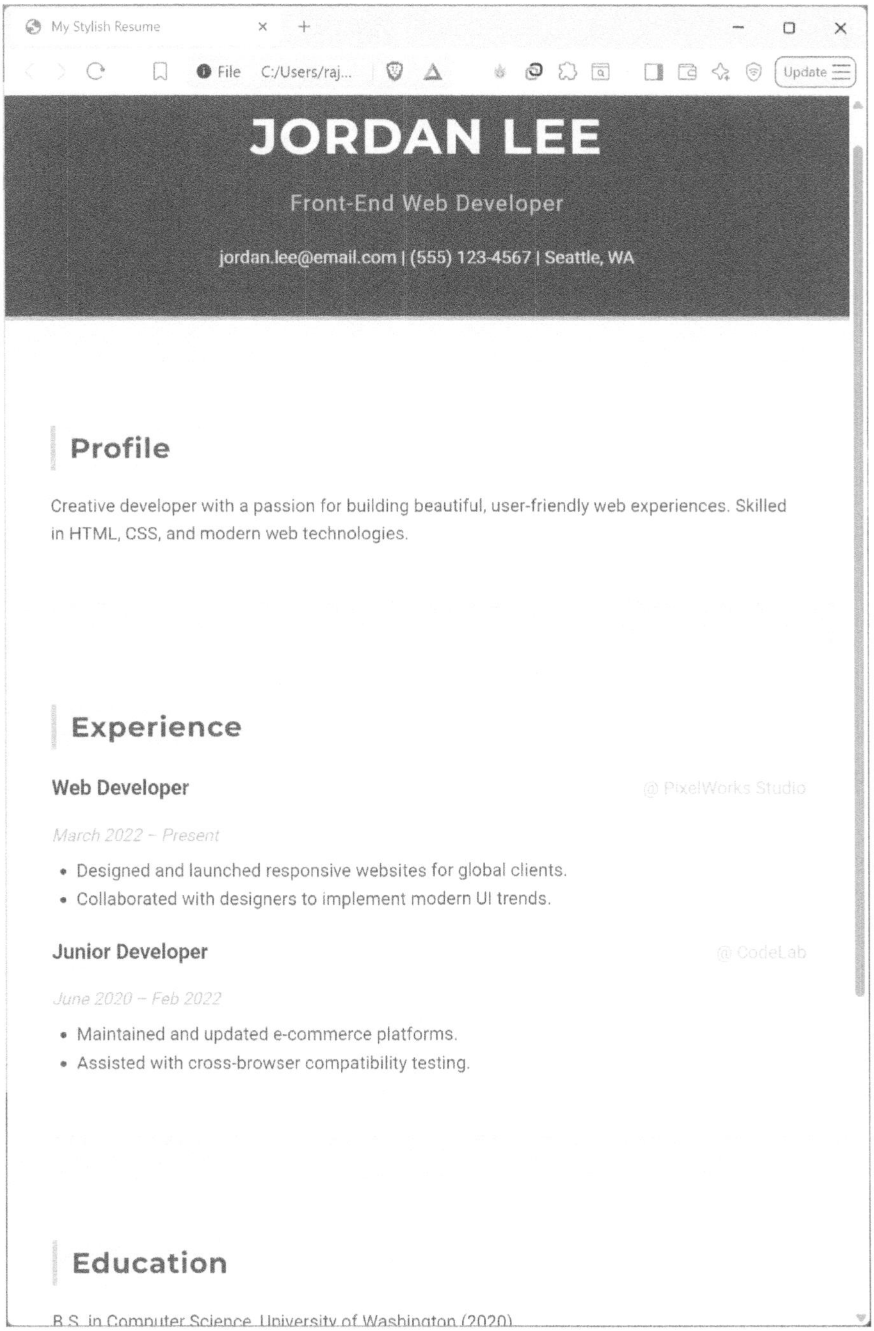

Responsive Design: Making Your Resume Look Great on Any Device

A professional resume should be easy to read on both desktop and mobile screens. Let's add a simple media query for responsiveness.

```
@media (max-width: 600px) {
  section {
    padding: 1.2rem 0.7rem;
  }

  h1 {
    font-size: 2rem;
  }

  h2 {
    font-size: 1.15rem;
    padding-left: 0.4rem;
  }

  .skills ul {
    flex-direction: column;
    gap: 0.4rem;
  }
}
```

This snippet adjusts spacing and font size for smaller screens, a key aspect of modern web development.

Real-World Scenario: Showcasing Your Work

Once styled, your resume page mirrors the "About Me" or bio sections found on portfolio websites and professional networking platforms. Many freelance designers and developers use a similar format to introduce themselves and list their skills, making this exercise directly relevant to your future web projects.

You've now combined semantic HTML, advanced text and font styling, modern layout techniques, and responsive design to create a standout resume page. You practiced using custom fonts, color accents, badges, and flexible layouts—skills that form the backbone of professional web design. Try expanding your page with more sections, links to projects, or additional media queries for further practice.

8. WORKING WITH IMAGES AND BACKGROUNDS IN CSS

Images and backgrounds have the power to transform a simple webpage into a visually engaging user experience. In this chapter, you'll unlock the creative tools that professional web designers use to make sites stand out—from eye-catching hero images and immersive background patterns to sleek gradients and modern blending effects. By mastering these techniques, you'll be able to shape the look and feel of your websites, making them not only functional but also visually memorable and appealing.

We'll start by exploring how to set images as backgrounds using CSS, giving you the ability to control exactly how visuals appear behind your content. You'll learn how to adjust the size and placement of these images so they look perfect on any screen, a skill that's essential for building responsive and user-friendly sites. Next, you'll dive into adding gradients, a popular design trend that brings depth and style to everything from portfolios and blogs to e-commerce landing pages.

But the creativity doesn't stop there. This chapter also introduces overlay effects and blending modes, empowering you to layer colors, images, and effects for extra visual impact. By the end, you'll put your new skills to the test by building a hands-on landing page project featuring a bold hero image—a feature seen on many professional websites worldwide.

Whether your goal is to impress potential clients, attract blog readers, or simply express your creative vision, the skills you develop here will give you the confidence to design with both beauty and purpose. Let's dive in and start bringing your web designs to life with images, backgrounds, and dynamic visual effects.

8.1. Setting Images as Backgrounds

Images play a powerful role in web design, often setting the tone and visual appeal of a website. While adding images directly to your content with the HTML `<img>` tag is common, there's another dynamic way to use images in your layouts: setting them as backgrounds with CSS. This technique allows you to control how images decorate sections, highlight areas, and even create engaging visual effects without affecting the HTML content structure.

Let's explore how to set images as backgrounds using CSS and apply these skills to real-world web projects.

Understanding the CSS background-image Property

The **background-image** property in CSS lets you assign an image as the background of any HTML element. Unlike the `<img>` tag, which places an image directly on the page, background images are part of an element's styling. This means you can use them for decorative effects, layer them with colors, and control their size, position, and repetition.

Here's the basic syntax:

```
selector {
  background-image: url( 'path-to-your-image.jpg');
}
```

selector targets the HTML element you want to style.
url('...') points to your image file. This can be a local file or a direct link to an online image.

Practical Example: Portfolio Header with a Background Image

Imagine you're building a portfolio website and want a striking header with a background image that showcases your personal style.

Create the HTML structure

```
<header class= "portfolio-header">
  <h1>Welcome to My Portfolio</h1>
  <p>Showcasing my web development projects</p>
</header>
```

Add CSS to set a background image for the header

```
.portfolio-header {
  background-image: url( 'images/portfolio-hero.jpg');
  color: #fff; /* Ensures text stands out */  padding: 80px 0;
  text-align: center;
}
```

The image `portfolio-hero.jpg` appears behind the header content.
The text remains readable because of the contrasting color.

Enhancing Background Images: Repeat, Position, and Size

By default, background images repeat horizontally and vertically to fill the element. Often, you'll want to customize how the image appears.

background-repeat controls how the image repeats.

- `no-repeat` stops the image from repeating.

background-position sets where the image is placed.

- `center center` centers the image.

background-size allows you to adjust the image's size.

- `cover` stretches the image to cover the entire element, cropping if necessary.
- `contain` ensures the whole image is visible, fitting inside the element.

Let's see these in practice.

```
.portfolio-header {
  background-image: url( 'images/portfolio-hero.jpg');
  background-repeat: no-repeat;
  background-position: center center;
  background-size: cover;
  color: #fff;
  padding: 80px 0;
  text-align: center;
}
```

Now the header displays a single, centered image that covers the entire area, creating a modern and professional look.

Exercise: Blog Section Banner with a Background Image

A blog site often uses banners with background images to highlight featured posts. Try this hands-on coding exercise to reinforce your skills.

HTML for a blog banner

```
<section class= "blog-banner">
  <h2>Latest Insights</h2>
  <p>Explore tips, tutorials, and stories from the world of web
development</p>
</section>
```

CSS to style the banner with a background image

```
.blog-banner {
  background-image: url( 'images/blog-banner.jpg');
  background-repeat: no-repeat;
  background-position: top center;
  background-size: cover;
  padding: 60px 20px;
  color: #222;
  border-radius: 10px;
```

```
  margin: 30px 0;
  box-shadow: 0 4px 12px rgba(0,0,0,0.1);
}
```

This approach creates an attention-grabbing section with a smooth, visually appealing background. Try changing the `background-position` and `background-size` properties to see how the image's appearance changes.

Real-World Scenario: E-commerce Product Highlight

On e-commerce websites, background images are often used to spotlight featured sales or products without cluttering the page with extra image tags. Here's a practical coding scenario you can apply.

HTML for a product highlight section

```
<div class= "product-highlight">
  <h3>Summer Collection</h3>
  <p>Shop the latest trends for the sunny season!</p>
  <a href= "#"class= "shop-now">Shop Now</a>
</div>
```

CSS to set a themed background

```
.product-highlight {
  background-image: url( 'images/summer-collection.jpg');
  background-size: cover;
  background-repeat: no-repeat;
  background-position: center;
  color: #fff;
  padding: 70px 40px;
  text-align: center;
  border-radius: 12px;
}
.shop-now {
  display: inline-block;
  margin-top: 18px;
  background: rgba(255,255,255,0.8);
  color: #333;
  padding: 10px 25px;
  border-radius: 5px;
  text-decoration: none;
  font-weight: bold;
}
```

Notice how the button uses a semi-transparent background to stand out against the image. This is a popular design pattern in modern e-commerce and landing page designs.

Tips for Using Background Images Effectively

Use high-quality, optimized images to ensure fast loading times.

Choose images that enhance (not overpower) your content for better readability.

Combine background images with gradients or overlays using `background-color` and `rgba` for additional effects.

Make sure your text remains readable by adjusting colors or adding overlays.

You've learned how to set images as backgrounds with CSS, control their placement, and apply them to headers, banners, and product highlights. Experiment with the `background-image`, `background-size`, and `background-position` properties in your own projects to create visually compelling sections that stand out.

Up next, you'll discover how to control the size and placement of images with even greater precision. Get ready to master techniques that let you tailor every detail of your backgrounds for a truly professional look!

8.2. Controlling Image Size and Placement

Images play a vital role in bringing life and personality to websites. Whether you are designing a captivating portfolio or showcasing products in an online shop, understanding how to control the size and placement of your images is essential. CSS empowers you to fine-tune images, making them fit perfectly into your layout and ensuring they look great on all devices.

Sizing Images with CSS

By default, when you add an image to your HTML using the **`<img>`** tag, the browser displays it at its natural size. However, images often need to be resized to fit a specific design or to keep your layout responsive. CSS offers several properties for this purpose.

Common CSS Properties for Image Sizing:

- **width**: Sets the width of the image.
- **height**: Sets the height of the image.
- **max-width**: Prevents the image from exceeding a certain width.
- **max-height**: Prevents the image from exceeding a certain height.

Let's walk through a practical example. Imagine you're building a modern blog and want your post images to always fit nicely within your content area.

Start by adding an image in your HTML:

```
<img src= "images/blog-photo.jpg"alt= "A cup of coffee on a
table"class= "blog-image">
```

In your CSS, set the image to fill the width of its container, but not exceed its original size:

```
.blog-image {
  width: 100%;
  max-width: 600px;
  height: auto;
  display: block;
  margin: 0 auto;
}
```

- **width: 100%;** makes the image scale to the width of its parent container.
- **max-width: 600px;** ensures the image never gets larger than 600 pixels wide.
- **height: auto;** keeps the image's original aspect ratio, preventing it from looking stretched.
- **display: block;** removes the small gap below images that are inline by default.
- **margin: 0 auto;** centers the image horizontally.

Try different **max-width** values to see how your image responds. This technique is widely used on blogs, portfolios, and e-commerce product galleries.

Placing Images with CSS

Besides sizing, you often want to control exactly where your images appear within your design. CSS provides various tools for image placement, including **margin**, **float**, **position**, and modern layout techniques like **Flexbox**.

Let's explore a portfolio scenario: You want to display your profile photo on the left side of your bio text.

HTML setup:

```
<div class= "profile">
  <img src= "images/profile.jpg"alt= "Your portrait"class= "profile-
photo">
  <div class= "bio">
    <h2>Alex Kim</h2>
    <p>Web developer and designer passionate about clean, user-
friendly interfaces.</p>
  </div>
</div>
```

CSS for side-by-side placement using Flexbox:

```
.profile {
  display: flex;
  align-items: center;
  gap: 20px;
  max-width: 700px;
  margin: 40px auto;
  padding: 24px;
  background: #f7f7f7;
  border-radius: 12px;
}

.profile-photo {
  width: 120px;
  height: 120px;
  object-fit: cover;
  border-radius: 50%;
  box-shadow: 0 2px 8px rgba(0,0,0,0.1);
}

.bio {
  flex: 1;
}
```

- **display: flex;** places the image and bio side by side.
- **align-items: center;** vertically centers the image with the text.
- **gap: 20px;** adds space between image and text.
- **object-fit: cover;** ensures the photo fills the circle without distortion.
- **border-radius: 50%;** makes the image perfectly round.

This approach keeps your image and bio aligned and visually balanced, a technique seen on many professional portfolio sites.

Responsive Image Placement

On mobile devices, large images or side-by-side layouts may not look good. To ensure your images and content adapt gracefully, use **media queries**.

Add a responsive tweak:

```
@media (max-width: 500px) {
  .profile {
    flex-direction: column;
    align-items: flex-start;
  }
  .profile-photo {
    margin-bottom: 12px;
  }
}
```

Now, when the screen width is 500 pixels or less, the profile photo stacks on top of the bio. This makes the layout much more user-friendly on phones and tablets.

Image Alignment in Content

Sometimes, you want an image to float to one side of text, as seen in magazine-style blogs or e-commerce sites. While Flexbox is great for layouts, the **float** property is still useful for simple image alignment within paragraphs.

HTML example for a product description:

```
    <div class= "product-description">
      <img src= "images/shoe.jpg"alt= "Running shoe"class= "float-left">
      <p>This lightweight running shoe is perfect for daily workouts and
weekend adventures. Its breathable material keeps your feet cool and
comfortable.</p>
    </div>
```

CSS for floating the image:

```
    .float-left {
      float: left;
      width: 180px;
      margin-right: 20px;
      margin-bottom: 10px;
      border-radius: 8px;
    }
```

This places the product image on the left, letting the text wrap around it. Use **margin-right** to add space between the image and the paragraph.

Hands-On Coding Summary

Use **width** and **max-width** to make images responsive and fit your design.
Experiment with **Flexbox** for side-by-side layouts, perfect for portfolios or team pages.
Apply **float** for classic text-wrapping effects in articles or product descriptions.
Combine these techniques with **media queries** to keep images and layouts attractive on all devices.

Try adapting these examples to a project you care about: center a banner image on your homepage, create a team section with aligned photos, or float images in a blog post. Practicing these skills will help you create visually engaging, professional websites.

Next, you'll explore how to add vibrant gradients as backgrounds, bringing a modern, stylish look to your designs. Get ready to blend colors smoothly and level up your web pages with the creative power of CSS gradients!

8.3. Adding Gradients for Modern Designs

Gradients are a powerful tool in modern web design, allowing you to create smooth transitions between colors for backgrounds, buttons, banners, and more. Unlike flat colors, gradients add depth and visual interest, making your site feel more dynamic and up to date. Whether you're building a personal portfolio or the homepage for a local café, learning how to apply CSS gradients will instantly give your projects a professional touch.

Understanding Gradients in CSS

Before we start coding, let's clarify what a **gradient** is in web development. A gradient is a gradual blend between two or more colors. CSS provides two main types of gradients:

Linear gradients: Colors blend along a straight line (vertical, horizontal, or at any angle).

Radial gradients: Colors radiate out from a central point, creating a circular or elliptical color transition.

Unlike images, gradients are generated by the browser, so they scale perfectly on any screen size and don't slow down your site.

Creating Your First Linear Gradient Background

Let's jump in with a simple linear gradient for a website header. Imagine you're designing a stylish landing page for a travel blog. You want the header to shift smoothly from deep blue to a soft turquoise, suggesting sky and water.

Start by setting up your HTML structure:

```
<!DOCTYPE html>
<html lang= "en">
<head>
  <meta charset= "UTF-8">
  <title>Travel Blog Header with Gradient</title>
  <link rel= "stylesheet"href= "styles.css">
</head>
<body>
  <header class= "hero-header">
    <h1>Explore the World</h1>
    <p>Your adventure starts here.</p>
  </header>
</body>
</html>
```

Next, open your `styles.css` file and style the header with a linear gradient:

```
.hero-header {
  height: 300px;
  color: white;
  display: flex;
  flex-direction: column;
  justify-content: center;
  align-items: center;
  /* Linear gradient background */  background: linear-gradient(135deg,
#0f2027, #2c5364, #20e3b2);
}
```

The `linear-gradient(135deg, #0f2027, #2c5364, #20e3b2)` function creates a color blend at a 135-degree angle, moving from a deep blue (`#0f2027`) to a lighter blue (`#2c5364`), and finally to turquoise (`#20e3b2`).

The gradient fills the entire header, giving it a modern, inviting feel.

Customizing Gradients: Direction, Colors, and Stops

You're not limited to just two colors or a single direction. CSS gradients are highly customizable, letting you control the angle, the number of colors, and exactly where each color appears.

For example, to create a sunrise effect for an e-commerce site's call-to-action banner:

```
.cta-banner {
  padding: 40px;
  text-align: center;
  background: linear-gradient(to right, #ffaf7b 20%, #d76d77 60%,
#3a1c71 100%);
  color: white;
  border-radius: 15px;
  font-size: 1.5rem;
}
```

`to right` means the gradient moves horizontally, left to right.

Color stops like `20%` and `60%` control where each color transition happens, so you can fine-tune the look.

Try adding this HTML inside your `<body>` to see it in action:

```
<div class= "cta-banner">
  <strong>Limited Time Offer!</strong> Save 20% on all travel gear.
</div>
```

Creating Radial Gradients for Unique Effects

Radial gradients start from a central point and spread outward, perfect for attention-grabbing backgrounds.

Suppose you want a vibrant, spotlight effect for your personal portfolio's hero section:

```
.portfolio-hero {
  min-height: 350px;
  display: flex;
  justify-content: center;
  align-items: center;
  background: radial-gradient(circle at 60% 40%, #ff5f6d 0%, #ffc371
80%);
  color: #222;
  font-size: 2rem;
}
```

`circle at 60% 40%` positions the center of the gradient slightly off-center for a more dynamic look.

The gradient moves from a vivid red-pink (`#ff5f6d`) at the center to a warm yellow (`#ffc371`) at the edges.

Use this HTML snippet:

```
<section class= "portfolio-hero">
  <div>
    <h2>Hello, I'm Maya!</h2>
    <p>Front-End Web Developer & Designer</p>
  </div>
</section>
```

Advanced: Repeating Gradients for Patterns

If you want more creative backgrounds, try repeating gradients. This technique can simulate stripes or patterns without images.

Here's a quick way to add subtle stripes to a background:

```
.striped-background {
  background: repeating-linear-gradient(
    45deg,
    #e0eafc,
    #e0eafc 10px,
    #cfdef3 10px,
    #cfdef3 20px
  );
  padding: 60px;
}
```

This code repeats a light blue (`#e0eafc`) and a slightly darker blue (`#cfdef3`) at a 45-degree angle, creating a crisp, lightweight stripe effect for banners or cards.

Test it with this HTML:

```
<div class= "striped-background">
  <h3>Modern Design, Effortless Style</h3>
  <p>Stripes created purely with CSS gradients!</p>
</div>
```

Practical Exercise: Gradient Button Styles

Let's apply what you've learned by making a gradient-styled button, a common element on modern sites.

Add this button HTML to your page:

```
<button class= "gradient-btn">Get Started</button>
```

Style it with a linear gradient in your CSS:

```
.gradient-btn {
  padding: 14px 32px;
  font-size: 1.1rem;
  color: white;
  border: none;
  border-radius: 25px;
  background: linear-gradient(90deg, #36d1c4 0%, #5b86e5 100%);
  cursor: pointer;
  transition: box-shadow 0.2s;
}

.gradient-btn:hover {
  box-shadow: 0 4px 20px rgba(91, 134, 229, 0.3);
}
```

Now your button will blend from teal to blue, standing out beautifully on the page.

You've just explored how to use CSS gradients to create modern, eye-catching designs, from full-page backgrounds to stylish buttons and patterned effects. Gradients are a versatile tool that instantly enhance your web layouts, scale perfectly on all devices, and require no extra image files. Experiment with different colors, angles, and gradient types to see how they transform your site's visual appeal.

Next, get ready to take your designs even further. In the upcoming section, you'll discover how to layer gradients and images using overlay effects and blending modes, unlocking even richer backgrounds and creative possibilities. Let's keep building your CSS skills with these advanced techniques!

8.4. Overlay Effects and Blending Modes

When you want to create visually stunning web pages, overlay effects and blending modes in CSS unlock a new level of design flexibility. These techniques allow you to layer colors, gradients, or even images over your background images, resulting in modern and dynamic visuals. Whether you're building a sleek portfolio, an engaging blog, or an eye-catching e-commerce homepage, understanding overlays and blending modes will help your site stand out.

Let's explore the essentials, see how popular sites use these effects, and then dive into practical, step-by-step coding.

Understanding Overlays: Why and When to Use Them

An **overlay** is a semi-transparent layer placed over an image or background. Overlays can:

Make text easier to read by increasing contrast
Add color branding to hero sections or banners
Create mood and visual depth for headers or cards

For example, many portfolio sites use a dark overlay over hero images to ensure the designer's name and tagline pop. E-commerce homepages often use overlays to highlight promotional text over vibrant product photos.

Basic Overlay with CSS: Step-by-Step

To add an overlay effect, you'll typically use a **container** with a background image, then layer a semi-transparent element (usually a `<div>`) on top. Here's how you can do it:

Structure your HTML: Create a container with an image background and an overlay div above it.

```
<div class= "hero-section">
  <div class= "overlay"></div>
  <div class= "hero-content">
    <h1>Stunning CSS Effects</h1>
    <p>Learn to layer and blend like a pro!</p>
  </div>
</div>
```

Style with CSS: Use `position: relative` for the container and `position: absolute` for the overlay, so it covers the background image.

```
.hero-section {
  position: relative;
  width: 100%;
  height: 350px;
  background-image: url( 'https://images.unsplash.com/photo-
1465101046530-73398c7f28ca?auto=format&fit=crop&w=800&q=80');
  background-size: cover;
  background-position: center;
  color: white;
  display: flex;
  align-items: center;
  justify-content: center;
}

.overlay {
  position: absolute;
  top: 0; left: 0; right: 0; bottom: 0;
  background: rgba(0,0,0,0.5); /* Black with 50% opacity */  z-index:
1;
}

.hero-content {
  position: relative;
  z-index: 2;
  text-align: center;
}
```

With this structure, the `.overlay` creates a subtle darkening effect, making the white text much easier to read.

Customizing Overlay Colors and Opacity

You're not limited to black overlays. Try branding your overlay by adjusting the color and transparency:

```
/* Blue overlay */.overlay {
  background: rgba(30, 144, 255, 0.4);
}
```

Adjust the fourth value (opacity) to control how much of the image remains visible.

Using Gradients as Overlays

Gradients can add even more visual interest to your overlays. Here's how to use a linear gradient instead of a solid color:

```
.overlay {
  background: linear-gradient(to bottom right, rgba(255, 0, 150, 0.6),
rgba(0, 204, 255, 0.4));
}
```

This produces a colorful overlay that shifts from pink to blue, which is popular in modern blog headers and creative portfolios.

Blending Modes: Blending Layers for Unique Effects

CSS **blending modes** allow you to control how an element's content blends with what's behind it. The `mix-blend-mode` property is especially useful for overlays.

Common blend modes include:

`multiply`: darkens underlying images, great for subtle overlays
`screen`: lightens, often used for dreamy effects
`overlay`: increases contrast, making both image and overlay pop

How to Apply a Blend Mode:

Add `mix-blend-mode` to your overlay CSS.

```
.overlay {
  background: rgba(255,255,255,0.5); /* semi-transparent white */  mix-
blend-mode: multiply;
}
```

This multiplies the overlay's color with the background, creating a richer, deeper effect.

Real-World Scenario: Blog Featured Image Card

Let's say you're designing a card for a blog's featured post. You want the image to have a dramatic color overlay using a blend mode.

```
<div class= "featured-card">
  <img src= "https://images.unsplash.com/photo-1519125323398-
675f0ddb6308?auto=format&fit=crop&w=800&q=80"alt= "Sunset over city"/>
  <div class= "card-overlay"></div>
  <div class= "text-content">
    <h2>Latest Design Trends</h2>
    <p>Discover how overlays and blending modes enhance your web
layouts.</p>
  </div>
</div>


.featured-card {
  position: relative;
  width: 400px;
  height: 250px;
  overflow: hidden;
  border-radius: 10px;
  margin: 2rem auto;
}
```

```
.featured-card img {
  width: 100%;
  height: 100%;
  object-fit: cover;
  display: block;
}

.card-overlay {
  position: absolute;
  top: 0; left: 0; right: 0; bottom: 0;
  background: linear-gradient(120deg, rgba(255, 140, 0, 0.7),
rgba(255,255,255,0.3));
  mix-blend-mode: overlay;
  z-index: 1;
}

.text-content {
  position: absolute;
  z-index: 2;
  bottom: 20px;
  left: 20px;
  color: #fff;
  text-shadow: 0 2px 5px rgba(0,0,0,0.6);
}
```

Now the overlay uses a gradient and the `overlay` blend mode, creating a vibrant, professional look that's both stylish and readable.

Tips for Effective Overlays and Blending Modes

Test your overlays for readability, especially if text sits on top.

Experiment with different blend modes and overlay colors to match your brand or theme.

Remember cross-browser support: while most modern browsers support blend modes, always check your design in Chrome, Firefox, and Edge.

Coding Practice Summary

By mastering overlay effects and blending modes, you add a professional polish to your web projects. Practice layering different colors, gradients, and blend modes over images to see how each combination changes the mood and clarity of your design. These techniques are widely used in real-world sites, from creative portfolios to online shops, so the skills you build here are directly transferable to your future projects.

In the next section, you'll bring these concepts together as you craft a visually striking landing page with hero images, combining overlays, blending modes, and creative text layouts. Get ready to build a modern, attention-grabbing page from scratch!

8.5. Hands-On Exercise: Crafting a Landing Page with Hero Images

A modern landing page often grabs attention immediately with a stunning hero image: a large, visually engaging photo or illustration at the top of the page, usually paired with bold text or a call-to-action button. Hero images are a defining feature of many contemporary websites, from tech startups to travel blogs and personal portfolios. In this hands-on exercise, you'll build your first landing page hero section using HTML and CSS, gaining skills that are essential for real-world web development.

Understanding Hero Images in Web Design

Before diving into the code, it's helpful to know what makes a hero image effective. Hero images are large banners, often full-width, that set the tone for a website and direct your visitor's attention to key content or actions. Sites like Airbnb and Unsplash use hero images to immediately immerse visitors in their brand and message.

In this exercise, you'll:

Structure a simple HTML layout for a landing page
Use CSS to set a responsive hero image as a background
Overlay text and a button for visual impact
Add a semi-transparent layer for readability
Explore blending modes for creative effects

Let's get started with the HTML foundation.

Step 1: Building the HTML Structure

Start with a basic HTML5 template. Focus on semantic elements that make your code easy to read and accessible.

```
<!DOCTYPE html>
<html lang= "en">
<head>
  <meta charset= "UTF-8">
  <title>My Landing Page</title>
  <link rel= "stylesheet"href= "styles.css">
</head>
<body>
  <header class= "hero">
    <div class= "hero-overlay">
      <h1>Welcome to Your Next Adventure</h1>
      <p>Discover unforgettable journeys, tailored for you.</p>
      <a href= "#"class= "cta-btn">Get Started</a>
    </div>
  </header>
  <main>
    <!-- Page content goes here -->  </main>
</body>
```

```
</html>
```

The `<header>` acts as the hero section, signaling its importance.

The `.hero-overlay` container holds your headline, description, and a call-to-action button.

The `<a>` element styled as a button encourages user interaction.

Step 2: Adding a Background Hero Image with CSS

Now, switch to your CSS file. Start by targeting the `.hero` class to make the header fill the top of the page with a background image.

```
.hero {
  min-height: 80vh;
  background-image: url( 'https://images.unsplash.com/photo-
1506744038136-46273834b3fb?auto=format&fit=crop&w=1200&q=80');
  background-size: cover;
  background-position: center;
  display: flex;
  align-items: center;
  justify-content: center;
  position: relative;
}
```

`min-height: 80vh;` ensures your hero section takes up 80% of the viewport height for a dramatic effect.

`background-image` uses a royalty-free Unsplash photo for immediate visual appeal.

`background-size: cover;` stretches the image to fill the space, cropping as needed.

`background-position: center;` keeps the image focused.

Flexbox properties align the content both vertically and horizontally.

Step 3: Creating a Readable Overlay

Text over images can be hard to read. To fix this, add a semi-transparent overlay using a background color with alpha transparency.

```
.hero-overlay {
  background: rgba(0, 0, 0, 0.5); /* Black with 50% opacity */
padding: 3rem 2rem;
  border-radius: 10px;
  text-align: center;
  color: #fff;
  max-width: 500px;
  width: 100%;
  box-shadow: 0 8px 24px rgba(0,0,0,0.2);
}
```

`background: rgba(0, 0, 0, 0.5);` ensures the text stands out against the image.

`padding` and `border-radius` create a visually pleasing container.
`box-shadow` adds subtle depth.
`color: #fff;` keeps your text crisp and readable.

Step 4: Styling the Call-to-Action Button

A landing page often features a clear call to action. Style the button to make it pop.

```
.cta-btn {
  display: inline-block;
  margin-top: 1.5rem;
  padding: 0.9rem 2rem;
  font-size: 1.2rem;
  background: #ff9800;
  color: #fff;
  border: none;
  border-radius: 30px;
  text-decoration: none;
  cursor: pointer;
  transition: background 0.3s;
}
.cta-btn:hover {
  background: #e65100;
}
```

`background: #ff9800;` gives the button a warm, inviting color.
Rounded corners and a subtle hover effect encourage interaction.

Step 5: Making the Hero Section Responsive

Websites should look great on any device. Adjust your CSS with a media query:

```
@media (max-width: 600px) {
  .hero {
    min-height: 60vh;
  }
  .hero-overlay {
    padding: 1.5rem 1rem;
    font-size: 0.95rem;
  }
  .cta-btn {
    width: 100%;
    font-size: 1rem;
  }
}
```

This ensures your hero section adapts to mobile screens, keeping content accessible and visually appealing.

Step 6: Adding a Creative Touch with Background Blending

For an extra professional touch, try a background blending mode. This technique overlays a color gradient on your image for a modern effect seen on trendy portfolio and startup sites.

```
.hero {
  background-image:
    linear-gradient(rgba(255, 152, 0, 0.4), rgba(0, 0, 0, 0.4)),
    url( 'https://images.unsplash.com/photo-1506744038136-
46273834b3fb?auto=format&fit=crop&w=1200&q=80');
  background-blend-mode: multiply;
}
```

`linear-gradient` adds a warm orange-to-black overlay.
`background-blend-mode: multiply;` blends the gradient and the image for richer color and readability.

Real-World Inspiration

Many professional sites use hero images to convey identity and engage users at first glance. For example, a personal portfolio might display a designer's portrait and tagline, while an e-commerce homepage could showcase featured products with a call-to-action. You can experiment with Unsplash, Pexels, or your own photos to make your landing page unique.

In this exercise, you crafted a visually impressive landing page hero section using HTML and CSS. You learned how to set a background image, add a transparent overlay, style text and buttons, and enhance your design with blending modes and responsive adjustments. Try swapping out the image, colors, or overlay styles to build confidence and see how small CSS changes can dramatically affect the look and feel of your page.

9. BUILDING TABLES AND TABULAR DATA

Tables are a powerful tool for organizing and presenting information clearly on the web. Whether you're comparing product features, displaying schedules, or summarizing data, well-structured tables transform complex details into easy-to-read formats that help users make informed decisions. In today's digital landscape, knowing how to create and style tables is an essential skill for anyone aiming to build useful, user-friendly websites—skills that employers and clients truly value.

In this chapter, you'll discover how to build tables from the ground up using essential HTML elements. You'll learn how to structure rows and columns, highlight headings for clarity, and add accessibility features that ensure your tables are helpful to everyone, including users with disabilities. Beyond structure, you'll dive into CSS techniques that bring your tables to life: customizing colors, borders, and spacing to create polished, professional designs that match your site's look and feel.

To put your new skills into practice, you'll design a product comparison table like those found on leading e-commerce sites—an exercise that blends real-world relevance with creative problem-solving. By the end of this chapter, you'll not only know how to code functional tables but also how to enhance them with styling and accessibility best practices. These are the building blocks for presenting data effectively on any website, boosting your confidence as a web developer and preparing you to tackle more advanced projects with ease. Let's get started by exploring the structure and purpose of tables in web development.

9.1. Creating Basic Tables with <table>, <tr>, <td>, and <th>

Tables are a powerful way to organize and display data on your web pages. Whether you want to show a schedule, product list, or comparison chart, understanding how to build tables with HTML is an essential skill for every aspiring web developer. In this section, you'll learn how to create well-structured tables using the core HTML elements: **<table>**, **<tr>**, **<td>**, and **<th>**. We'll walk through practical examples and hands-on exercises so you can see immediate results in your browser.

Why Use Tables on the Web?

Tables are best for presenting tabular data—information that's naturally organized in rows and columns. Think about a restaurant menu, a sports team roster, or a list of products with prices. Many real-world websites, like online stores and business directories, rely on tables to keep their data neat and accessible.

Let's dive into the basics and build your first table from scratch.

Understanding Table Structure: The Building Blocks

Before you start coding, it's important to know what each table tag does:

<table>: This tag creates the overall table structure.
<tr>: Short for "table row," each **<tr>** defines a row in your table.
<td>: "Table data" cell, used for standard data in the table.
<th>: "Table header" cell, used for headings. These are bold and centered by default, helping users understand what each column means.

Tables are built by nesting **<tr>** (rows) inside **<table>**, and then placing **<td>** or **<th>** inside each row.

Hands-On: Build a Simple Data Table

Let's create a basic table that could appear on a portfolio or small business website. Imagine you want to display your skills and experience in a clear, professional way.

Open your code editor and create a new HTML file called `skills-table.html`.
Add the following code inside the `<body>` element:

```
<table>
  <tr>
    <th>Skill</th>
    <th>Level</th>
    <th>Years of Experience</th>
  </tr>
  <tr>
    <td>HTML</td>
    <td>Advanced</td>
    <td>3</td>
  </tr>
  <tr>
    <td>CSS</td>
    <td>Intermediate</td>
    <td>2</td>
  </tr>
  <tr>
    <td>JavaScript</td>
    <td>Beginner</td>
    <td>1</td>
  </tr>
</table>
```

Open the file in your browser. You'll see a simple, unstyled table with three rows of data and a header row. This kind of table is perfect for summarizing your skills on a portfolio site.

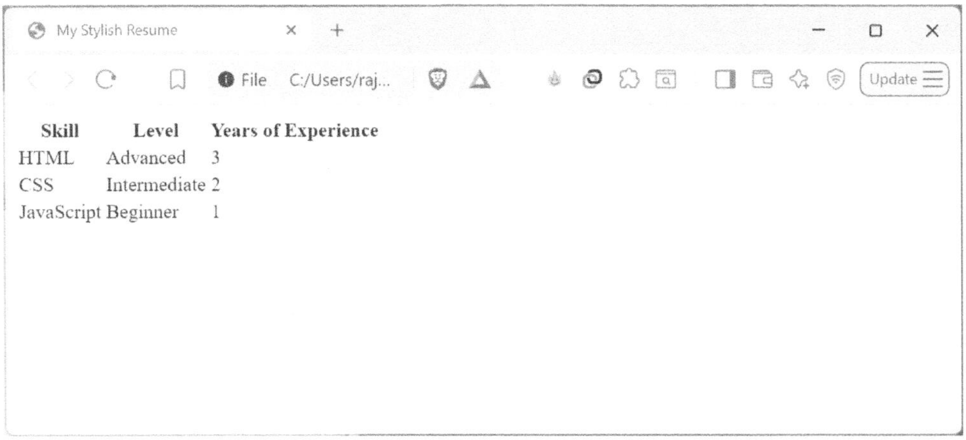

Breaking Down the Example

Let's look at how the table is structured:

The `<table>` tag wraps around everything, defining the table area.
The first `<tr>` contains three `<th>` header cells: "Skill," "Level," and "Years of Experience."
Each following `<tr>` is a new row, with `<td>` cells holding the actual data (for example, "HTML," "Advanced," "3").

This structure makes it easy for both browsers and people to read your data.

Real-World Scenario: A Product Price List

Imagine you're building a simple e-commerce landing page. You want to list products, their prices, and availability. Let's create a table for this scenario.

In your HTML file, below the previous table, add:

```
<table>
  <tr>
    <th>Product</th>
    <th>Price</th>
    <th>Availability</th>
  </tr>
  <tr>
    <td>Coffee Mug</td>
    <td>$12.99</td>
    <td>In Stock</td>
  </tr>
  <tr>
    <td>Notebook</td>
    <td>$4.50</td>
    <td>Out of Stock</td>
  </tr>
  <tr>
    <td>T-Shirt</td>
```

```
    <td>$15.00</td>
    <td>In Stock</td>
  </tr>
</table>
```

Refresh your browser to see the product list appear in a neat table. This approach is widely used on online stores and restaurant menus.

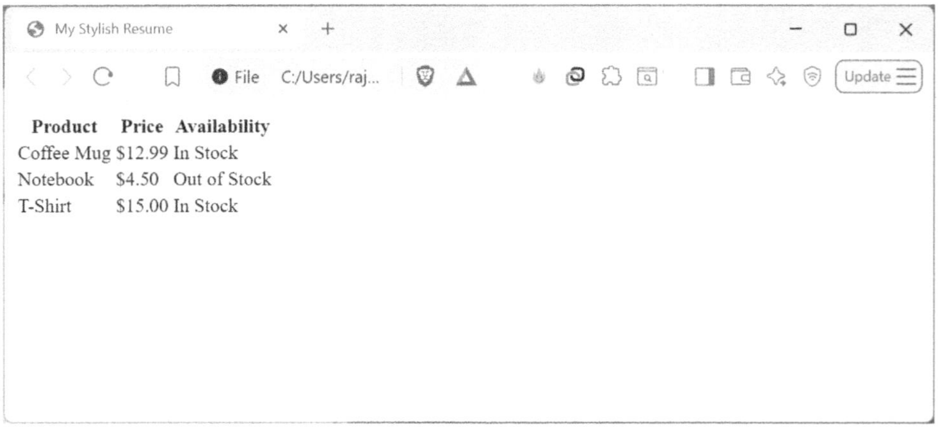

Key Points for Creating Tables

Always use **<th>** for headings at the top of columns. This improves accessibility and makes your table easier to read.

Every row starts with **<tr>** and ends with **</tr>**.

Place your data inside **<td>** cells.

You can have as many rows and columns as you need by repeating **<tr>** and **<td>** elements.

Practice Exercise: Create a Simple Schedule Table

Try building a weekly class schedule using a table. Follow these steps:

After your previous tables, add the following code:

```
<table>
  <tr>
    <th>Day</th>
    <th>9:00 AM</th>
    <th>11:00 AM</th>
    <th>1:00 PM</th>
  </tr>
  <tr>
    <td>Monday</td>
    <td>Math</td>
    <td>English</td>
    <td>Science</td>
  </tr>
  <tr>
```

```
    <td>Tuesday</td>
    <td>Art</td>
    <td>History</td>
    <td>Physical Ed</td>
  </tr>
  <tr>
    <td>Wednesday</td>
    <td>Computer</td>
    <td>Music</td>
    <td>Math</td>
  </tr>
</table>
```

Try changing the subjects or times to make it your own. This exercise models the type of tabular data used by schools or event planners on their websites.

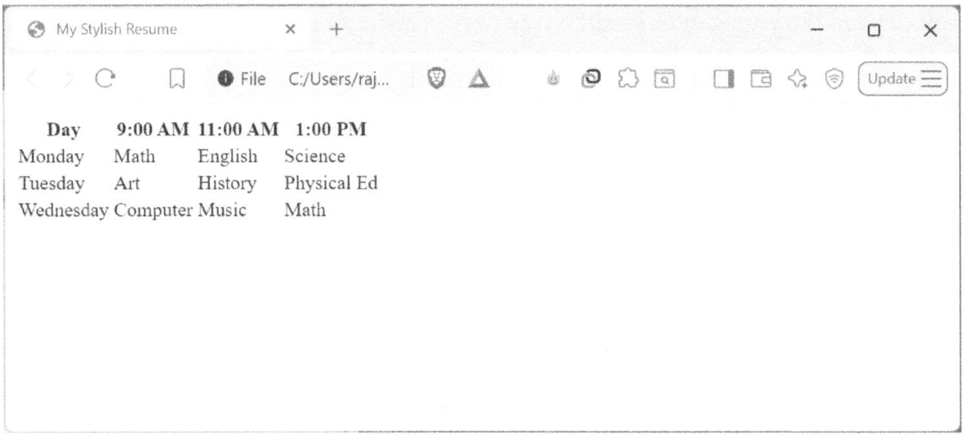

Table Tags in Action: Accessibility and Semantics

Using the correct table tags isn't just about looks. **<th>** elements signal to screen readers and search engines that these are headings, making your content more accessible and meaningful. For best practice:

Use **<th>** for header cells, not just bolded **<td>**.
Keep your table content concise and relevant.

Combining Tables and Web Trends

Modern web design often uses tables for data presentation, not for layouts. For example, a blog might display recent posts in a table, or a travel website could show flight schedules. By mastering tables, you're building skills that are directly useful for real-world projects.

Quick Tips for Table Structure

The number of **<td>** or **<th>** cells should match across rows to keep columns aligned.
You can mix **<th>** and **<td>** in a row, but typically, the first row uses **<th>** for clarity.

Summary Exercise: Build Your Own Table

Create a table for any scenario you find interesting, such as a favorite movies list, a workout plan, or a comparison of online courses. Use at least three columns and three rows. Preview your table in a browser and tweak the data as you like.

In this section, you learned how to construct basic tables using **<table>**, **<tr>**, **<td>**, and **<th>**. You practiced building tables for skills, product listings, and schedules—just like real websites do. With these HTML building blocks, you can now organize and display information in a way that's clear and accessible.

Up next, you'll discover how to style your tables with CSS. Get ready to turn plain tables into visually striking, easy-to-read designs that match your website's look and feel. Let's make your tables stand out with professional polish!

9.2. Styling Tables with CSS

Tables are a powerful way to organize and display data on your website, but their default appearance can look plain and outdated. With CSS, you can completely transform the look and feel of your tables, making them more inviting, readable, and professional. In this section, you'll discover how to take control of table presentation using CSS, starting with basic styling and progressing to more advanced layout techniques. You'll get hands-on experience by working through real-world scenarios that you can use in your own web projects, such as designing a simple product list for an online shop or showcasing a class schedule for a student portfolio.

Why Style Tables with CSS?

HTML gives you the structure for your table, but CSS provides the creativity and polish. Stylish tables help users scan information quickly, highlight important data, and keep your site consistent with your brand or theme. Whether you're creating a pricing table for a business website or a list of recent blog posts, thoughtful CSS styling can elevate the entire user experience.

Let's dive into practical steps and see how you can give your tables a modern, professional look using CSS.

Getting Started: The Basic Table Markup

Before applying CSS, you need a well-structured table. Here's a sample HTML table for a fictional online bookstore inventory:

```
<table class= "book-table">
  <thead>
    <tr>
      <th>Title</th>
```

```
      <th>Author</th>
      <th>Genre</th>
      <th>Price</th>
    </tr>
  </thead>
  <tbody>
    <tr>
      <td>The Smarter Way to Learn HTML & CSS</td>
      <td>Mark Myers</td>
      <td>Web Development</td>
      <td>$25</td>
    </tr>
    <tr>
      <td>Intro to CSS Design</td>
      <td>Jane Doe</td>
      <td>Design</td>
      <td>$20</td>
    </tr>
    <tr>
      <td>Modern JavaScript Basics</td>
      <td>John Smith</td>
      <td>Programming</td>
      <td>$30</td>
    </tr>
  </tbody>
</table>
```

You'll style this table step by step, learning practical CSS techniques along the way.

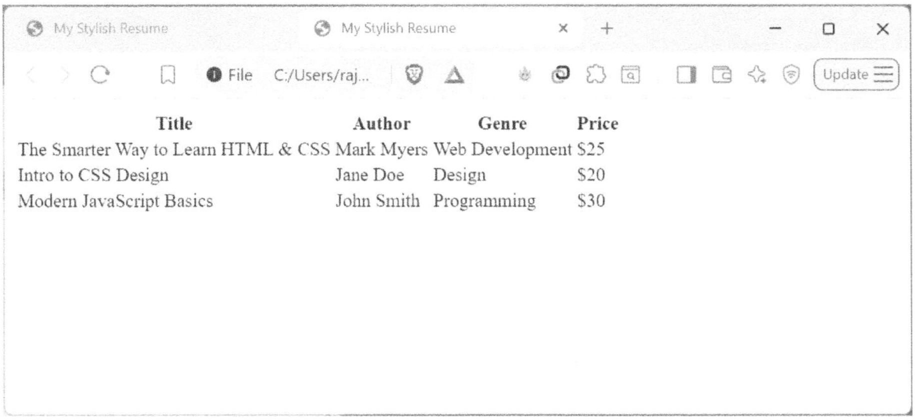

Adding Borders and Spacing

By default, table borders can be hard to see or may have gaps between cells. CSS allows you to control borders and spacing for a cleaner look.

Add a border to your table, header, and data cells:

```
.book-table,
.book-table th,
.book-table td {
```

```
  border: 1px solid #cccccc;
}
```

Remove unwanted space between borders using `border-collapse`:

```
.book-table {
  border-collapse: collapse;
}
```

Add padding for better readability:

```
.book-table th,
.book-table td {
  padding: 12px 20px;
}
```

Now, your table has consistent lines and enough space inside each cell, making the data easier to read.

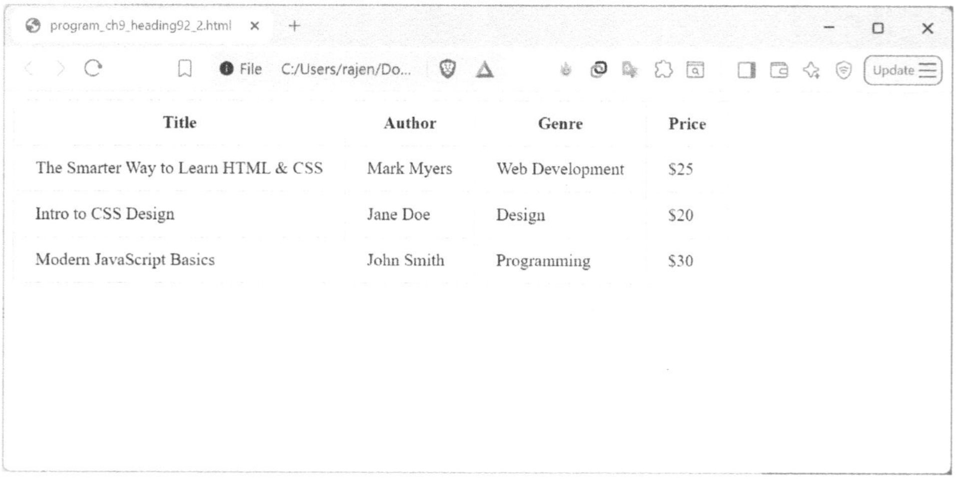

Giving Your Table a Modern Look with Colors

Colors can improve visual hierarchy and help users distinguish between headers and rows.

Style the table header with a background color and white text:

```
.book-table th {
  background-color: #4a90e2;
  color: #fff;
  font-weight: bold;
  text-align: left;
}
```

Add zebra striping for alternate rows using `nth-child`:

```
.book-table tbody tr:nth-child(even) {
  background-color: #f6f8fa;
}
```

This alternating row color effect (often called "zebra striping") makes long tables easier to scan.

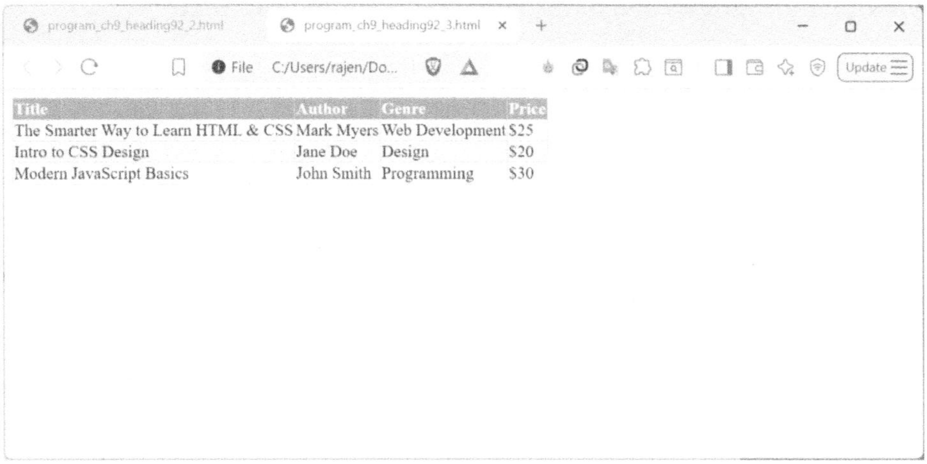

Customizing Font and Alignment

Typography plays a key role in table readability. Let's adjust the font and align the content for a more polished result.

Set a clean, modern font and adjust font size:

```
.book-table {
  font-family: 'Segoe UI', Arial, sans-serif;
  font-size: 16px;
}
```

Align price data to the right for clarity:

```
.book-table td:last-child {
  text-align: right;
}
```

This ensures price values line up neatly, making them easier to compare at a glance.

Highlighting Important Data

Sometimes, you want to draw attention to specific cells, such as highlighting the highest price or a special genre.

Add a custom class to highlight a cell in the HTML:

```
<td class= "highlight">$30</td>
```

Style the highlight with a background color and bold text:

```
.book-table .highlight {
  background-color: #ffe082;
  font-weight: bold;
  color: #444;
}
```

This technique is perfect for showcasing best sellers, discounts, or other key data points in your tables.

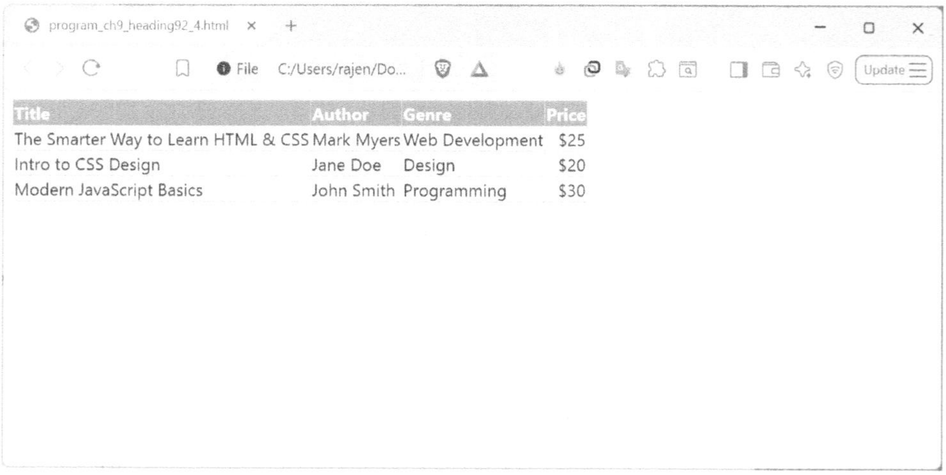

Real-World Scenario: Student Class Schedule

Let's apply what you've learned to a different context: a student portfolio site featuring a class schedule.

HTML:

```
<table class= "schedule-table">
  <thead>
    <tr>
      <th>Time</th>
      <th>Monday</th>
      <th>Tuesday</th>
      <th>Wednesday</th>
```

```
      <th>Thursday</th>
      <th>Friday</th>
    </tr>
  </thead>
  <tbody>
    <tr>
      <td>9-10am</td>
      <td>Math</td>
      <td>History</td>
      <td>Math</td>
      <td>Science</td>
      <td>English</td>
    </tr>
    <tr>
      <td>10-11am</td>
      <td>English</td>
      <td>Math</td>
      <td>Science</td>
      <td>Math</td>
      <td>History</td>
    </tr>
    <tr>
      <td>11-12pm</td>
      <td>Science</td>
      <td>English</td>
      <td>History</td>
      <td>English</td>
      <td>Math</td>
    </tr>
  </tbody>
</table>
```

CSS:

```
.schedule-table {
  border-collapse: collapse;
  width: 100%;
  font-family: 'Roboto', Arial, sans-serif;
}

.schedule-table th,
.schedule-table td {
  border: 1px solid #9e9e9e;
  padding: 10px 12px;
  text-align: center;
}

.schedule-table th {
  background-color: #0288d1;
  color: #fff;
  letter-spacing: 1px;
}

.schedule-table tbody tr:nth-child(odd) {
  background-color: #e3f2fd;
}
```

This table is now visually appealing and clearly communicates the class schedule, making it perfect for a student's personal site or school project.

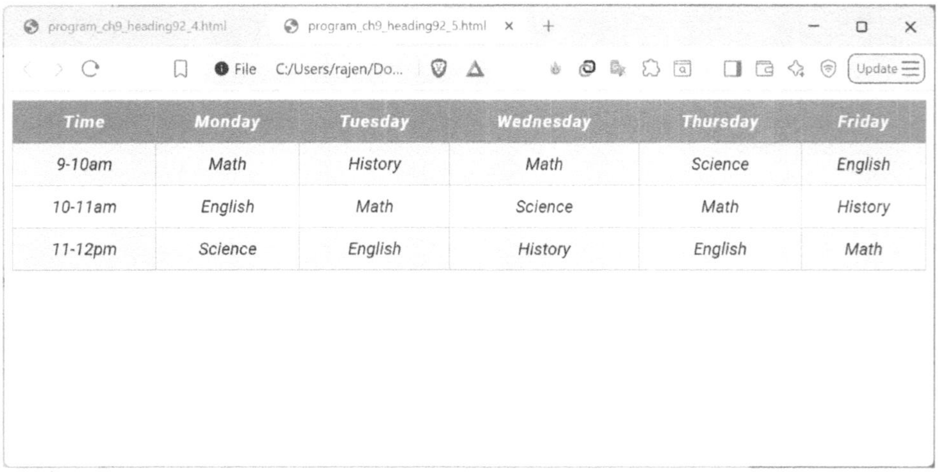

Responsive Tables: Keeping Data Accessible on All Devices

Tables can be challenging to view on small screens. Modern CSS allows you to make tables more mobile-friendly.

Enable horizontal scrolling for wide tables:

```
.table-responsive {
  overflow-x: auto;
  display: block;
}

.book-table,
.schedule-table {
  min-width: 600px;
}
```

Wrap your table in a responsive container:

```
<div class= "table-responsive">
  <table class= "book-table">
    <!-- table content -->  </table>
</div>
```

With this approach, users can scroll horizontally if the table overflows on smaller screens, ensuring your data remains accessible on smartphones and tablets.

Hands-On Exercise: Style Your Own Table

Now it's your turn! Create a table for a topic of your choice, such as a product price list, movie schedule, or travel itinerary. Apply the following:

- Add borders and padding for clarity
- Style the header with a bold background color
- Use alternating row colors for readability
- Highlight one or more cells using a custom class
- Make your table responsive for mobile devices

Experiment with different color schemes and fonts to match your site's personality.

In this section, you learned how to use CSS to transform plain HTML tables into eye-catching, user-friendly designs. You practiced styling borders, adding padding, using color for headers and rows, aligning and highlighting data, and making tables responsive for all devices. These skills are essential for presenting tabular data professionally and can be adapted for many types of websites, from portfolio pages to e-commerce product lists.

Next, you'll discover how to make your tables even more accessible by adding captions and summaries. These HTML features not only improve usability for everyone, but also enhance your site's searchability and inclusivity. Get ready to take your tables to the next level with accessibility best practices!

9.3. Adding Captions and Summaries for Accessibility

When you create tables for your websites, it's easy to focus on just the visual layout. However, making your tables accessible is just as important as making them look good. Accessibility means everyone, including people who use screen readers or assistive technologies, can understand your data. Two essential HTML features for accessible tables are **captions** and **summaries**. Let's explore how to use them, step by step, with practical coding exercises you can immediately apply to real-world projects.

Why Captions and Summaries Matter

Before we dive into the code, it's important to understand *why* these elements are crucial:

Captions give all users a clear, concise description of the table's purpose.

Summaries (using the `summary` attribute or a visually hidden section) help users with screen readers understand the table's structure, especially if the data is complex.

If you're building a portfolio site, a blog, or an e-commerce platform, captions and summaries make your tables more usable and professional, while also improving your site's SEO and compliance with web standards.

Adding a Caption to Your Table

A **caption** is a simple way to add a descriptive title to your table. It appears directly above the table and summarizes what the table contains. This helps all users quickly grasp the purpose of the data.

Let's walk through the steps to add a caption to a basic HTML table.

Create a simple table with headers and data.
Place the **caption** element immediately after the opening `<table>` tag.

```
<table>
  <caption>Monthly Website Traffic by Source</caption>
  <thead>
    <tr>
      <th>Source</th>
      <th>Visits</th>
      <th>Bounce Rate</th>
    </tr>
  </thead>
  <tbody>
    <tr>
      <td>Organic Search</td>
      <td>5,320</td>
      <td>34%</td>
    </tr>
    <tr>
      <td>Social Media</td>
      <td>2,410</td>
      <td>50%</td>
    </tr>
    <tr>
      <td>Email Campaigns</td>
      <td>1,890</td>
      <td>28%</td>
    </tr>
  </tbody>
</table>
```

Key points:

The **<caption>** tag must be the first child of the **<table>** element.
Keep captions short, clear, and descriptive.

Try adding a caption to a table you've already built, or create a new one to practice.

Styling Your Table Caption with CSS

By default, captions are plain and left-aligned. You can improve readability and match your site's design by styling captions with CSS. Let's make the caption stand out:

```
caption {
  font-weight: bold;
  font-size: 1.1em;
  color: #2a7ae2;
  margin-bottom: 0.5em;
  text-align: center;
  letter-spacing: 1px;
}
```

You can add this CSS to your `<style>` section or external stylesheet. Now, the caption above your table will be centered and more visually appealing.

Providing Table Summaries for Screen Readers

While the HTML5 specification no longer supports the `summary` attribute directly on the `<table>` tag, you can still provide a summary for screen readers using visually hidden text. A summary should explain the table's organization, what kind of data it displays, and any special relationships.

Let's see how you can add a summary that only screen readers will access, without displaying it visually on the page.

Create a visually hidden class in your CSS:

```
.sr-only {
  position: absolute;
  width: 1px;
  height: 1px;
  padding: 0;
  margin: -1px;
  overflow: hidden;
  clip: rect(0, 0, 0, 0);
  border: 0;
}
```

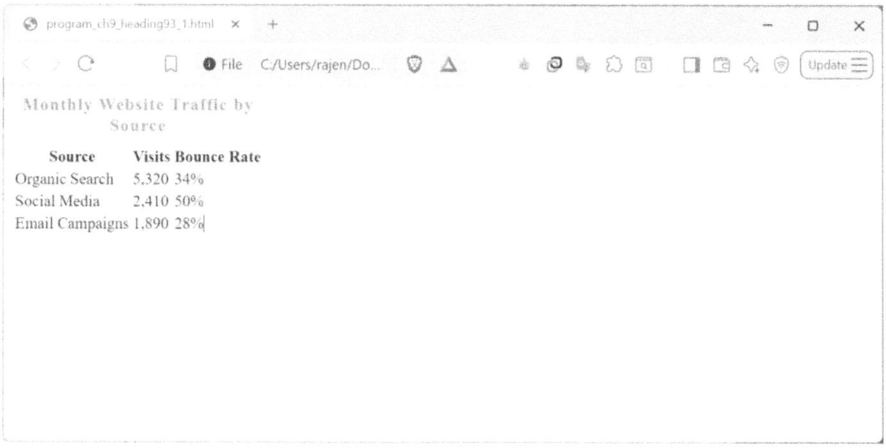

Use a `<div>` or `<p>` with the `sr-only` class before your table:

```
<div class= "sr-only"id= "traffic-table-summary">
  This table shows monthly website traffic segmented by source,
including the number of visits from each channel and their respective
bounce rates. Use the column headers to compare traffic performance.
</div>
<table aria-describedby= "traffic-table-summary">
  <caption>Monthly Website Traffic by Source</caption>
  <thead>
    <tr>
      <th>Source</th>
      <th>Visits</th>
      <th>Bounce Rate</th>
    </tr>
  </thead>
  <tbody>
    <!-- data rows here -->  </tbody>
</table>
```

What's happening here?

The `aria-describedby` attribute on the table links to the hidden summary, so screen readers announce the summary before reading the table's content.

The **.sr-only** class ensures the summary is accessible to assistive technology but not visible to sighted users.

Real-World Example: E-Commerce Product Specification Table

Imagine you're building an e-commerce site that compares product features. Adding captions and summaries helps all shoppers quickly understand the comparison and supports users with visual impairments.

HTML Example:

```
<div class= "sr-only"id= "specs-summary">
  This table compares key specifications of three smartphone models,
including display size, battery life, and camera resolution, to help
customers make informed decisions.
</div>
<table aria-describedby= "specs-summary">
  <caption>Smartphone Features Comparison</caption>
  <thead>
    <tr>
      <th>Feature</th>
      <th>Model A</th>
      <th>Model B</th>
      <th>Model C</th>
    </tr>
  </thead>
  <tbody>
    <tr>
      <td>Display</td>
```

```
      <td>6.1" OLED</td>
      <td>6.5" LCD</td>
      <td>5.8" AMOLED</td>
    </tr>
    <tr>
      <td>Battery Life</td>
      <td>18 hrs</td>
      <td>20 hrs</td>
      <td>15 hrs</td>
    </tr>
    <tr>
      <td>Camera</td>
      <td>12 MP</td>
      <td>16 MP</td>
      <td>12 MP</td>
    </tr>
  </tbody>
</table>
```

Enhanced CSS Styling:

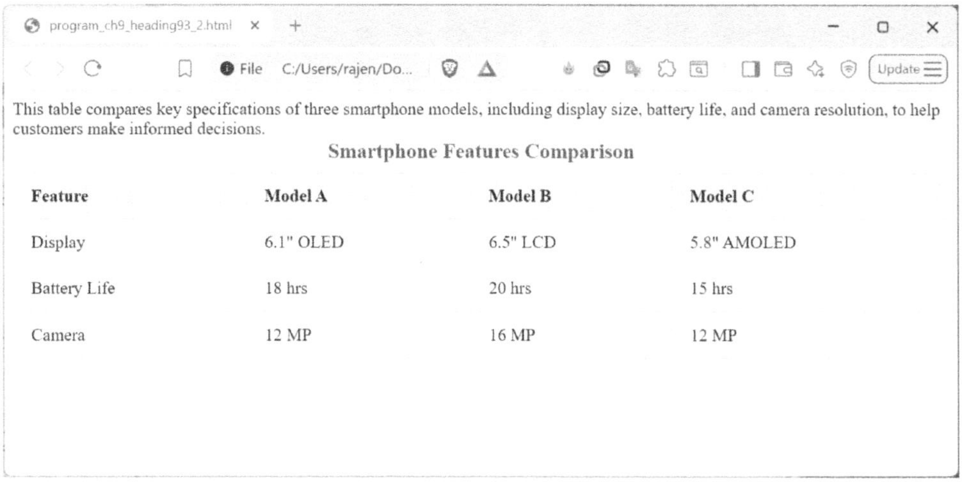


```
caption {
  font-size: 1.2em;
  font-weight: 600;
  color: #3b3b3b;
  padding-bottom: 0.5em;
}

table {
  border-collapse: collapse;
  width: 100%;
  margin-bottom: 2em;
}

th, td {
  border: 1px solid #ddd;
  padding: 0.75em 1em;
  text-align: left;
```

```
}

th {
  background-color: #f5f7fa;
}
```

With these enhancements, your table is not only visually appealing but also accessible to everyone.

Real-World Example: Blog Post List Table

Suppose you run a blog and want to display a list of recent posts in a table. Let's use captions and summaries to make this table accessible.

HTML Example:

```
<p class= "sr-only"id= "blog-table-summary">
  This table lists the five most recent blog posts, including the
title, publication date, and author for each entry.
</p>
<table aria-describedby= "blog-table-summary">
  <caption>Recent Blog Posts</caption>
  <thead>
    <tr>
      <th>Title</th>
      <th>Date</th>
      <th>Author</th>
    </tr>
  </thead>
  <tbody>
    <tr>
      <td>Getting Started with HTML</td>
      <td>June 10, 2024</td>
      <td>Sam Lee</td>
    </tr>
    <tr>
      <td>Mastering CSS Layouts</td>
      <td>June 8, 2024</td>
      <td>Rina Patel</td>
    </tr>
    <tr>
      <td>Responsive Web Design Basics</td>
      <td>June 6, 2024</td>
      <td>Juan Torres</td>
    </tr>
    <tr>
      <td>SEO for Beginners</td>
      <td>June 3, 2024</td>
      <td>Emily Chen</td>
    </tr>
    <tr>
      <td>Building Accessible Tables</td>
      <td>June 1, 2024</td>
      <td>Alex Kim</td>
    </tr>
```

```
  </tbody>
</table>
```

CSS Styling for Clarity:

```
caption {
  color: #1d3557;
  font-size: 1.1em;
  font-style: italic;
  margin-bottom: 0.3em;
}
```

This approach makes your blog's data friendly for all users, regardless of how they access your site.

In this section, you learned how to use the **caption** element and visually hidden summaries to make tables accessible and informative. By combining semantic HTML with targeted CSS styling, you create tables that are both attractive and inclusive. Practice by updating your own tables with captions and summaries, or try building a new table from scratch using these techniques.

Up next, you'll put your skills to the test by designing a feature-rich product comparison table for an e-commerce site. Get ready to combine accessibility, layout, and modern design practices to build a table that truly stands out!

9.4. Hands-On Exercise: Designing a Product Comparison Table

Creating a visually appealing and easy-to-navigate product comparison table is a common task for modern e-commerce websites. These tables help users quickly evaluate features, prices, and differences between products, making their shopping experience smoother and more informed. In this hands-on exercise, you'll build and style a professional product comparison table from scratch, gaining practical experience with both HTML table elements and modern CSS techniques.

Understanding Product Comparison Tables in the Real World

Before we dive into code, let's look at how product comparison tables are used on sites you may know:

Amazon: Side-by-side comparisons help shoppers understand product specs, price differences, and customer ratings.

Best Buy: Tech products are compared based on features, reviews, and available options, enhancing buyer confidence.

Our goal is to recreate a similar experience for a fictional e-commerce site, focusing on clarity, readability, and responsive design.

Step 1: Structuring the Table with Semantic HTML

A well-structured HTML table uses the following tags:

`<table>`: The main container for tabular data.
`<thead>`: Groups the header content.
`<tbody>`: Contains the main data rows.
`<tr>`: Table row.
`<th>`: Header cell, usually bold and centered.
`<td>`: Standard data cell.

Let's create the basic structure for a product comparison table featuring three smartphones.

```
<table>
  <thead>
    <tr>
      <th>Feature</th>
      <th>Smartphone A</th>
      <th>Smartphone B</th>
      <th>Smartphone C</th>
    </tr>
  </thead>
  <tbody>
    <tr>
      <td>Price</td>
      <td>$499</td>
      <td>$599</td>
      <td>$699</td>
    </tr>
    <tr>
      <td>Screen Size</td>
      <td>6.1"</td>
      <td>6.5"</td>
      <td>6.7"</td>
    </tr>
    <tr>
      <td>Battery Life</td>
      <td>18 hrs</td>
      <td>20 hrs</td>
      <td>22 hrs</td>
    </tr>
    <tr>
      <td>Camera</td>
      <td>12MP</td>
      <td>48MP</td>
      <td>108MP</td>
    </tr>
    <tr>
      <td>5G Support</td>
```

```
      <td>Yes</td>
      <td>Yes</td>
      <td>Yes</td>
    </tr>
  </tbody>
</table>
```

This markup provides a clear framework for comparison. Notice how the first column defines the feature, and each subsequent column represents a product.

Step 2: Adding Basic CSS for Readability

Next, let's style the table to make it visually appealing and easy to read. We'll use CSS to add spacing, borders, and background colors.

```
table {
  width: 100%;
  border-collapse: collapse;
  margin: 32px 0;
  font-family: Arial, sans-serif;
  font-size: 1rem;
}

th, td {
  border: 1px solid #ddd;
  padding: 16px 12px;
  text-align: center;
}

thead th {
  background-color: #f8f9fa;
  color: #333;
  font-size: 1.1rem;
}

tbody tr:nth-child(even) {
  background-color: #f5f7fa;
}
```

Here's what these styles accomplish:

`border-collapse: collapse;` removes double borders between cells.
Padding creates breathing room for each cell.
Alternating row colors enhance readability, a technique often seen on sites like Shopify and Alibaba.

Step 3: Highlighting Key Features and Best Values

It's common to help users spot the best deal or a standout feature. Let's emphasize the most affordable smartphone by adding a special class and styling it.

Update your HTML for the price cell of Smartphone A:

```
<td class= "best-value">$499</td>
```

Now, add the CSS:

```
.best-value {
  background-color: #d4edda;
  color: #155724;
  font-weight: bold;
  border: 2px solid #28a745;
}
```

This subtle highlight draws attention to the best value, a strategy widely used by e-commerce leaders like Newegg and Flipkart.

Step 4: Improving Accessibility

Building accessible tables helps all users, including those with screen readers. Add a **caption** and **scope** attributes to your table.

```
<table>
  <caption>Compare Smartphone Features</caption>
  <thead>
    <tr>
      <th scope= "col">Feature</th>
      <th scope= "col">Smartphone A</th>
      <th scope= "col">Smartphone B</th>
      <th scope= "col">Smartphone C</th>
    </tr>
  </thead>
  <tbody>
    <tr>
      <td>Price</td>
      <td class= "best-value">$499</td>
      <td>$599</td>
      <td>$699</td>
    </tr>
    <!-- Remaining rows as before -->  </tbody>
</table>
```

`<caption>` provides a summary for screen readers.
`scope="col"` ensures header cells are correctly associated with data cells.

Step 5: Making the Table Responsive

Modern websites must look good on all devices. Let's add CSS to ensure your table scrolls horizontally on smaller screens, a mobile-friendly pattern used by Apple and Samsung.

```
.table-responsive {
  overflow-x: auto;
  -webkit-overflow-scrolling: touch;
  margin-bottom: 24px;
}

table {
  min-width: 600px;
}
```

Wrap your table in a div:

```
<div class= "table-responsive">
  <!-- Table goes here --></div>
```

On small screens, users can now scroll horizontally to view all columns, keeping your comparison accessible and usable.

Step 6: Adding Interactive Hover Effects

Small interactive touches make your table feel modern and engaging. Let's add a hover effect to highlight rows when the user moves their mouse over them.

```
tbody tr:hover {
  background-color: #ffeeba;
  transition: background 0.2s;
  cursor: pointer;
}
```

This effect guides the user's eye, especially when comparing products row by row.

Step 7: Practical Coding Challenge – Customize Your Table

Try expanding your comparison with one of these real-world scenarios:

Add a **"Buy Now"** button to each product column using a `<button>` inside a `<td>`.
Compare additional features like "Storage Options" or "Warranty."
Swap product types, such as comparing laptops, headphones, or even subscription plans.

For example, let's add a "Buy Now" button:

```
<td>
  <button class= "buy-btn">Buy Now</button>
</td>
```

With CSS:

```
.buy-btn {
  background: #007bff;
  color: #fff;
  border: none;
  padding: 8px 16px;
  border-radius: 5px;
  font-size: 1rem;
  cursor: pointer;
  transition: background 0.2s;
}
.buy-btn:hover {
  background: #0056b3;
}
```

This approach closely mirrors interactive product tables found on top e-commerce platforms, increasing both usability and visual appeal.

Coding Summary

In this exercise, you learned how to build a structured, accessible, and beautifully styled product comparison table using HTML and CSS. You practiced using table elements, semantic markup, responsive design, and interactive styling, all skills essential for aspiring web developers. Continue experimenting by adding new features, adjusting styles, and testing your table on different devices to master real-world web development techniques.

10. HTML FORMS AND USER INPUT

Imagine visiting your favorite online store, submitting a job application, or leaving a review on a travel blog—all of these interactions rely on forms. Forms are the silent workhorses of the web, powering everything from newsletter signups to checkout pages. In this chapter, you'll unlock the essential skills to create interactive, user-friendly forms that invite visitors to connect, share, and take action on your website.

You'll dive into the practical use of HTML elements designed for collecting user input, including **form**, **input**, **textarea**, and **button**. Step by step, you'll learn how to create dropdown menus, radio buttons, and checkboxes, making your forms intuitive and accessible. Discover how to clearly label and group form fields, ensuring your layout is both organized and easy to navigate. We'll also explore HTML5's built-in validation features, empowering you to guide users and prevent common input mistakes—no extra coding required.

To bring your new knowledge to life, you'll build a professional contact or feedback form, a hands-on project that mirrors what real-world websites use every day. By the end of this chapter, you'll have the confidence to design forms that not only gather information but also enhance your site's functionality and professionalism. Whether you're dreaming of launching a portfolio, managing a blog, or creating an e-commerce experience, mastering forms will set you apart as a capable and versatile web developer. Get ready to transform your web pages into interactive destinations that engage and empower your users.

10.1. Building Forms with <form>, <input>, <textarea>, and <button>

Forms are the gateway between your website and its users. Whether you're signing up for a newsletter, leaving a comment on a blog, or placing an order in an online shop, forms make it all possible. In this section, you'll discover how to create your own interactive forms using essential HTML elements: **<form>**, **<input>**, **<textarea>**, and **<button>**. By the end, you'll have built a practical contact form with styled inputs, ready for real-world use.

Understanding the Building Blocks of Forms

Before we dive into coding, let's clarify what these elements do:

<form>: Wraps all form elements and defines how data is sent.
<input>: Creates various types of fields for user input (like text, email, or password).
<textarea>: Lets users enter longer blocks of text, perfect for comments or messages.

<button>: Triggers form actions, such as submitting the form.

These are the core tools every web developer uses to gather information from visitors.

Step-by-Step: Building a Contact Form

Let's build a simple, user-friendly contact form — a must-have feature for portfolios, blogs, and business sites worldwide.

Start with the <form> element

The **<form>** tag wraps all your input fields. It needs an **action** attribute (where to send the data) and a **method** attribute (how to send it). For now, use `"#"` for action (so it won't go anywhere) and `"post"` as the method.

```
<form action= "#"method= "post">
  <!-- Form fields will go here --></form>
```

Add Input Fields

Inside the form, add fields for a name and an email address. Use **<input>** with different **type** attributes.

```
<form action= "#"method= "post">
  <label for= "name">Name:</label>
  <input type= "text"id= "name"name= "name"required>

  <label for= "email">Email:</label>
  <input type= "email"id= "email"name= "email"required>
</form>
```

The **label** tag is linked to the input by the **for** and **id** attributes, which improves accessibility.

Include a <textarea> for Messages

Add a **<textarea>** so users can write a message.

```
<label for= "message">Message:</label>
<textarea id= "message"name= "message"rows= "5"required></textarea>
```

The **rows** attribute sets the visible height (number of lines).

Add a Submit Button

A **<button>** lets users send their input.

```
<button type= "submit">Send Message</button>
```

Combine Everything

Here's the full HTML for your contact form:

```
<form action= "#"method= "post">
  <label for= "name">Name:</label>
  <input type= "text"id= "name"name= "name"required>

  <label for= "email">Email:</label>
  <input type= "email"id= "email"name= "email"required>

  <label for= "message">Message:</label>
  <textarea id= "message"name= "message"rows= "5"required></textarea>

  <button type= "submit">Send Message</button>
</form>
```

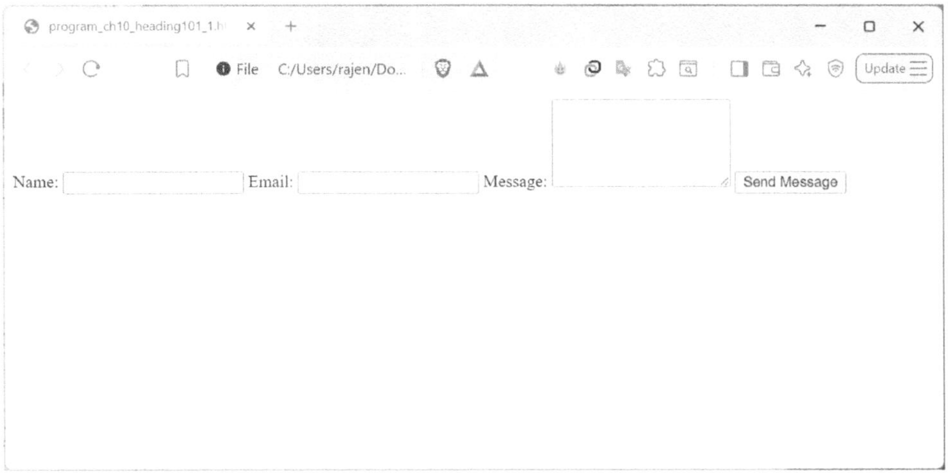

Styling the Form with CSS

A well-designed form is easier to use and more visually appealing. Let's add some CSS to transform our basic form into a professional-looking feature.

```
form {
  max-width: 400px;
  margin: 40px auto;
  padding: 24px;
  background-color: #fafafa;
  border-radius: 8px;
  box-shadow: 0 2px 8px rgba(0,0,0,0.05);
}

label {
  display: block;
  margin-top: 16px;
```

```
  font-weight: bold;
  color: #333;
}

input[type= "text"],
input[type= "email"],
textarea {
  width: 100%;
  padding: 10px;
  margin-top: 6px;
  border: 1px solid #ccc;
  border-radius: 4px;
  font-size: 1em;
  box-sizing: border-box;
  transition: border-color 0.3s;
}

input[type= "text"]:focus,
input[type= "email"]:focus,
textarea:focus {
  border-color: #0077cc;
  outline: none;
}

button {
  display: block;
  width: 100%;
  padding: 12px;
  margin-top: 22px;
  background-color: #0077cc;
  color: #fff;
  font-size: 1.1em;
  border: none;
  border-radius: 4px;
  cursor: pointer;
  transition: background 0.3s;
}

button:hover {
  background-color: #005fa3;
}
```

The form is centered and has a soft background with rounded corners for a modern look.

Inputs and textarea fields are styled for readability and user comfort.

The button stands out with color and expands on hover for better interaction.

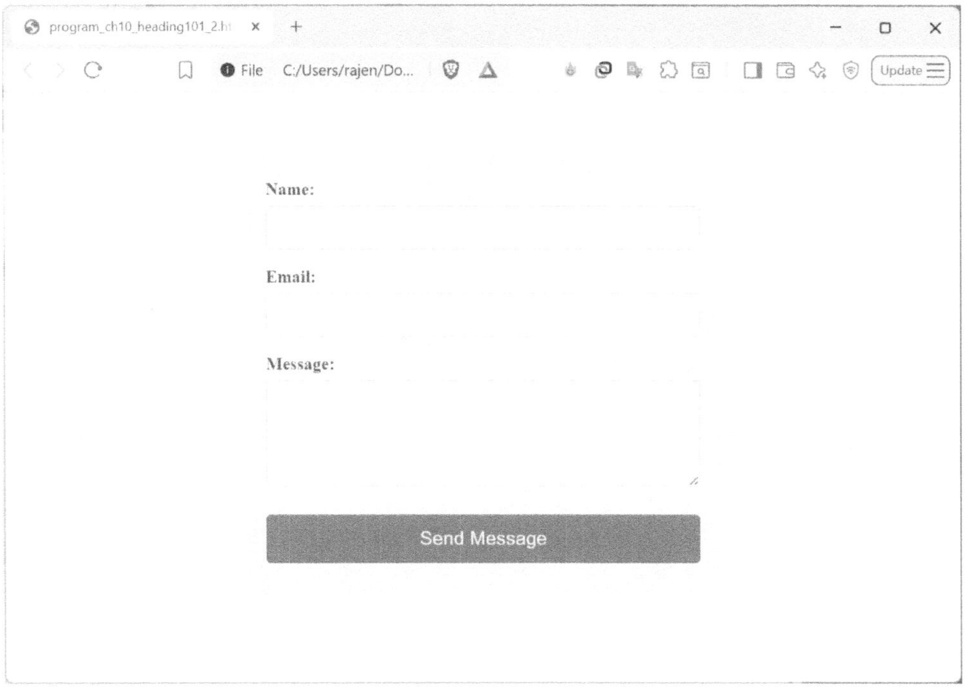

Real-World Scenario: Portfolio Contact Form

Imagine you're building your first web developer portfolio. A contact form lets potential clients reach out. Here's how your form might look on a personal site:

```
<section id= "contact">
  <h2>Contact Me</h2>
  <form action= "#"method= "post">
    <label for= "name">Your Name:</label>
    <input type= "text"id= "name"name= "name"required>

    <label for= "email">Your Email:</label>
    <input type= "email"id= "email"name= "email"required>

    <label for= "message">Your Message:</label>
    <textarea id= "message"name= "message"rows= "5"required></textarea>

    <button type= "submit">Send</button>
  </form>
</section>
```

Apply the same CSS as above and your portfolio instantly feels more professional and approachable.

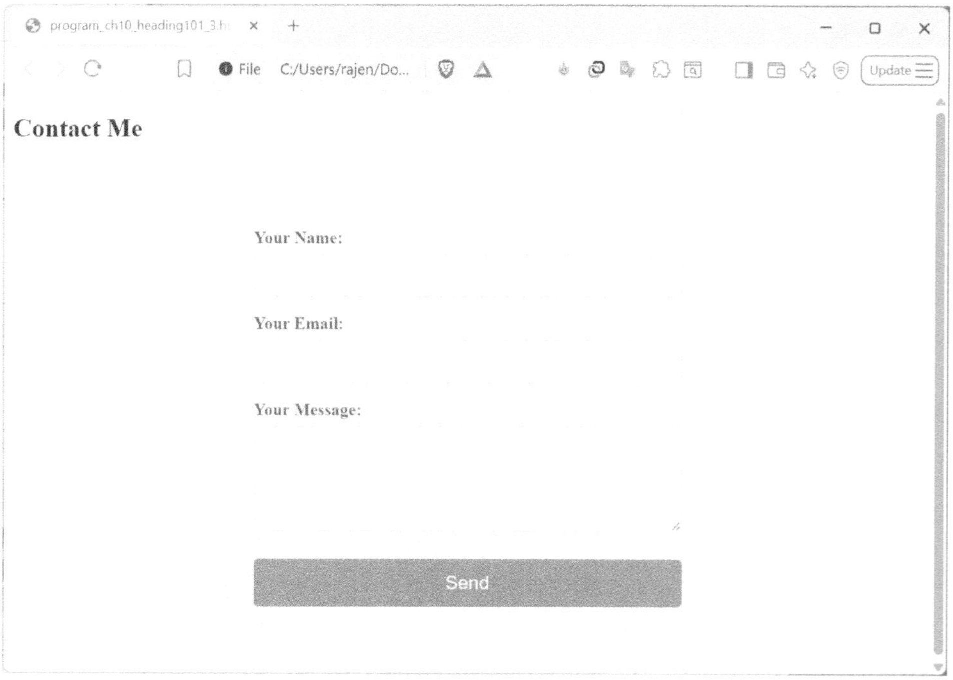

Hands-On Exercise: Create a Simple Blog Comment Form

Try building a comment form for a blog post. Include fields for name, email, and a comment. Use the **<form>**, **<input>**, **<textarea>**, and **<button>** elements you've just learned, and style them using the CSS techniques above. This exercise will reinforce your understanding of form structures and styling.

You've learned how to build basic web forms with **<form>**, **<input>**, **<textarea>**, and **<button>**. You practiced structuring forms for real-world uses and enhanced them with CSS for a polished, user-friendly appearance. Keep experimenting with layouts, colors, and input types to deepen your understanding.

In the next section, you'll unlock even more interactive form features by learning how to add dropdown menus, radio buttons, and checkboxes. These controls give users more ways to interact, making your forms even more dynamic and engaging. Get ready to take your forms to the next level!

10.2. Creating Dropdowns, Radio Buttons, and Checkboxes

Forms become truly interactive and useful when you provide options for users to select from, such as dropdown lists, radio buttons, and checkboxes. These controls are everywhere on the web, from registration forms and surveys to online shops and job applications. In this section, you'll learn how to create these essential elements using

HTML, then style and lay them out with CSS for a professional, user-friendly appearance.

Understanding Dropdown Lists

A dropdown list, built with the **<select>** element, lets users pick one choice from a list. Dropdowns save space and keep forms tidy, making them perfect for selecting countries, categories, or quantities.

Let's start with a practical scenario: imagine you're building a sign-up form for a global travel blog where users must select their country.

Open your project's HTML file and add a simple form section:

```
<form>
  <label for= "country">Choose your country:</label>
  <select id= "country"name= "country">
    <option value= "">--Please choose an option--</option>
    <option value= "usa">United States</option>
    <option value= "uk">United Kingdom</option>
    <option value= "india">India</option>
    <option value= "japan">Japan</option>
  </select>
</form>
```

<select> defines the dropdown menu.
<option> elements provide the possible choices.
The **value** attribute is what gets submitted with the form.

Now, let's add some style to make the dropdown visually appealing and easy to use:

```
select {
  padding: 8px 12px;
  border-radius: 4px;
  border: 1px solid #aaa;
  background-color: #f9f9f9;
  font-size: 1rem;
}

label {
  font-weight: bold;
  margin-right: 10px;
}
```

This CSS gives your dropdown a modern, friendly look, matching styles seen on popular websites.

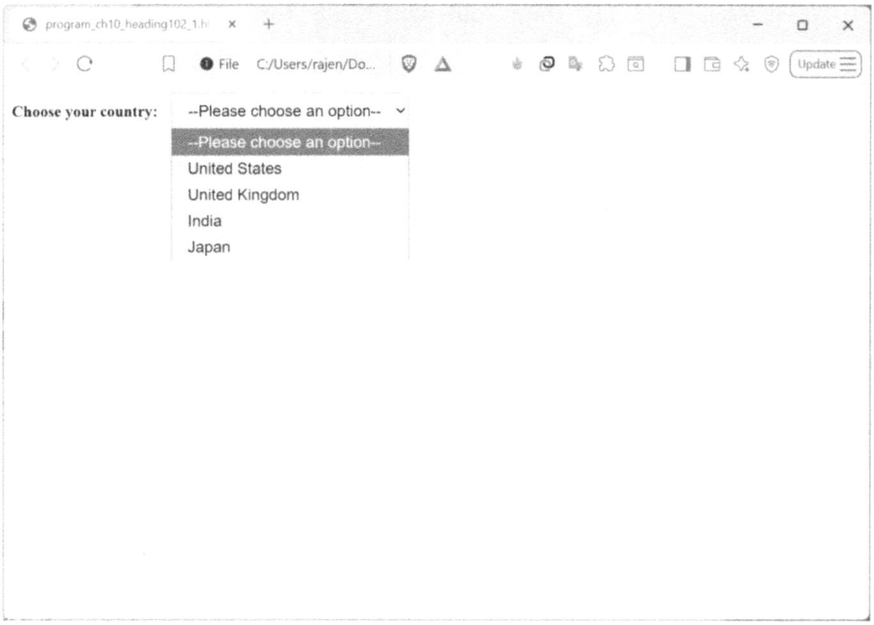

Building Radio Button Groups

Radio buttons let users select one option from a set—perfect for single-choice questions like payment methods or yes/no answers.

Suppose you're creating a checkout form for an online art shop, and you want users to choose a delivery method.

In your HTML file, add a radio button group:

```
<form>
  <p>Choose a delivery method:</p>
  <label>
    <input type= "radio"name= "delivery"value= "standard"checked>
    Standard Shipping
  </label>
  <label>
    <input type= "radio"name= "delivery"value= "express">
    Express Delivery
  </label>
  <label>
    <input type= "radio"name= "delivery"value= "pickup">
    In-Store Pickup
  </label>
</form>
```

All **input** elements have **type="radio"** and share the same **name** ("delivery"), linking them as a group so only one can be selected.
The **checked** attribute pre-selects an option.

Let's style these for clarity and accessibility:

```
form p {
  font-size: 1.1rem;
  margin-bottom: 6px;
}

input[type= "radio"] {
  accent-color: #0077cc;
  margin-right: 6px;
}

label {
  display: block;
  margin-bottom: 8px;
  cursor: pointer;
}
```

This approach uses the **accent-color** property, which changes the color of the radio button to match your brand or design theme. Spacing and bold labels ensure readability.

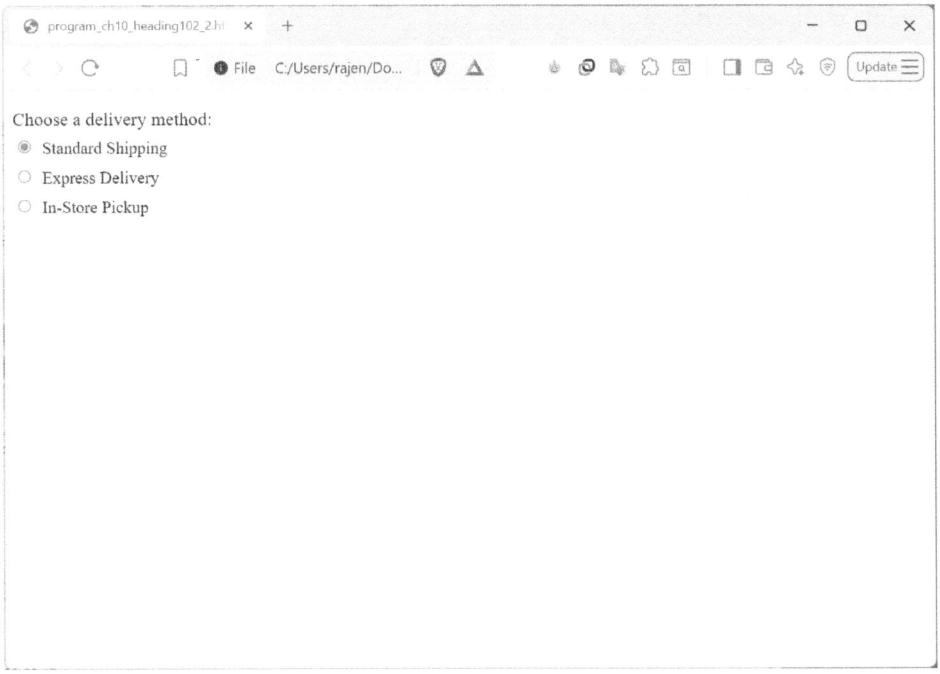

Adding Checkboxes for Multiple Selections

Checkboxes allow users to select multiple options from a list, making them ideal for things like newsletter subscriptions, agreeing to terms, or choosing product features.

Let's practice with a blog registration form where users can subscribe to different content categories.

Insert these checkboxes into your form:

```
<form>
  <p>Subscribe to topics:</p>
  <label>
    <input type= "checkbox"name= "topics"value= "html">
    HTML Tips
  </label>
  <label>
    <input type= "checkbox"name= "topics"value= "css">
    CSS Tricks
  </label>
  <label>
    <input type= "checkbox"name= "topics"value= "javascript">
    JavaScript News
  </label>
</form>
```

Each **input** uses **type="checkbox"**, and the **name** can be the same if you want to collect all selected topics as a group.

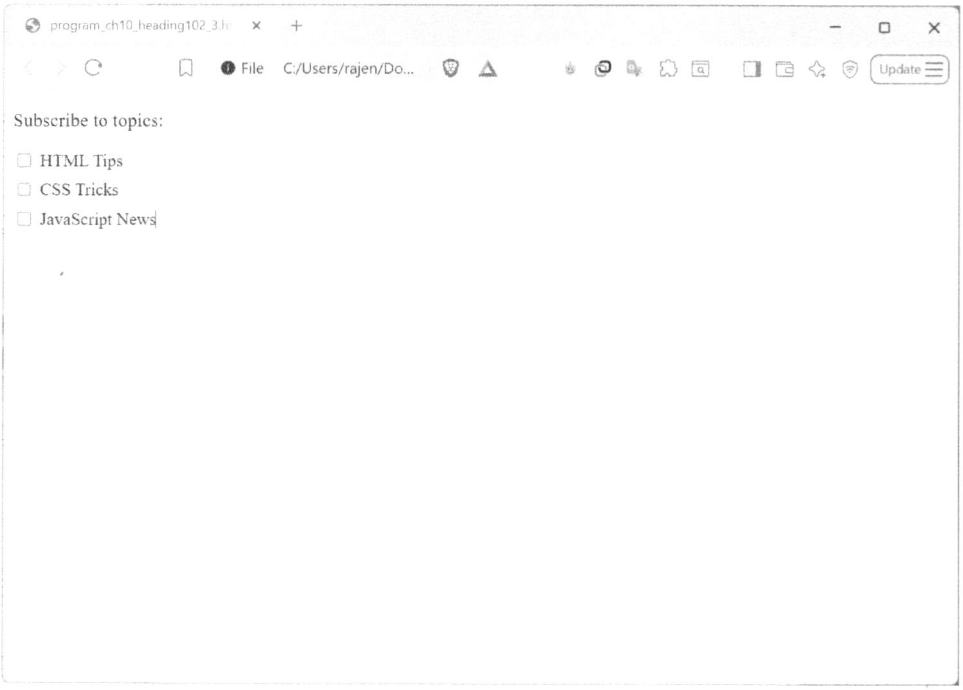

Add this CSS to enhance usability and appearance:

```
input[type= "checkbox"] {
  accent-color: #28a745;
  margin-right: 8px;
}

form p {
  font-size: 1.1rem;
}

label {
  display: flex;
  align-items: center;
  margin-bottom: 8px;
  cursor: pointer;
}
```

Using **display: flex** aligns the checkbox and label text neatly, while **accent-color** gives a visual cue when boxes are checked.

Hands-On Exercise: Build a Mini Signup Form

It's time to put your skills into practice. Combine all three elements in a single form, suitable for a real-world scenario like registering for a coding workshop.

```
<form>
  <label for= "role">Registering as:</label>
  <select id= "role"name= "role">
    <option value= "">--Select your role--</option>
    <option value= "student">Student</option>
    <option value= "teacher">Teacher</option>
    <option value= "professional">Professional</option>
  </select>

  <p>Preferred session:</p>
  <label>
    <input type= "radio"name= "session"value= "morning"checked>
    Morning
  </label>
  <label>
    <input type= "radio"name= "session"value= "afternoon">
    Afternoon
  </label>

  <p>Topics of Interest:</p>
  <label>
    <input type= "checkbox"name= "interest"value= "html">
    HTML Basics
  </label>
  <label>
    <input type= "checkbox"name= "interest"value= "css">
    CSS Design
  </label>
  <label>
    <input type= "checkbox"name= "interest"value= "js">
    JavaScript Essentials
  </label>
```

```
</form>
```

Enhance the form with the CSS techniques you've just learned to create a clean, inviting layout. Try tweaking **accent-color** or add a background for the form container to further personalize the design.

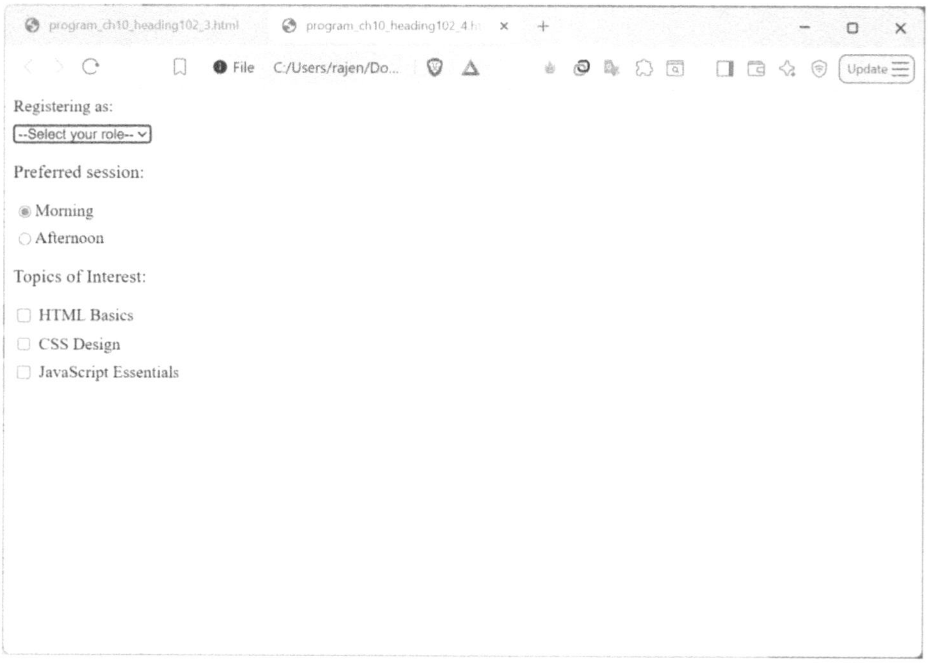

Coding Summary and Practice

You've now created and styled dropdown menus, radio button groups, and checkboxes. These controls are the backbone of interactive forms across millions of websites, from e-commerce checkouts to blog registration pages. Practice by building your own custom forms or recreating those you see on your favorite sites. Experiment with different styles and layouts to see how small changes in CSS can make a big difference.

Next, you'll discover how to organize your forms for clarity and accessibility. In the upcoming section, 'Labeling and Grouping Form Elements', you'll learn how to associate labels with inputs and use grouping to guide users through complex forms—unlocking even more powerful and user-friendly form designs!

10.3. Labeling and Grouping Form Elements

Designing forms that are easy for users to understand and navigate is a core skill for every web developer. Well-labeled and logically grouped form elements make your forms more accessible, professional, and enjoyable to use. In this section, you'll explore how to effectively associate labels with form fields and organize related inputs using fieldsets and legends. With hands-on practice, you'll see how these techniques enhance both usability and design in real-world scenarios.

Why Labels and Groups Matter

Imagine you're creating a sign-up form for a portfolio website or a checkout page for a small e-commerce store. Clear labels help users know exactly what information is needed, while grouping related fields (like shipping address or payment info) reduces confusion. Semantic HTML and proper CSS styling also make your forms easier to navigate for users with assistive technologies.

Associating Labels with Inputs

The **label** element is used to provide a descriptive caption for an input field. Associating a label with an input improves accessibility, enabling users to click the label to focus the input and screen readers to announce the label's text.

To link a label to an input, use the **for** attribute on the label and the matching **id** attribute on the input.

Let's walk through a practical example:

```
<form>
  <label for= "fullname">Full Name:</label>
  <input type= "text"id= "fullname"name= "fullname"/>

  <label for= "email">Email Address:</label>
  <input type= "email"id= "email"name= "email"/>
</form>
```

In this code:

The first label is connected to the "fullname" input. Clicking "Full Name:" focuses the text box.

The second label works the same way for the "email" input.

Let's add some basic CSS to make the form easier to read and more visually appealing:

```
form {
  max-width: 400px;
  margin: 2em auto;
  padding: 1em;
  background: #f7f7f7;
  border-radius: 8px;
  box-shadow: 0 2px 8px rgba(0,0,0,0.1);
}
```

```
label {
  display: block;
  margin-top: 1em;
  font-weight: bold;
  color: #333;
}

input[type= "text"],
input[type= "email"] {
  width: 100%;
  padding: 0.5em;
  margin-top: 0.25em;
  border: 1px solid #ccc;
  border-radius: 4px;
  font-size: 1em;
}
```

Now, your form is not only accessible but also looks professional and inviting.

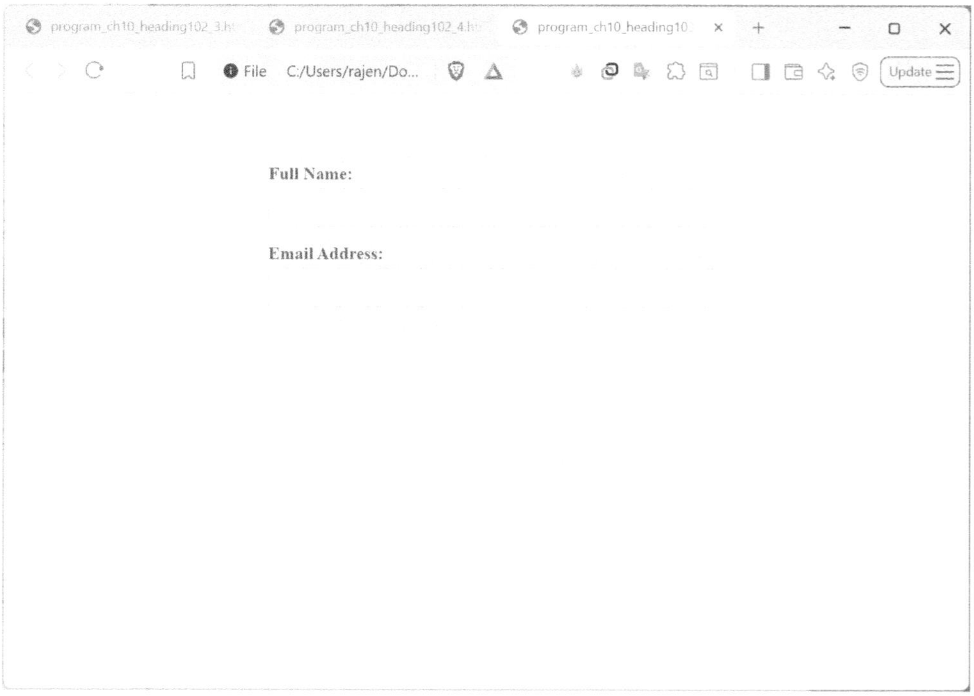

Grouping Related Form Elements

When forms include multiple related questions, grouping them makes the form clearer and easier to scan. This is especially useful in more complex scenarios like registration forms or e-commerce checkout pages.

The **fieldset** element is used to group related inputs, while the **legend** element provides a caption for the group. Here's how you can use them:

```
<form>
  <fieldset>
    <legend>Personal Information</legend>

    <label for= "firstName">First Name:</label>
    <input type= "text"id= "firstName"name= "firstName"/>

    <label for= "lastName">Last Name:</label>
    <input type= "text"id= "lastName"name= "lastName"/>
  </fieldset>

  <fieldset>
    <legend>Account Details</legend>

    <label for= "username">Username:</label>
    <input type= "text"id= "username"name= "username"/>

    <label for= "password">Password:</label>
    <input type= "password"id= "password"name= "password"/>
  </fieldset>
</form>
```

Notice how each set of related fields is grouped and introduced by a legend. This mirrors real-world web forms you might find on job application or e-commerce sites.

Styling Fieldsets and Legends

Fieldsets and legends can be styled to match your site's branding and improve the visual flow of forms. Here's an example of CSS that enhances the grouping:

```
fieldset {
  border: 2px solid #4CAF50;
  border-radius: 6px;
  padding: 1em;
  margin-bottom: 1.5em;
  background-color: #fff;
}

legend {
  font-size: 1.1em;
  font-weight: bold;
  color: #4CAF50;
  padding: 0 0.5em;
}
```

With this styling, your form sections stand out, making it easier for users to focus on one group at a time.

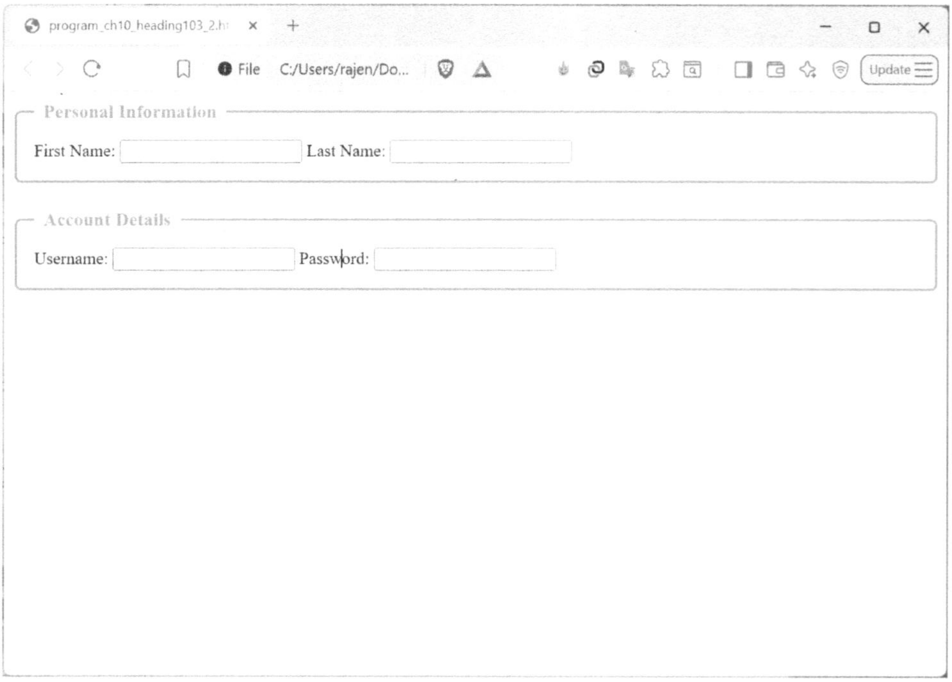

Real-World Coding Scenario: User Profile Form

Suppose you're building a user profile form for a blog platform. You want to group personal details separately from communication preferences. Here's how your HTML and CSS might look:

```
<form>
  <fieldset>
    <legend>Profile Info</legend>
    <label for= "displayName">Display Name:</label>
    <input type= "text"id= "displayName"name= "displayName"/>
    <label for= "bio">Short Bio:</label>
    <input type= "text"id= "bio"name= "bio"/>
  </fieldset>
  <fieldset>
    <legend>Preferences</legend>
    <label for= "newsletter">
      <input type= "checkbox"id= "newsletter"name= "newsletter"/>
      Receive monthly updates
    </label>
    <label for= "theme">Site Theme:</label>
    <select id= "theme"name= "theme">
      <option value= "light">Light</option>
      <option value= "dark">Dark</option>
    </select>
  </fieldset>
</form>
```

Add CSS for clarity and emphasis:

```
input[type= "checkbox"] {
  margin-right: 0.5em;
}
select {
  margin-top: 0.25em;
  width: 100%;
  padding: 0.5em;
  border-radius: 4px;
}
```

This approach not only organizes your form logically but also makes it visually appealing and user-friendly.

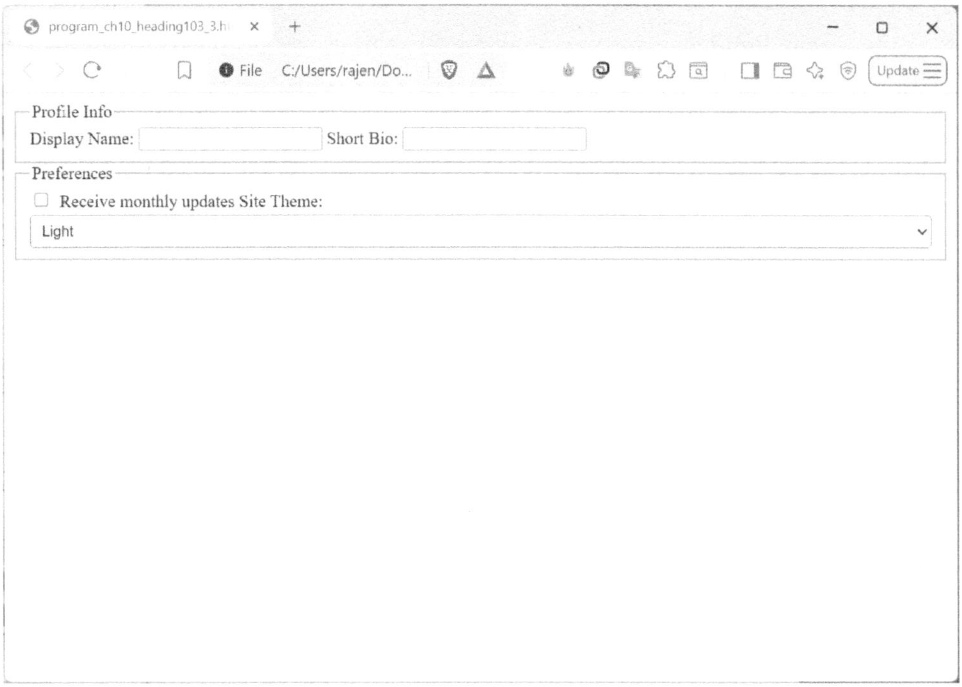

Hands-On Exercise: Build a Contact Form

Now it's your turn! Create a contact form for a portfolio website with these requirements:

- Group user details (name and email) in one fieldset.
- Group message details (subject and message box) in another.
- Use clear labels for each input.
- Style your fieldsets and labels for readability.

Experiment with different CSS colors, borders, and spacing to match your style.

You've learned how to use the **label** element to connect descriptive text with form inputs, improving both accessibility and user experience. By grouping related fields with **fieldset** and **legend**, you keep forms organized and easy to navigate. Styling these elements with CSS brings clarity and professionalism to your forms, whether for a blog, portfolio, or online store.

Up next, you'll discover how to make your forms even smarter and user-friendly by validating user input with built-in HTML5 attributes. Get ready to learn how to prevent mistakes and guide users to submit perfect forms every time!

10.4. Validating User Input with HTML5 Attributes

When you build web forms, making sure users enter the right information is critical. HTML5 introduced a set of powerful **form validation attributes** that help you control the kind of data users submit, without needing complex JavaScript. These built-in features make your forms friendlier and more reliable, so visitors are less likely to make mistakes or get frustrated.

Why Validate User Input?

Validating input means checking that the data entered fits your requirements. For example, on a registration form, you might want to ensure the email field contains a valid email address, the password is long enough, or a required field isn't left empty. Without validation, your website might receive incomplete, incorrect, or even harmful data.

Let's explore how HTML5 attributes make this job easier, and then style the validation messages for a seamless user experience.

Common HTML5 Validation Attributes

HTML5 gives you several attributes you can add to input elements to enforce rules:

- **required**: Makes the field mandatory.
- **type**: Sets the data type, such as "email", "url", or "number".
- **minlength** and **maxlength**: Set the minimum and maximum number of characters.
- **pattern**: Defines a regular expression the input must match.
- **min** and **max**: Restrict the range of numbers or dates.
- **step**: Specifies the legal number intervals for numeric fields.

Let's bring these attributes to life with a practical form example you might see on a modern blog or portfolio site.

Real-World Example: Newsletter Signup Form

Imagine you're adding a newsletter signup section to your personal portfolio or blog. You want to collect a user's name (at least 2 characters), email (must be valid), and a short message (optional, up to 150 characters).

Here's how you can use HTML5 validation attributes:

```
<form class= "newsletter-signup"autocomplete= "on">
  <label for= "name">Name:</label>
  <input type= "text"id= "name"name= "name"required minlength=
"2"maxlength= "30"placeholder= "Your Name">

  <label for= "email">Email:</label>
  <input type= "email"id= "email"name= "email"required placeholder=
"you@example.com">

  <label for= "message">Message (optional):</label>
  <textarea id= "message"name= "message"maxlength= "150"placeholder=
"Say hello! (max 150 characters)"></textarea>

  <button type= "submit">Subscribe</button>
</form>
```

Let's break down what each attribute does:

- **required** on the name and email fields ensures users can't submit the form without filling them out.
- **minlength="2"** and **maxlength="30"** on the name input control how short or long the name can be.
- **type="email"** on the email field checks for a valid email format.
- **maxlength="150"** on the message limits the text area input.

Try submitting the form with missing or invalid data—your browser will highlight errors and display helpful messages automatically.

Styling Validation States with CSS

While browsers provide default styles for invalid fields, customizing these with CSS improves user experience and keeps your design consistent.

You can style form fields based on their validity using CSS pseudo-classes:

:valid targets inputs that pass validation.
:invalid targets inputs that fail validation.
:focus targets the field the user is currently typing in.

Let's add some CSS to highlight fields and give users clear, immediate feedback:

```
.newsletter-signup input:invalid,
.newsletter-signup textarea:invalid {
  border: 2px solid #e74c3c; /* Red border for errors */  background-
color: #fff0f0;
}

.newsletter-signup input:valid,
.newsletter-signup textarea:valid {
  border: 2px solid #27ae60; /* Green border when valid */  background-
color: #f0fff0;
}

.newsletter-signup input:focus,
.newsletter-signup textarea:focus {
  outline: none;
  border-color: #2980b9; /* Blue border on focus */  background-color:
#eaf6ff;
}

.newsletter-signup label {
  display: block;
  margin-top: 15px;
  font-weight: bold;
}

.newsletter-signup button {
  margin-top: 20px;
  padding: 10px 16px;
  background-color: #2980b9;
  color: #fff;
  border: none;
  cursor: pointer;
  border-radius: 4px;
  font-size: 1rem;
}

.newsletter-signup button:hover {
  background-color: #3498db;
}
```

With this CSS, users immediately see which fields need attention, making your form both stylish and intuitive.

Using Pattern Attribute for Custom Validation

Sometimes you need to enforce a specific format that isn't covered by the default types. The **pattern** attribute lets you set a regular expression for custom validation.

Suppose you want users to enter a username that starts with a letter and is 4-12 characters long, using only letters and numbers. Here's how:

```
<label for= "username">Username:</label>
<input
  type= "text"
  id= "username"
```

```
  name= "username"
  required
  pattern= "^[A-Za-z][A-Za-z0-9]{3,11}$"
  title= "Username must start with a letter and be 4-12 characters
long, using only letters and numbers."
  placeholder= "Choose a username">
```

The **pattern** enforces the format.
The **title** provides a helpful tooltip when validation fails.

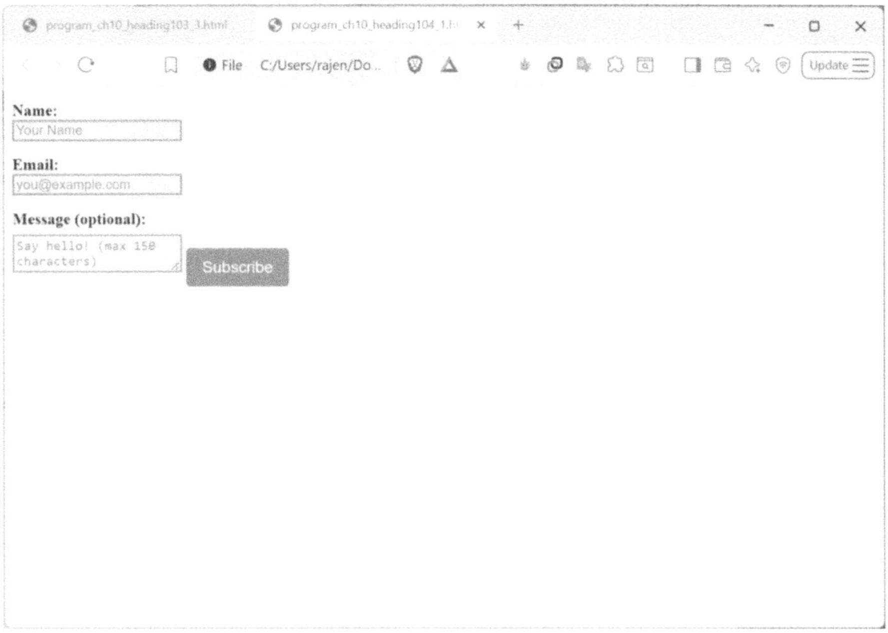

Real-World Scenario: E-commerce Checkout

On e-commerce sites, validating user input ensures smooth checkouts. For example, you can use:

type="email" for billing email.
pattern for phone numbers.
required for address fields.
min and **max** for quantity selectors.

This reduces errors and speeds up the buying process, improving both user satisfaction and your site's professionalism.

Coding Summary

By combining HTML5 validation attributes and custom CSS, you've learned to guide users toward correct input, reduce errors, and create visually engaging forms.

These skills are essential for any web developer, whether you're building a simple blog signup or a sophisticated e-commerce checkout.

Experiment with different validation attributes and styles in your own forms. Try tweaking the rules, such as requiring stronger passwords or validating phone numbers, and see how the browser responds.

Next, you'll put these concepts into action by building a fully functional contact or feedback form. Get ready to combine everything you've learned so far to create a professional, user-friendly form that shines with both functionality and style!

10.5. Hands-On Exercise: Making a Contact or Feedback Form

A well-designed contact or feedback form is a core feature on countless modern websites, from personal portfolios to business pages and online shops. Forms invite users to connect, ask questions, or provide valuable feedback. In this hands-on exercise, you'll build your own contact form from scratch, style it with CSS for visual appeal, and learn practical layout skills you'll use in real-world projects.

Before diving in, consider two real-world examples:

Personal Portfolio: A contact form lets visitors connect with you for freelance opportunities or collaborations.
Small Business Website: A feedback form helps customers reach out for support or to make suggestions, boosting trust and engagement.

Let's get started by creating the HTML structure, then styling your form to make it accessible, attractive, and user-friendly.

Structuring Your Contact Form with HTML

A contact form typically includes fields for a user's name, email, message, and a submit button. Here's how you can create this form step by step:

Open your code editor and create a new HTML file named `contact.html`.
Add the basic HTML structure. Inside the `<body>`, include a `<form>` element with relevant input fields.

```
<!DOCTYPE html>
<html lang= "en">
<head>
  <meta charset= "UTF-8">
  <title>Contact Us</title>
  <link rel= "stylesheet"href= "styles.css">
</head>
<body>
```

```
  <section class= "contact-section">
    <h2>Contact Us</h2>
    <form class= "contact-form"action= "#"method= "post">
      <label for= "name">Name</label>
      <input type= "text"id= "name"name= "name"required>

      <label for= "email">Email</label>
      <input type= "email"id= "email"name= "email"required>

      <label for= "message">Message</label>
      <textarea id= "message"name= "message"rows=
"5"required></textarea>

      <button type= "submit">Send Message</button>
    </form>
  </section>
</body>
</html>
```

Labels are connected to inputs using the `for` attribute, which improves accessibility. The `required` attribute ensures fields must be filled before submission.

Adding CSS for Form Layout and Design

Now it's time to make your form visually engaging and easy to use. You'll style the form using CSS, focusing on layout, spacing, and user interface cues.

Create a new CSS file named `styles.css` in the same folder as your HTML file.
Start by setting up some basic styles for the body and section to center your form and add padding.

```
body {
  font-family: Arial, sans-serif;
  background-color: #f5f7fa;
  margin: 0;
  padding: 40px 0;
}

.contact-section {
  max-width: 400px;
  margin: 0 auto;
  background: #fff;
  padding: 32px 24px;
  border-radius: 8px;
  box-shadow: 0 2px 12px rgba(0,0,0,0.09);
}
```

The `.contact-section` class centers the form and creates a card-like appearance with padding and a subtle shadow.
Next, style the form elements for clarity and usability.

```
.contact-form label {
  display: block;
  margin-bottom: 6px;
  font-weight: bold;
  color: #333;
}

.contact-form input,
.contact-form textarea {
  width: 100%;
  padding: 10px;
  border: 1px solid #c8c8c8;
  border-radius: 4px;
  margin-bottom: 18px;
  font-size: 1em;
  transition: border-color 0.2s;
}

.contact-form input:focus,
.contact-form textarea:focus {
  border-color: #5ab4f8;
  outline: none;
}

.contact-form button {
  background-color: #5ab4f8;
  color: #fff;
  border: none;
  padding: 12px 0;
  width: 100%;
  border-radius: 4px;
  font-size: 1.1em;
  font-weight: bold;
  cursor: pointer;
  transition: background 0.2s;
}

.contact-form button:hover {
  background: #488ac7;
}
```

Labels are bold and clear, guiding users through the form.

Inputs and **textareas** expand to fill the form width, with subtle borders and rounded corners for a friendly look.

Focus states (activated when users click inside a field) use a colored border to provide feedback.

The **button** uses a blue background for emphasis and changes shade on hover for interactivity.

Improving Responsiveness for All Devices

A key part of modern web design is ensuring your form looks great on any screen size. You can accomplish this with a simple responsive CSS rule.

```
@media (max-width: 500px) {
  .contact-section {
    padding: 18px 8px;
  }

  .contact-form button {
    font-size: 1em;
    padding: 10px 0;
  }
}
```

The **media query** adjusts padding and button size for smaller screens, such as smartphones, ensuring your form remains usable and visually balanced.

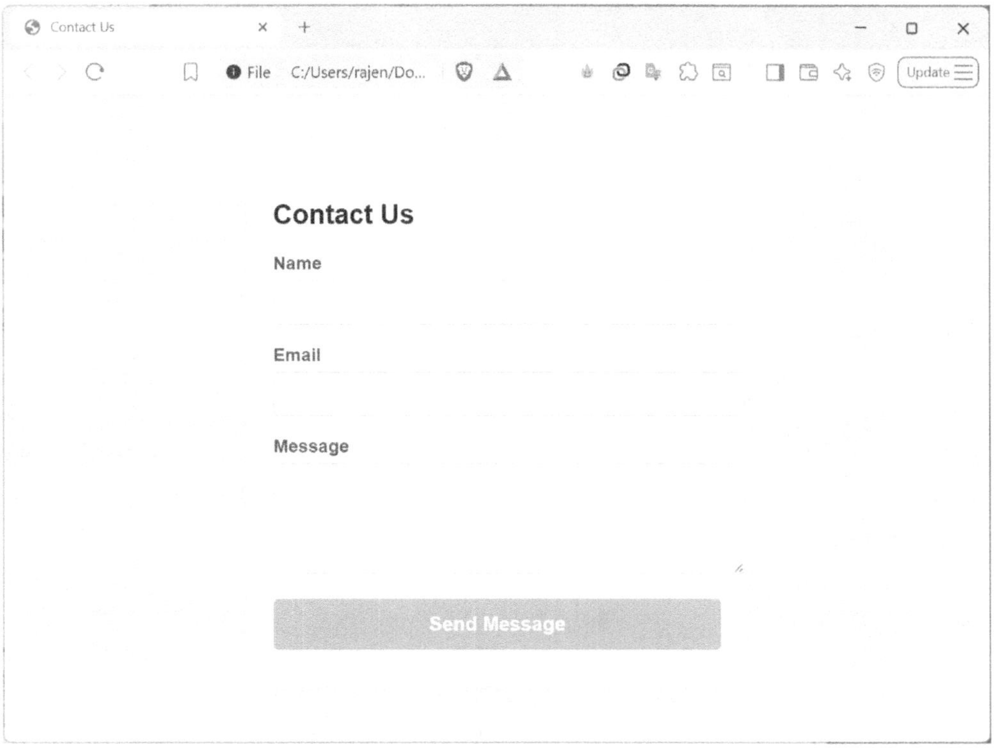

Real-World Application Scenario

Imagine you're building a **travel blog**. Visitors may want to share travel tips or ask about destinations. Embedding a styled, responsive contact form in your blog gives readers an easy way to interact and builds a sense of community.

You can use the exact techniques from this exercise for:

Portfolio contact pages
Feedback forms on small e-commerce sites
Newsletter sign-up forms (with slight modifications)

Practice Exercise: Personalize Your Contact Form

Try these enhancements to reinforce your learning:

Add a field for “Subject” to your form.
Change the color scheme to match your personal style or website branding.
Add a short introductory paragraph above the form inviting users to get in touch.

For example, to add a subject field:

```
<label for= "subject">Subject</label>
<input type= "text"id= "subject"name= "subject"required>
```

And update your CSS if you want a new accent color:

```
.contact-form button {
  background-color: #28a745;
}
.contact-form button:hover {
  background-color: #218838;
}
```

Coding Summary

You’ve just created a fully functional, stylish contact form and learned essential HTML and CSS techniques for building real-world user input features. This hands-on exercise covered form structure, labels, interactive styling, and responsive design—skills you’ll use in virtually every web project. Keep refining and experimenting to make your forms even better!

11. CSS LAYOUT FUNDAMENTALS

Unlocking the secrets of professional web design begins with mastering how content is arranged and displayed on a page. In this chapter, you'll dive deep into the essential skills that bring structure and order to your website layouts. Whether you've admired sleek portfolios, engaging blogs, or dynamic online stores, their polished appearance is no accident—it's powered by smart use of CSS layout techniques.

Here, you'll revisit and strengthen your understanding of the **CSS Box Model**, the core principle that shapes every element on a web page. You'll learn how to control the way content is stacked, aligned, and layered using the **display** and **position** properties, unlocking the ability to transform simple blocks of text and images into sophisticated, multi-column layouts. With hands-on guidance, you'll explore how to use **z-index** for layering elements, giving your designs depth and visual hierarchy.

The skills you build in this chapter are practical and immediately applicable. By the end, you'll not only understand how to construct a clean, professional two-column blog layout—a classic feature found in countless real-world websites—but you'll also gain the confidence to tackle more advanced designs as your coding journey continues. Each exercise is designed to empower you with techniques that modern web developers use every day, ensuring you're ready to apply them to your own projects or future career opportunities.

Get ready to see your web pages come alive with structure and style as you take the next step in becoming a confident, capable web developer.

11.1. Understanding the CSS Box Model (Revisited)

When you're building websites that look good and work well on any device, understanding how elements are sized and spaced is essential. This is where the **CSS box model** comes into play. Whether you're styling a blog post, a portfolio, or a product card for an e-commerce site, mastering the box model gives you precise control over your layouts.

The Box Model Refresher: What Makes Up Every Element

Every visible element on a webpage is a rectangular box, even if it looks round or has fancy borders. The CSS box model describes how the size of these boxes is calculated. Each box is made of four parts:

Content: The actual text, image, or other media inside the element.

Padding: The space between the content and the border. It pushes the border away from the content.
Border: The line surrounding the padding (and content).
Margin: The outermost space, separating one element from another.

Visualize it like a gift box: the **content** is your present, **padding** is bubble wrap, the **border** is the box itself, and the **margin** is the space between this box and others on the shelf.

Hands-On: Inspecting and Tweaking the Box Model

Let's practice by building a simple product card, a common feature on e-commerce websites. You'll see how changing the padding, border, and margin affects the layout.

Create a new HTML file and add the following markup inside your `<body>`:

```
<div class= "product-card">
  <h2>Wireless Headphones</h2>
  <p>Crystal clear sound. 30 hours of battery life.</p>
  <button>Add to Cart</button>
</div>
```

Add CSS to style your product card and explore the box model:

```
.product-card {
  width: 320px;
  background-color: #f3f6fa;
  padding: 24px;
  border: 2px solid #3a4d63;
  margin: 40px auto;
  border-radius: 12px;
  box-shadow: 0 4px 16px rgba(58, 77, 99, 0.08);
}

.product-card h2 {
  margin: 0 0 12px 0;
  font-size: 1.5rem;
}

.product-card p {
  margin: 0 0 18px 0;
  color: #5c6b77;
}

.product-card button {
  padding: 10px 32px;
  border: none;
  background: #3273dc;
  color: #fff;
  border-radius: 6px;
  cursor: pointer;
  font-size: 1rem;
```

```
}
```

Open your page in the browser and use your browser's developer tools (right-click the card and choose "Inspect") to see the box model in action. You'll notice:

Padding: Creates space inside the card around the text.

Border: The solid colored line outlining the card.

Margin: The space outside the card, centering it on the page.

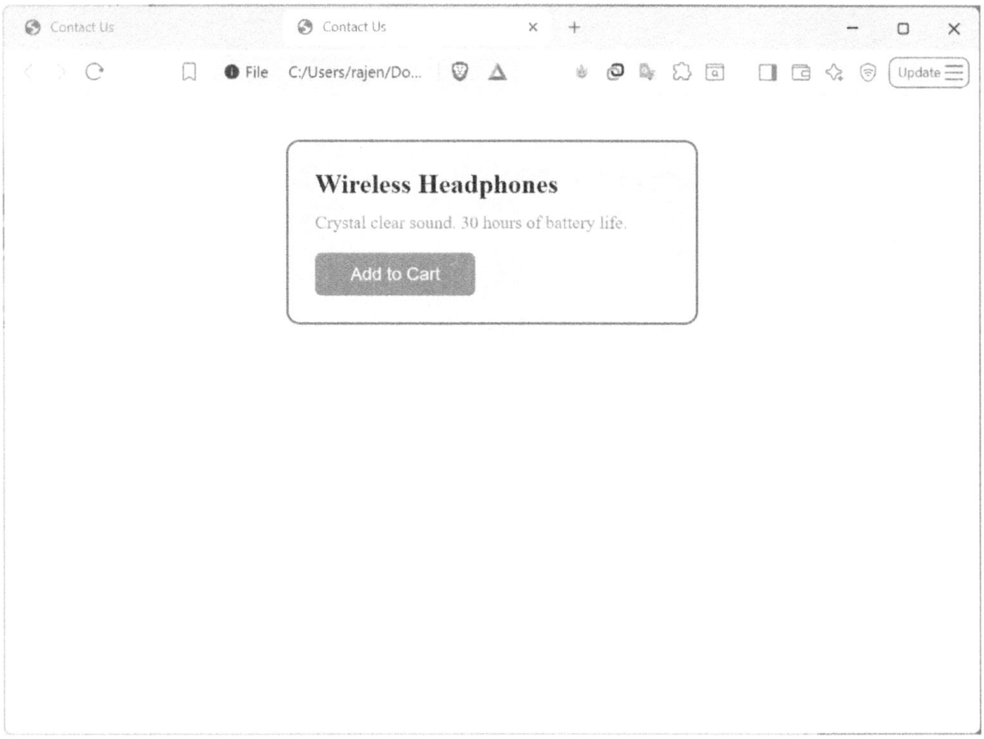

How the Box Model Influences Responsive Design

When making layouts that adapt to different devices, understanding how each part of the box model affects the overall size of your elements is vital. For example, if you set a **width** on an element, by default that width only applies to the content area, not including padding and border. This can trip up beginners when layouts break or overflow.

Let's see this with a blog post preview box, a common feature on blogs and news sites.

Add this HTML for a blog preview card:

```
<div class= "blog-preview">
  <h3>10 Tips for Learning CSS Fast</h3>
  <p>Master the basics of CSS with these practical tips and hands-on
```

```
exercises. Perfect for beginners!</p>
</div>
```

Style it to highlight the box model effect:

```
.blog-preview {
  width: 300px;
  padding: 30px;
  border: 3px solid #ffb347;
  margin: 32px;
  background: #fff8e1;
  box-sizing: content-box;
}
```

With **box-sizing: content-box** (the default), the width (300px) applies only to the content area. The padding and border add extra width, making the total element much wider than 300px.

Now, add this line to your CSS for `.blog-preview`:

```
box-sizing: border-box;
```

With **box-sizing: border-box**, the width (300px) includes the content, padding, and border, making layouts much easier to manage, especially for responsive designs.

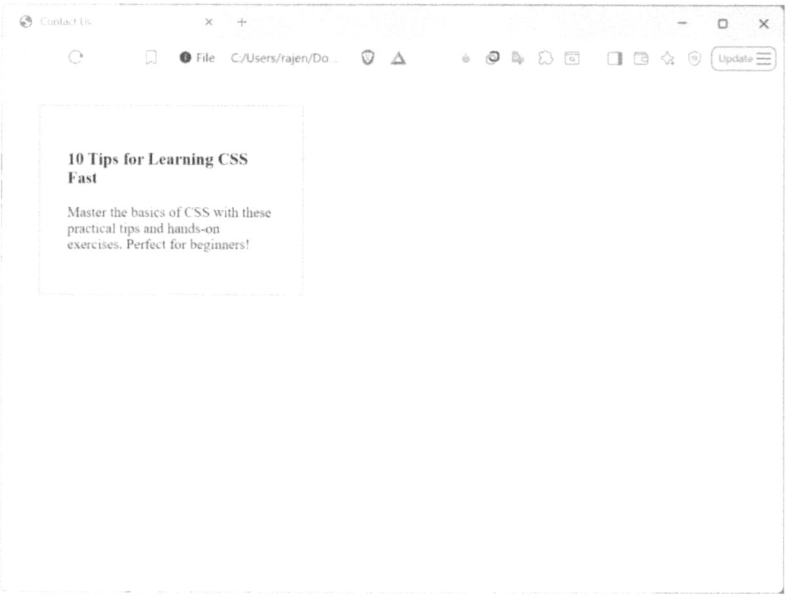

Coding Exercise: Practice the Box Model

Try this hands-on exercise to reinforce your skills:

Build a simple navigation bar with three links.

```
<nav class= "main-nav">
  <a href= "#">Home</a>
  <a href= "#">About</a>
  <a href= "#">Contact</a>
</nav>
```

Style the navigation bar using box model properties:

```
.main-nav {
  background: #e6f2ff;
  padding: 18px 0;
  border-bottom: 2px solid #3399ff;
  margin-bottom: 28px;
  text-align: center;
}

.main-nav a {
  color: #205081;
  text-decoration: none;
  display: inline-block;
  padding: 10px 18px;
  margin: 0 8px;
  border-radius: 4px;
  transition: background 0.2s;
}

.main-nav a:hover {
  background: #3399ff;
  color: #fff;
}
```

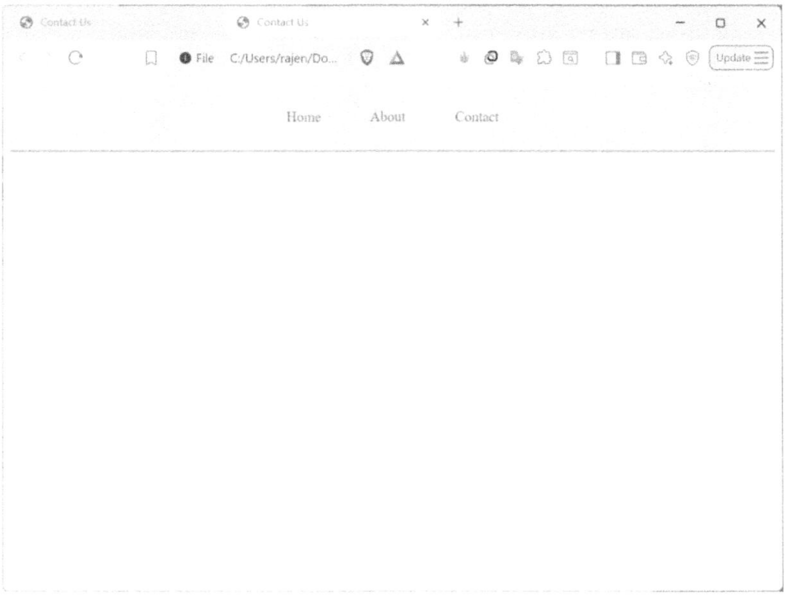

Experiment by adjusting the **padding**, **border**, and **margin** values for `.main-nav` and `.main-nav a`. Observe how these changes affect the size and spacing of your navigation links.

Real-World Application: Portfolio Layout

Imagine you're designing a personal portfolio website. You want your project cards to be evenly spaced and visually appealing on all devices. By properly applying the box model and using **box-sizing: border-box**, you can ensure your cards stay aligned and don't break the layout, no matter the padding or border you add for style.

You've now explored how the CSS box model controls every element's size and spacing, directly impacting your site's layout and responsiveness. By practicing with product cards, blog previews, and navigation bars, you've seen how **margin**, **padding**, **border**, and **box-sizing** work together. Remember, small changes in these properties can transform the look and feel of your website.

Keep experimenting with different box model settings to see their effects. Up next, you'll unlock even more layout power by learning about **display** and **position** properties, which help you move and organize elements exactly where you want them on the page. Get ready to take your layouts to the next level!

11.2. Display and Position Properties

CSS gives you the power to transform a basic HTML document into a visually appealing and organized web page. Two of the most important tools for layout are the **display** and **position** properties. These CSS properties let you control how elements appear, stack, and move on your page, forming the foundation for any modern, responsive design.

Understanding the Display Property

Every HTML element has a default way it behaves on the page, known as its **display** value. Some elements, like `<div>`, are **block-level** and take up the full width available, stacking vertically. Others, such as `<span>`, are **inline** and only take up as much space as they need, flowing horizontally with text.

To build creative layouts, you often need to override these defaults:

display: block; Forces an element to start on a new line and stretch to fill the container.

display: inline; Makes the element flow with surrounding text, taking up only its content's width.

display: inline-block; Allows the element to sit inline with text but lets you set width and height.

display: none; Hides the element completely from view.

Let's take a look at a real-world scenario: imagine you're designing a simple navigation bar for a portfolio website. You want the navigation links to appear side by side instead of stacking.

Create the HTML for your navigation:

```
<nav>
  <a href= "#">Home</a>
  <a href= "#">Projects</a>
  <a href= "#">Contact</a>
</nav>
```

Add CSS to display the links inline:

```
nav a {
  display: inline-block;
  padding: 10px 20px;
  text-decoration: none;
  background: #0077cc;
  color: white;
  border-radius: 4px;
  margin-right: 8px;
}
```

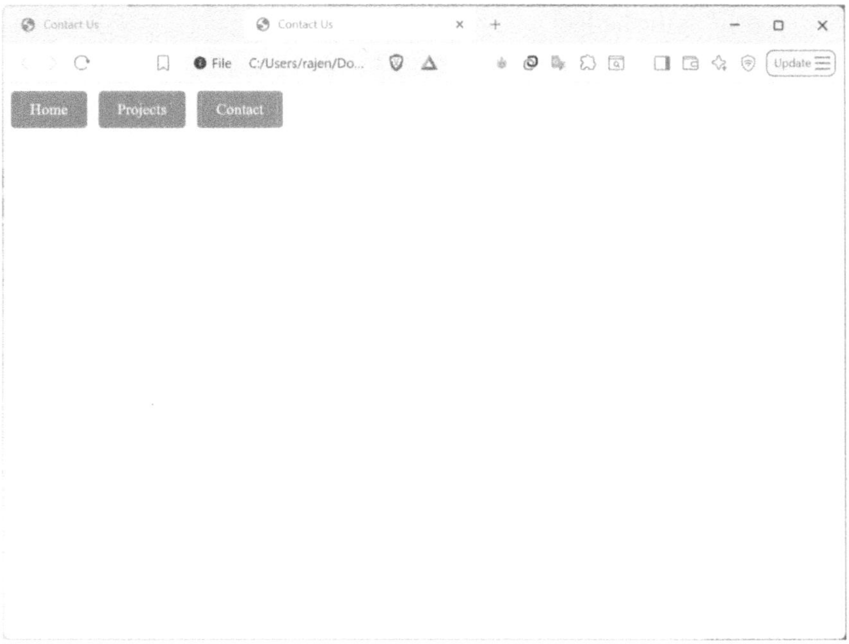

Now, each link sits neatly beside the others, just as you'd expect in a professional navigation bar.

Mastering the Position Property

The **position** property gives you precise control over where elements appear within their containing blocks. There are several useful position values:

static: The default; elements appear in the normal document flow.
relative: Elements can be nudged from their original spot without affecting others.
absolute: Positions the element relative to its nearest ancestor with a position other than static, or to the page if none exists.
fixed: Pins the element to a specific spot in the browser window, even when scrolling.
sticky: The element toggles between relative and fixed as you scroll.

Let's say you want to create a "sale" badge on a product image for an e-commerce site, positioned at the top-right corner.

Write the HTML for a product card:

```
<div class= "product-card">
  <img src= "images/shirt.jpg"alt= "Stylish Shirt"/>
  <span class= "badge">Sale!</span>
  <h2>Stylish Shirt</h2>
  <p>$29.99</p>
</div>
```

Use CSS to position the badge:

```
.product-card {
  position: relative;
  width: 220px;
  padding: 16px;
  border: 1px solid #ddd;
  border-radius: 8px;
  background: #fafafa;
}

.product-card img {
  width: 100%;
  border-radius: 4px;
}

.badge {
  position: absolute;
  top: 12px;
  right: 12px;
  background: #ff3366;
  color: white;
  padding: 6px 14px;
  border-radius: 16px;
```

```
  font-weight: bold;
  font-size: 0.9em;
}
```

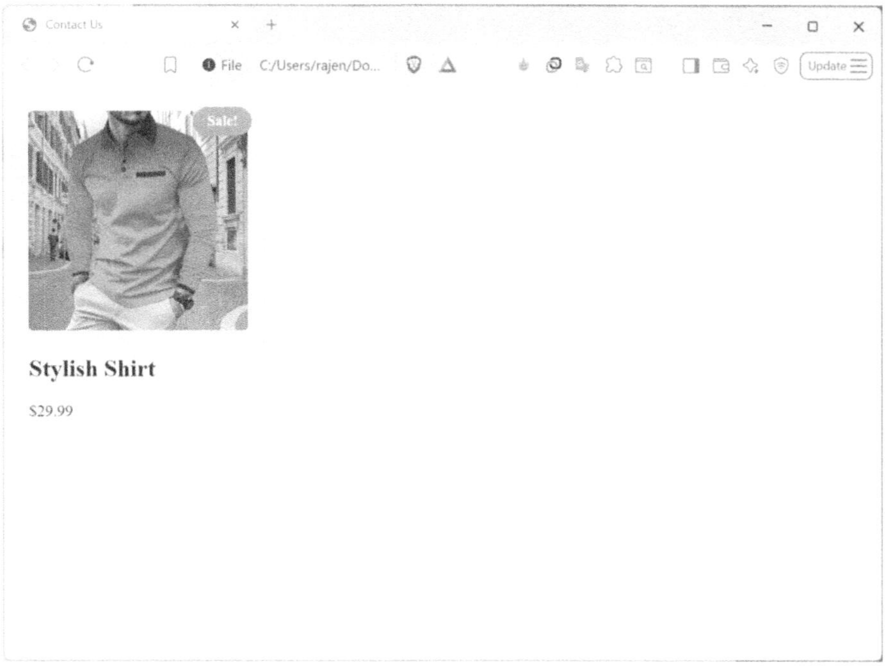

Here's what's happening:

.product-card is set to `position: relative;`, making it the reference point for absolutely positioned children.

.badge uses `position: absolute;`, placing it 12 pixels from the top and right of the card.

This simple technique is widely used for notification dots, price tags, or featured labels in online stores.

Combining Display and Position for Dynamic Layouts

Sometimes you'll need to mix both properties for creative effects. For instance, consider a call-to-action button that stays fixed at the bottom of a blog post, encouraging readers to subscribe.

Add the button to your HTML:

```
<article>
  <h1>5 Travel Photography Tips</h1>
  <p>...blog content...</p>
  <button class= "subscribe-btn">Subscribe</button>
</article>
```

Style it with CSS:

```
.subscribe-btn {
  position: fixed;
  bottom: 24px;
  right: 24px;
  background: #22b573;
  color: white;
  padding: 16px 32px;
  border: none;
  border-radius: 30px;
  font-size: 1.1em;
  box-shadow: 0 2px 8px rgba(0,0,0,0.18);
  cursor: pointer;
  z-index: 100;
  display: inline-block;
}
```

The button uses `position: fixed;` to anchor it to the viewport's bottom right. It remains visible as the user scrolls.

Adding `display: inline-block;` ensures the button's width and height remain under your control and don't stretch undesirably.

Practical Exercise: Try It Yourself

Let's reinforce your learning with a hands-on challenge. Create a simple alert box that appears centered at the top of your page and disappears when a close button is clicked.

HTML structure:

```
<div class= "alert">
  <span>This is an important announcement!</span>
  <button class= "close-btn">×</button>
</div>
```

CSS for layout:

```
.alert {
  position: fixed;
  top: 30px;
  left: 50%;
  transform: translateX(-50%);
  background: #f7b731;
  color: #222;
  padding: 14px 36px;
  border-radius: 24px;
  font-size: 1.1em;
  box-shadow: 0 4px 16px rgba(0,0,0,0.09);
  display: flex;
  align-items: center;
  z-index: 200;
```

```
}

.close-btn {
  margin-left: 20px;
  background: none;
  border: none;
  font-size: 1.3em;
  cursor: pointer;
  color: #222;
}
```

This alert uses `position: fixed;` and a transform to perfectly center itself horizontally at the top of the page, while `display: flex;` arranges the text and close button neatly side by side.

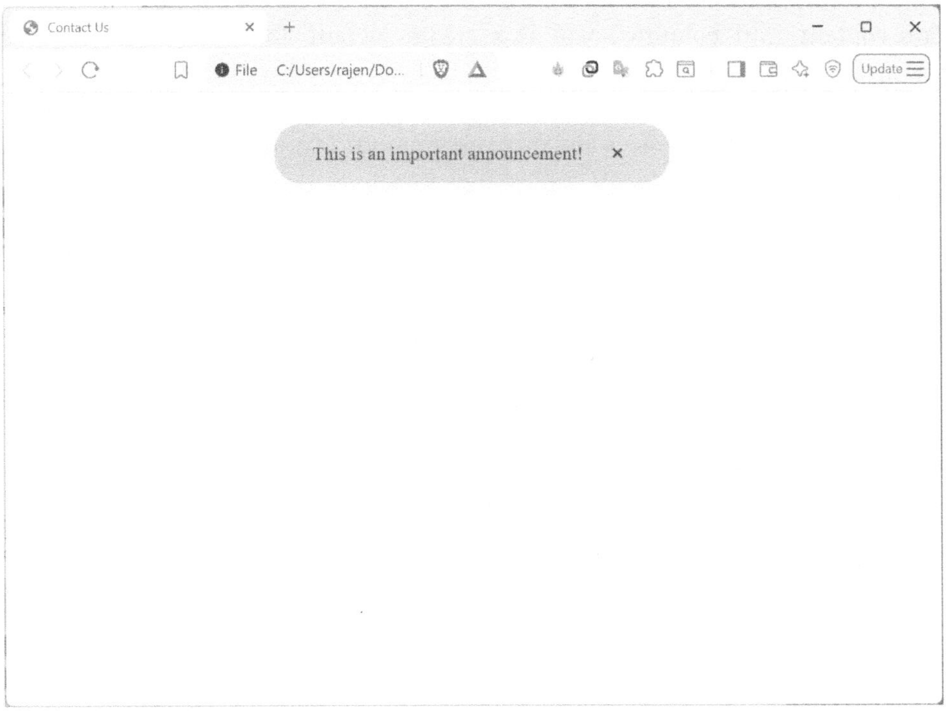

You've now discovered how the **display** and **position** properties form the backbone of modern CSS layout. Whether you're building navigation menus, product cards, or notification banners, these tools let you control how elements appear and interact on the page. Practice these patterns on your own web projects to master their flexibility and power.

Next, you'll learn how to combine these fundamental layout skills to create simple, attractive multi-column designs using only HTML and CSS. Get ready to build dynamic page layouts that look great on any screen size!

11.3. Creating Simple Multi-Column Layouts

Multi-column layouts are a hallmark of modern web design, allowing you to organize content in visually appealing and user-friendly ways. Whether you're building a personal portfolio, a blog, or an online shop, mastering columns unlocks endless possibilities for your sites. In this section, you'll learn the practical steps to create simple, flexible multi-column layouts using only HTML and CSS—no frameworks or complex tools required.

Understanding Columns in Web Design

Columns help break up content, making it easier for users to scan and interact with your webpages. Think of a typical news website, where articles might appear side by side, or an e-commerce homepage with products arranged in neat grids. By structuring your content into columns, you can create layouts that are both organized and responsive to different device sizes.

Let's start with the basics and progress to more advanced techniques.

Building a Two-Column Layout with Flexbox

CSS **Flexbox** is a powerful layout model that makes it straightforward to create columns that adapt to different screen sizes. Unlike older techniques (like using floats), Flexbox keeps your code cleaner and more responsive.

Let's build a simple two-column layout for a portfolio website:

Create the HTML structure:

```
<div class= "container">
  <div class= "sidebar">
    <h2>About Me</h2>
    <p>Short introduction goes here.</p>
  </div>
  <div class= "main-content">
    <h1>My Portfolio</h1>
    <p>Welcome to my work showcase! Check out my latest projects
below.</p>
  </div>
</div>
```

Add the Flexbox CSS:

```
.container {
  display: flex;
  gap: 24px; /* Space between columns */  max-width: 900px;
  margin: 0 auto;
  padding: 32px;
}
```

```
.sidebar {
  flex: 1;
  background-color: #f5f5fa;
  padding: 20px;
  border-radius: 8px;
}

.main-content {
  flex: 2;
  background-color: #fff;
  padding: 20px;
  border-radius: 8px;
  box-shadow: 0 2px 8px rgba(0,0,0,0.05);
}
```

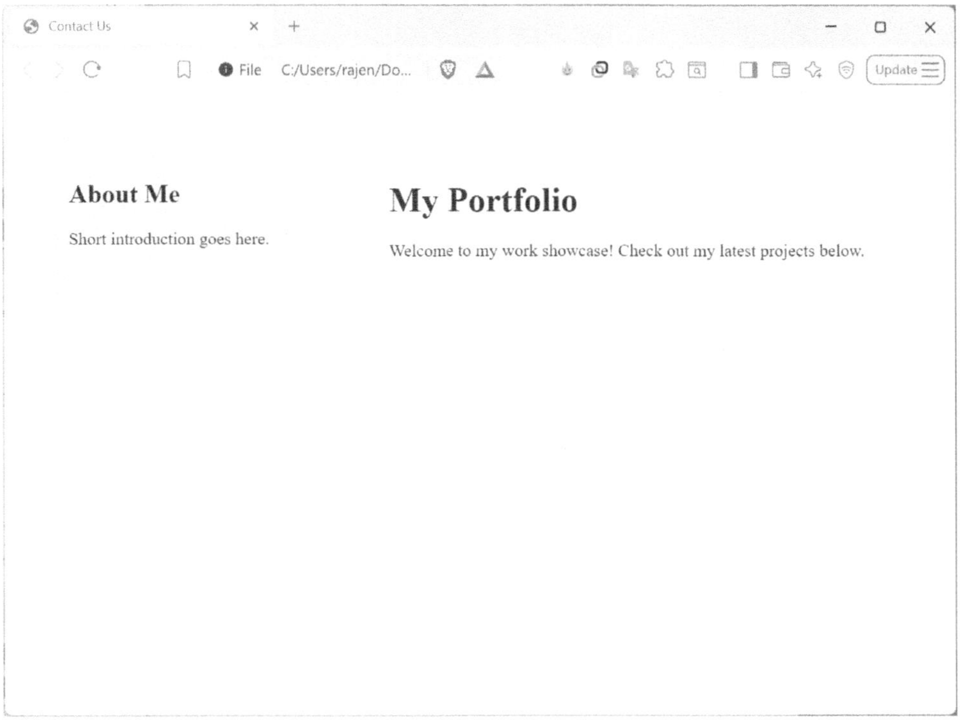

How does this work?

The **container** uses `display: flex` to lay out its children (the sidebar and main content) in a row.

The **sidebar** takes up 1 part of the available space, while **main-content** takes up 2, creating a 1:2 column ratio.

`gap` adds spacing between columns for a clean look.

Responsive Tip: To make your columns stack on smaller screens, add a media query:

```
@media (max-width: 700px) {
  .container {
    flex-direction: column;
  }
}
```

Now, when the browser is narrow, your columns will stack vertically—perfect for mobile devices.

Building a Three-Column Layout for a Blog

Let's apply these skills to a real-world scenario: a simple three-column layout for a blog homepage. This pattern is common in news sites or blogs, featuring a main article area flanked by navigation and extra content.

Write the HTML markup:

```
<div class= "blog-layout">
  <nav class= "left-nav">
    <ul>
      <li><a href= "#">Home</a></li>
      <li><a href= "#">Categories</a></li>
      <li><a href= "#">About</a></li>
    </ul>
  </nav>
  <main class= "blog-main">
    <h1>Latest Articles</h1>
    <article>
      <h2>How to Start Coding</h2>
      <p>Tips and resources for beginners learning web development.</p>
    </article>
    <!-- Add more articles here -->  </main>
  <aside class= "sidebar-right">
    <h3>Trending Topics</h3>
    <ul>
      <li>HTML & CSS</li>
      <li>Responsive Design</li>
      <li>JavaScript Basics</li>
    </ul>
  </aside>
</div>
```

Style with CSS Flexbox:

```
.blog-layout {
  display: flex;
  gap: 16px;
  max-width: 1200px;
  margin: 40px auto;
}

.left-nav,
.sidebar-right {
```

```
  flex: 1;
  background: #eef2f6;
  padding: 20px;
  border-radius: 6px;
}

.blog-main {
  flex: 2.5;
  background: #fff;
  padding: 24px;
  border-radius: 6px;
  box-shadow: 0 1px 6px rgba(0,0,0,0.05);
}
```

The **left-nav** and **sidebar-right** take up equal, smaller widths.
blog-main is larger, giving the articles more space and focus.

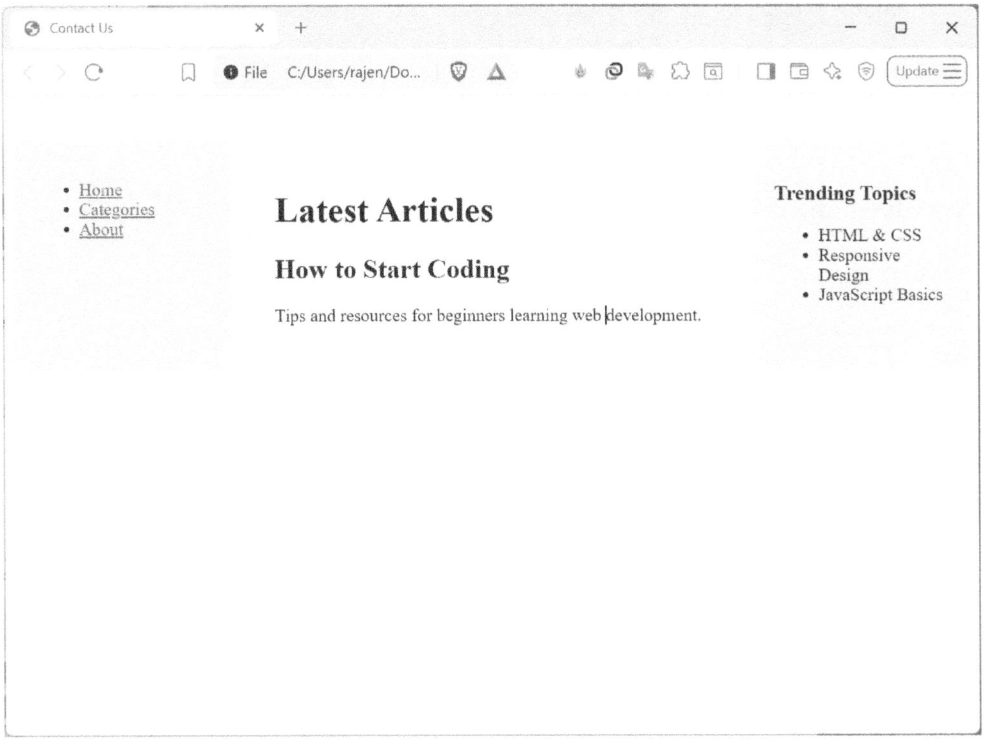

Making It Responsive:

To ensure your three-column layout adapts to phones and tablets, add:

```
@media (max-width: 900px) {
  .blog-layout {
    flex-direction: column;
  }
  .left-nav,
  .sidebar-right,
  .blog-main {
```

```
    margin-bottom: 16px;
  }
}
```

This stacks all three columns vertically, making content easy to read on any device.

CSS Grid: An Alternative for Multi-Column Layouts

While Flexbox is excellent for single-dimensional layouts, **CSS Grid** offers even more control for complex, two-dimensional layouts. Let's build a simple e-commerce product grid as an example.

HTML for a product grid:

```
<div class= "product-grid">
  <div class= "product-card">Product 1</div>
  <div class= "product-card">Product 2</div>
  <div class= "product-card">Product 3</div>
  <div class= "product-card">Product 4</div>
</div>
```

CSS with Grid:

```
.product-grid {
  display: grid;
  grid-template-columns: repeat(2, 1fr);
  gap: 20px;
  max-width: 600px;
  margin: 32px auto;
}

.product-card {
  background: #f8fafb;
  padding: 24px;
  border-radius: 8px;
  text-align: center;
  font-weight: bold;
}
```

`grid-template-columns: repeat(2, 1fr);` creates two equal-width columns.
`gap` provides spacing between products.

Responsive Grid:

Add this for a single-column layout on small screens:

```
@media (max-width: 500px) {
  .product-grid {
    grid-template-columns: 1fr;
  }
```

```
}
```

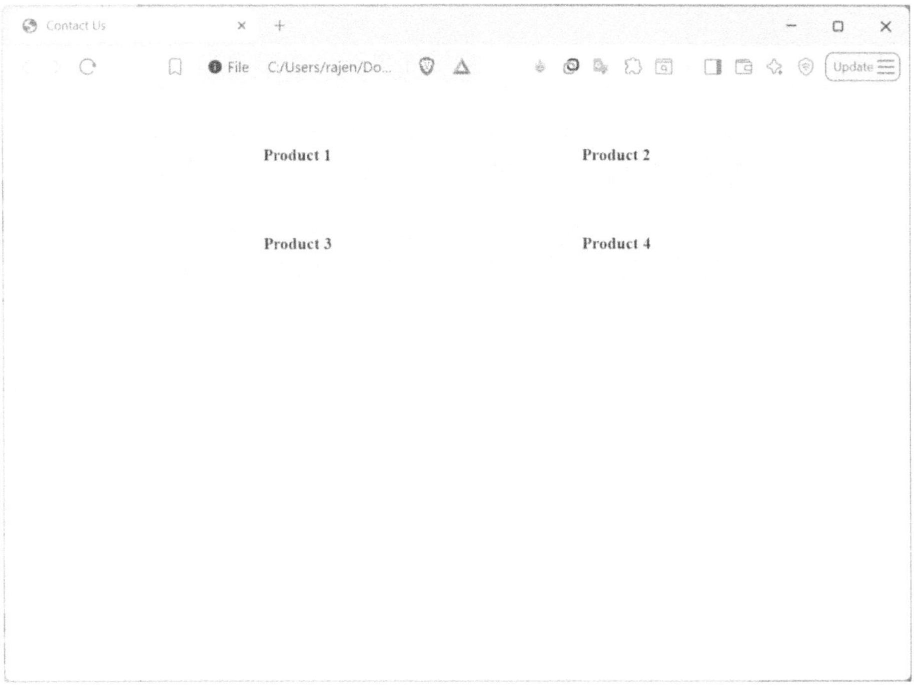

You've now practiced creating multi-column layouts using both Flexbox and Grid—two essential tools for responsive, modern web design. Try customizing the layouts above: experiment with column ratios, colors, and content. The best way to master layouts is to build different types of pages, such as a portfolio, blog, or product showcase, using these techniques.

Next, you'll unlock the power of **z-index** for layering content, giving you precise control over which elements sit on top in your layouts. Get ready to add depth and creativity to your web pages with this essential CSS tool!

11.4. Using z-index for Layering

Layering is a key concept in modern web design, allowing you to control which elements appear on top of others within your layout. The **z-index** property in CSS is the tool that makes this possible. By assigning different **z-index** values to elements with a positioned context, you can create interactive menus, stylish banners, pop-up notifications, and visually rich effects that bring your websites to life.

Before diving into practical code, let's clarify what **z-index** does. The **z-index** property controls the stacking order of elements along the z-axis (think of it as an invisible vertical axis that rises above the web page). Elements with a higher **z-index** value appear in front of those with a lower value, but only if their position is set to **relative**, **absolute**, **fixed**, or **sticky**.

Let's see how this works in real-world scenarios.

Understanding Positioning and z-index

To use **z-index**, you must first apply a positioning property to your CSS rule. If you try adding **z-index** to a static element, it won't have any effect. Here are the positioning properties that work with **z-index**:

- **relative**: Positions the element relative to its normal place.
- **absolute**: Positions the element relative to the nearest positioned ancestor.
- **fixed**: Positions the element relative to the browser window.
- **sticky**: Switches between relative and fixed, depending on scroll.

Let's set up a simple example to illustrate how layering works.

Basic Layering Example: Overlapping Cards

Suppose you want to create an overlapping card effect for a portfolio homepage, where one card slightly sits on top of another to highlight a featured project.

Start with your HTML structure:

```
<div class= "card-container">
  <div class= "card card-one">Project Alpha</div>
  <div class= "card card-two">Project Beta</div>
  <div class= "card card-three">Project Gamma</div>
</div>
```

Now, add CSS to create the card layout and set up overlapping:

```
.card-container {
  position: relative;
  width: 360px;
  height: 220px;
  margin: 40px auto;
}

.card {
  position: absolute;
  width: 200px;
  height: 120px;
  border-radius: 8px;
  box-shadow: 0 6px 20px rgba(0,0,0,0.15);
  background: #f3f3fb;
  display: flex;
  align-items: center;
  justify-content: center;
```

```
  font-size: 1.25rem;
  transition: box-shadow 0.2s;
}

.card-one {
  left: 0;
  top: 40px;
  background: #cce3f8;
  z-index: 1;
}

.card-two {
  left: 50px;
  top: 0;
  background: #a6dcef;
  z-index: 3; /* This card will appear on top */}

.card-three {
  left: 100px;
  top: 60px;
  background: #ffd6e0;
  z-index: 2;
}
```

Here's what's happening:

Each card uses **position: absolute** to enable stacking and precise placement.

z-index is set differently for each card. **card-two** has the highest **z-index** (3), so it will visually appear above the other two cards.

Changing the **z-index** values immediately alters which card is on top.

Try swapping the **z-index** values to see how the stacking changes.

Real-World Scenario: Pop-up Modal on an E-Commerce Page

A practical example where **z-index** is essential is when displaying a pop-up modal over a dimmed background, such as a product quick view on an e-commerce site. You want the modal to always appear above the overlay, and both need to be above the main content.

Write the HTML structure for a modal and overlay:

```
<div class= "page-content">
  <h2>Welcome to Our Store!</h2>
  <button id= "show-modal">Show Product Details</button>
</div>
<div class= "modal-overlay"></div>
<div class= "modal-window">
  <h3>Product Details</h3>
  <p>High-quality headphones with rich sound and deep bass.</p>
  <button id= "close-modal">Close</button>
```

```
</div>
```

Style the overlay and modal with CSS using **z-index**:

```
.page-content {
  padding: 40px;
}

.modal-overlay {
  position: fixed;
  top: 0;
  left: 0;
  width: 100vw;
  height: 100vh;
  background: rgba(0,0,0,0.4);
  display: none; /* Hidden by default */  z-index: 10;
}

.modal-window {
  position: fixed;
  top: 50%;
  left: 50%;
  width: 300px;
  padding: 30px;
  background: #fff;
  border-radius: 10px;
  transform: translate(-50%, -50%);
  box-shadow: 0 8px 32px rgba(0,0,0,0.25);
  display: none; /* Hidden by default */  z-index: 20; /* Higher than
overlay */}
```

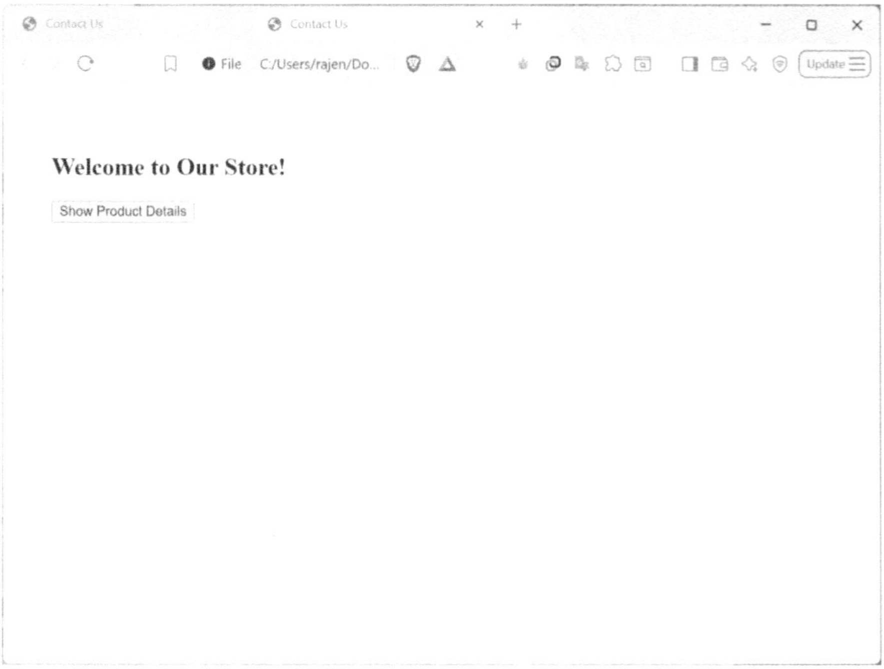

Now add a bit of JavaScript (optional for this exercise) to show and hide the modal, or simply toggle the **display** property in your browser's developer tools to see the effect.

Notice how the **modal-window** uses a higher **z-index** (20) than the **modal-overlay** (10), ensuring the modal appears above the dimmed background. Without the correct **z-index** setup, the modal could be hidden behind the overlay or other page elements.

Common Pitfalls and Best Practices

Positioning Context: Remember, **z-index** only works when the element's position is not static. Always check that your target element uses **relative**, **absolute**, **fixed**, or **sticky**.

Layering Hierarchy: Higher **z-index** values bring elements closer to the viewer. Elements with the same **z-index** (and sharing the same stacking context) will stack in the order they appear in the HTML.

Stacking Context: If you set **position** and **z-index** on a parent, children with **z-index** are layered within that parent, not globally across the page. This is important when working on complex layouts.

Practical Use Cases: Use **z-index** for navigation menus, tooltips, modals, dropdowns, banners, and any component that needs to float above the main content.

Quick Practice: Highlighting a Featured Blog Post

Let's reinforce your understanding with a hands-on mini-exercise. Imagine a blog homepage where you want to highlight a featured post by having its card float above the others.

Create a container with three blog post cards.
Use **position: relative** on the container, **position: absolute** on each card.
Assign a higher **z-index** to the featured card.
Experiment by switching the **z-index** values to see the change.

You should now have a solid grasp of how to use **z-index** to layer elements and control the visual depth of your web layouts. This skill is crucial for designing compelling, interactive interfaces.

Next, you'll put your new layering expertise to the test as we build a practical two-column blog layout. Get ready to combine positioning, **z-index**, and responsive techniques to create a polished and modern blog homepage!

11.5. Hands-On Exercise: Building a Two-Column Blog Layout

Creating a clean, modern two-column blog layout is a cornerstone skill for any aspiring web developer. It's a practical challenge found in countless real-world websites, from personal blogs to news portals and e-commerce homepages. In this hands-on exercise, you'll learn how to structure your HTML, apply advanced CSS layout techniques, and build a fully responsive design that adapts beautifully across screens. By the end, you'll have a stylish blog homepage template and a deeper understanding of CSS layouts, ready for use in your own web projects.

Understanding the Two-Column Layout

Before we dive into code, let's break down what a two-column layout looks like in a typical blog:

A **main content area** (where blog posts or articles appear)
A **sidebar** (for recent posts, author info, or ads)

On wider screens, these areas sit side by side. On smaller screens, like smartphones, the sidebar usually drops below the main content for easy reading. This approach is called **responsive design**, which ensures your website looks great everywhere.

Step 1: Structuring the HTML

Let's begin by creating the basic HTML structure for our blog layout. You'll use familiar elements, but organized for layout flexibility.

Create an HTML file named `blog.html`.
Insert the following HTML markup:

```
<!DOCTYPE html>
<html lang="en">
<head>
  <meta charset="UTF-8">
  <title>My Two-Column Blog</title>
  <link rel="stylesheet" href="styles.css">
</head>
<body>
  <header>
    <h1>My Web Dev Journey</h1>
  </header>
  <div class="container">
    <main class="main-content">
      <article>
        <h2>Launching My First Website</h2>
        <p>Today I launched my first website using HTML and CSS! It was
a fantastic learning experience...</p>
      </article>
```

```
      <article>
        <h2>Tips for New Developers</h2>
        <p>Here are some practical tips for anyone starting their
coding journey...</p>
      </article>
    </main>
    <aside class="sidebar">
      <section>
        <h3>About Me</h3>
        <p>Budding developer, passionate about clean code and
design.</p>
      </section>
      <section>
        <h3>Recent Posts</h3>
        <ul>
          <li>How to Style Your First Webpage</li>
          <li>Understanding CSS Flexbox</li>
          <li>Setting Up Your Coding Environment</li>
        </ul>
      </section>
    </aside>
  </div>
  <footer>
    <p>&copy; 2024 My Web Dev Journey</p>
  </footer>
</body>
</html>
```

Notice the use of semantic elements like **main**, **aside**, **header**, and **footer**. The posts appear in the **main-content** area. The sidebar holds profile and recent posts info. The **container** class wraps both columns for styling.

Step 2: Styling the Layout with CSS Flexbox

Now, let's use **CSS Flexbox** to arrange the two columns side by side, and add some basic styling for a clean, appealing look.

Create a CSS file named `styles.css` in the same folder.
Add the following CSS code:

```
body {
  font-family: Arial, sans-serif;
  margin: 0;
  background: #f9f9f9;
}

header, footer {
  background: #222;
  color: #fff;
  text-align: center;
  padding: 1.5em 0;
}

.container {
  display: flex;
```

```
  max-width: 1040px;
  margin: 2em auto;
  gap: 2em;
  padding: 0 1em;
}

.main-content {
  flex: 2;
  background: #fff;
  padding: 2em;
  border-radius: 8px;
  box-shadow: 0 2px 8px rgba(0,0,0,0.05);
}

.sidebar {
  flex: 1;
  background: #f1f3f6;
  padding: 2em 1.5em;
  border-radius: 8px;
  box-shadow: 0 2px 8px rgba(0,0,0,0.03);
}

h2, h3 {
  color: #222;
}

ul {
  padding-left: 1.2em;
}

@media (max-width: 800px) {
  .container {
    flex-direction: column;
    gap: 1em;
  }
  .sidebar {
    order: 2;
  }
  .main-content {
    order: 1;
  }
}
```

Here's what's happening:

display: flex on `.container` creates a flexible row for the two columns.
flex: 2 and **flex: 1** assign more space to the main content than the sidebar.
gap adds space between columns.
Responsive design is achieved with a **media query**: when the screen is 800px wide or less, columns stack vertically, perfect for mobile devices.

Real-World Example: News Portal Homepages

Many major news and magazine websites, such as [The Guardian](https://www.theguardian.com) or [Medium](https://medium.com), use a similar two-column layout. They feature a main content area for latest stories and a sidebar for trending articles or author profiles. This layout is flexible, widely recognized, and user-friendly, making it a go-to choice for blogs and content-rich sites globally.

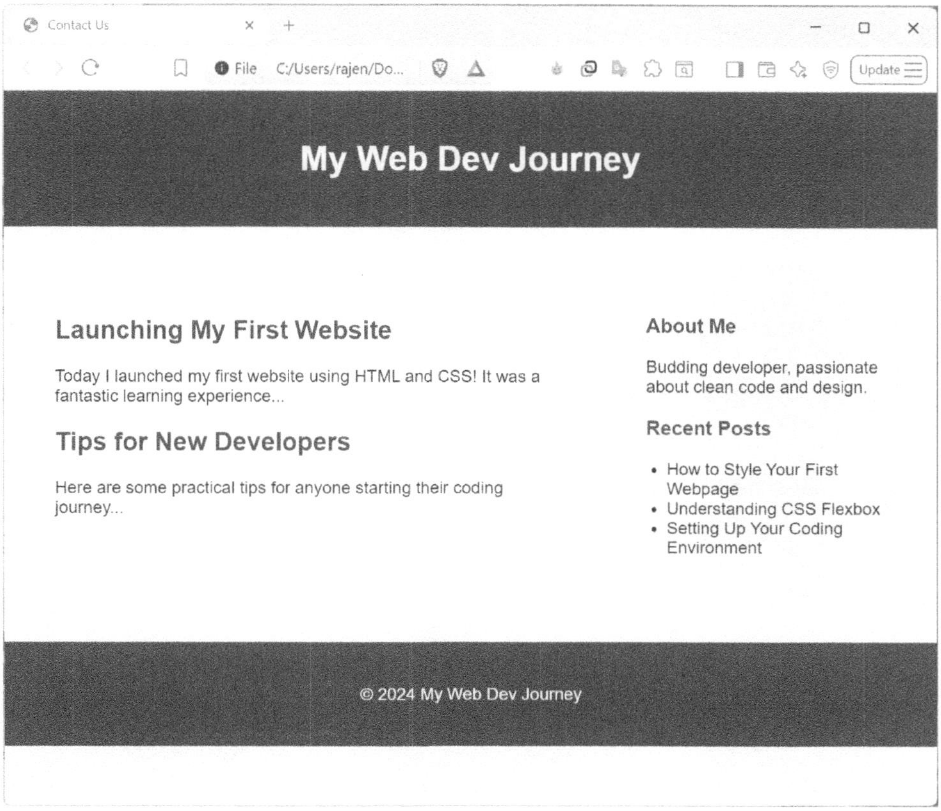

Step 3: Enhancing with Visual Details

To make the layout even more engaging:

Add small accent colors and visual hierarchy:

```
.main-content article {
  margin-bottom: 2em;
  border-bottom: 1px solid #eee;
  padding-bottom: 1em;
}

.sidebar h3 {
  border-bottom: 2px solid #3498db;
  padding-bottom: 0.5em;
  margin-bottom: 1em;
  color: #3498db;
}

.sidebar ul li {
```

```
  margin-bottom: 0.5em;
}
```

This helps separate articles, highlight sidebar sections, and improve readability.

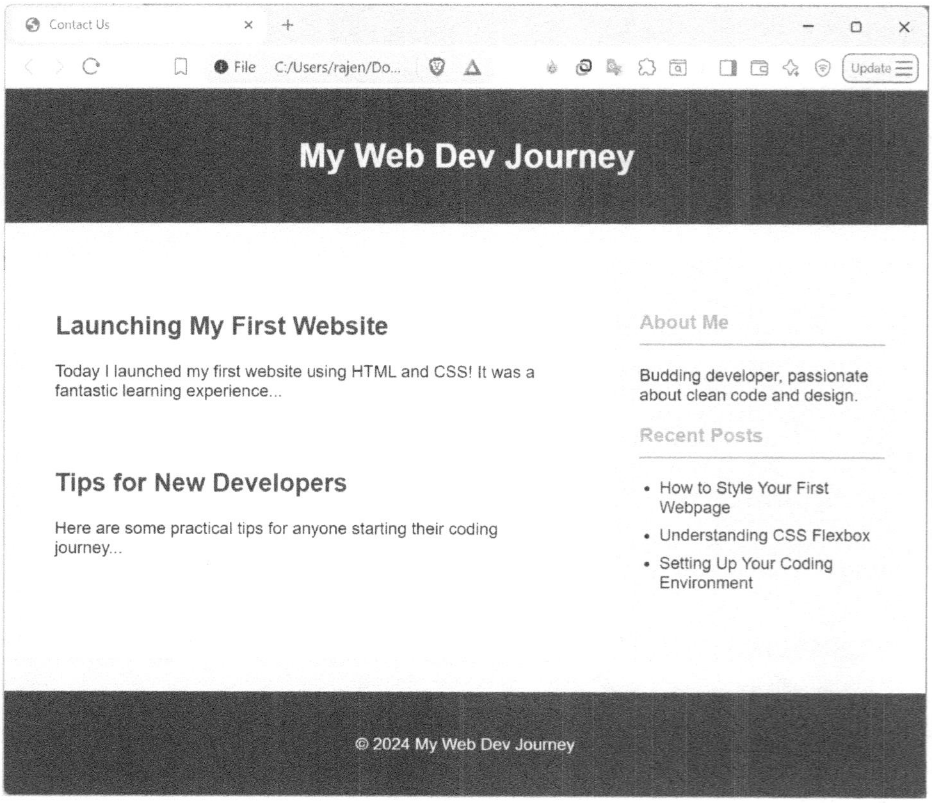

Step 4: Make It Your Own

Experiment by:

Changing **background colors** to match your personal style.
Adding images to posts or the sidebar.
Replacing sidebar content with categories or social media links.

Try using a different color scheme or typography to give your blog a unique feel.

In this hands-on exercise, you built a two-column blog layout by combining semantic HTML structure and modern CSS Flexbox techniques. You practiced responsive design with media queries, making your layout adapt to all screen sizes. Mastering this pattern gives you a powerful template for blogs, portfolios, and countless real-world projects.

12. MODERN LAYOUTS WITH FLEXBOX

If you've ever wondered how professional websites create those beautifully organized layouts, perfectly spaced navigation bars, and flexible rows of content that look great on any device, you're about to uncover the secret: Flexbox. In this chapter, you'll unlock the power of the Flexible Box Layout—a modern CSS tool designed to make page structure not only easier to create but also more dynamic and adaptable than ever before.

Mastering Flexbox will transform how you approach web design. Instead of struggling with floats or unpredictable spacing, you'll learn to align, distribute, and size content with precision and ease. These skills are essential for building layouts that adjust smoothly across phones, tablets, and desktops—a must-have for any website in today's mobile-first world.

Throughout this chapter, you'll explore practical ways to use Flexbox to arrange navigation menus, content blocks, and interactive galleries. By working through hands-on exercises, you'll create a responsive portfolio grid, giving you real experience with the same techniques used in modern portfolio sites, blogs, and e-commerce pages around the globe. Each section will guide you step by step, building your confidence and coding ability as you see your layouts come to life.

Whether you dream of launching your own online presence or aim to stand out as a future web developer, Flexbox is a vital tool in your coding toolkit. Let's dive in and discover how flexible, responsive layouts can elevate your web projects to professional quality—starting right now.

12.1. Introducing Flexbox: Flexible Box Layout

Modern websites need to look great and work smoothly on any device, whether it's a wide desktop monitor or a small phone screen. This challenge makes layout skills essential for every aspiring web developer. One of the most powerful tools you'll use is **Flexbox** (short for Flexible Box Layout), a CSS layout model designed to easily arrange, align, and distribute space among items in a container—even when their size is unknown or changes dynamically.

Flexbox makes creating modern, responsive page structures much simpler than older layout techniques like floats or inline-block. By learning Flexbox, you'll unlock the ability to build dynamic navigation bars, responsive galleries, product listings, and more—all with minimal code.

What is Flexbox and Why Should You Use It?

Flexbox is a one-dimensional layout system that helps you control the direction, alignment, and spacing of items within a container. Unlike traditional layouts, Flexbox adapts content automatically, making your designs flexible and responsive with less effort.

Key benefits of Flexbox:

Effortlessly create horizontal or vertical layouts
Align items vertically and horizontally without hacks
Distribute space between items evenly
Build layouts that adapt to different screen sizes

Let's see how Flexbox works through hands-on practice.

Setting Up a Flexbox Container

To use Flexbox, you first designate a parent element as a **flex container**. All direct child elements become **flex items**. This gives you powerful layout controls using just a few CSS properties.

Let's start by building a simple navigation bar using Flexbox. This is a common feature on portfolio sites and blogs.

Step-by-Step: Creating a Flexbox Navigation Bar

Write the HTML Structure

Create a new file named `flexbox-navbar.html`. Add the following markup:

```
<!DOCTYPE html>
<html lang= "en">
<head>
  <meta charset= "UTF-8">
  <title>Flexbox Navigation Example</title>
  <link rel= "stylesheet"href= "styles.css">
</head>
<body>
  <nav class= "navbar">
    <a href= "#">Home</a>
    <a href= "#">Portfolio</a>
    <a href= "#">Blog</a>
    <a href= "#">Contact</a>
  </nav>
</body>
</html>
```

Add Flexbox to the CSS

Open `styles.css` and add the following:

```
.navbar {
  display: flex; /* Activates Flexbox */  background: #2d3e50;
  padding: 10px 20px;
}

.navbar a {
  color: #fff;
  text-decoration: none;
  padding: 10px 15px;
  font-size: 1.1rem;
  transition: background 0.2s;
}

.navbar a:hover {
  background: #1abc9c;
  border-radius: 4px;
}
```

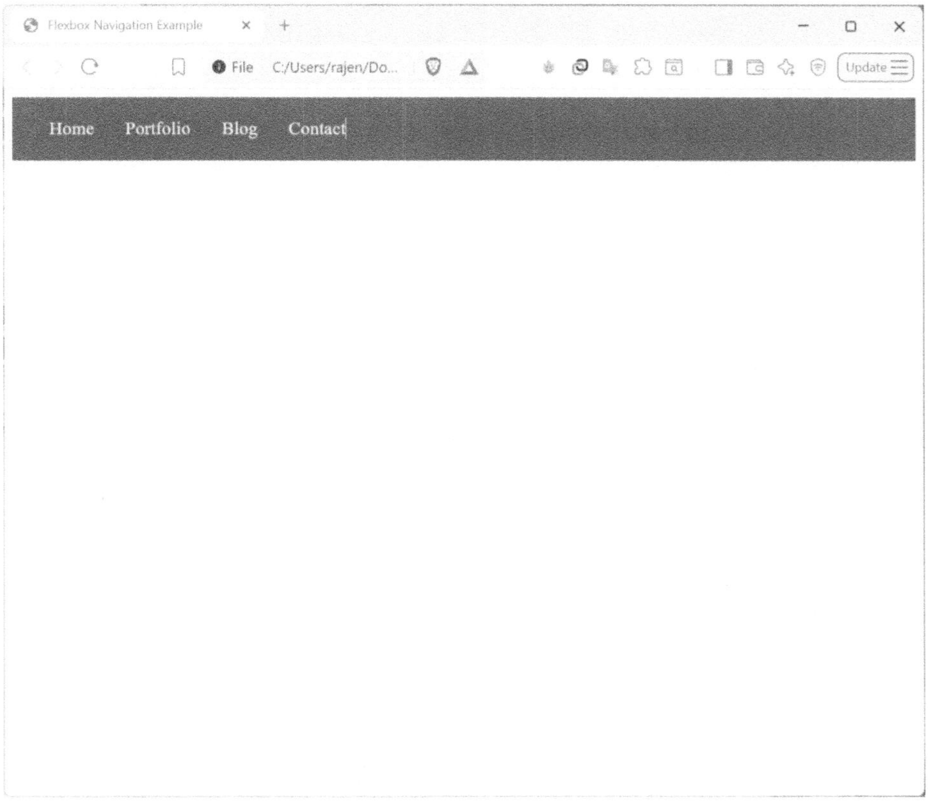

Let's break down what's happening:

display: flex; turns `.navbar` into a flex container, making its child `<a>` elements flex items.

The navigation links now line up horizontally, automatically adjusting to the available space.

Open your HTML file in a browser. Notice how easily the links align side by side, and how the layout adjusts if you add, remove, or rename links.

Real-World Example: Responsive Blog Post Previews

Imagine you're building a blog homepage that displays article previews in a row. On smaller screens, you want the posts to wrap to a new line. Flexbox makes this easy.

Add the HTML for Blog Previews

Replace your `<nav>` section with a blog layout:

```
<section class= "blog-list">
  <article class= "blog-post">
    <h2>How I Built My First Website</h2>
    <p>Tips and lessons from my journey as a beginner web
developer.</p>
  </article>
  <article class= "blog-post">
    <h2>Understanding CSS Selectors</h2>
    <p>A beginner-friendly guide to mastering CSS selectors for
styling.</p>
  </article>
  <article class= "blog-post">
    <h2>Responsive Design Basics</h2>
    <p>Learn how to make your site look great on any device.</p>
  </article>
</section>
```

Apply Flexbox Styling for a Responsive Layout

Add the following to your CSS:

```
.blog-list {
  display: flex; /* Make the section a flex container */  gap: 20px; /*
Space between blog posts */  flex-wrap: wrap; /* Allow posts to wrap on
small screens */  margin: 30px 0;
}

.blog-post {
  background: #f9f9f9;
  border-radius: 8px;
  padding: 20px;
  box-shadow: 0 2px 5px rgba(0,0,0,0.07);
  flex: 1 1 250px; /* Grow, shrink, and set a base width */  min-width:
200px;
  max-width: 400px;
}
```

gap: sets space between flex items (more modern and easier than using margins).

flex-wrap: wrap; allows the flex items to move to the next line if there isn't enough space, making your blog layout responsive.

flex: 1 1 250px; is a shorthand for `flex-grow`, `flex-shrink`, and `flex-basis`. It means each blog post can grow and shrink, but starts at 250px wide.

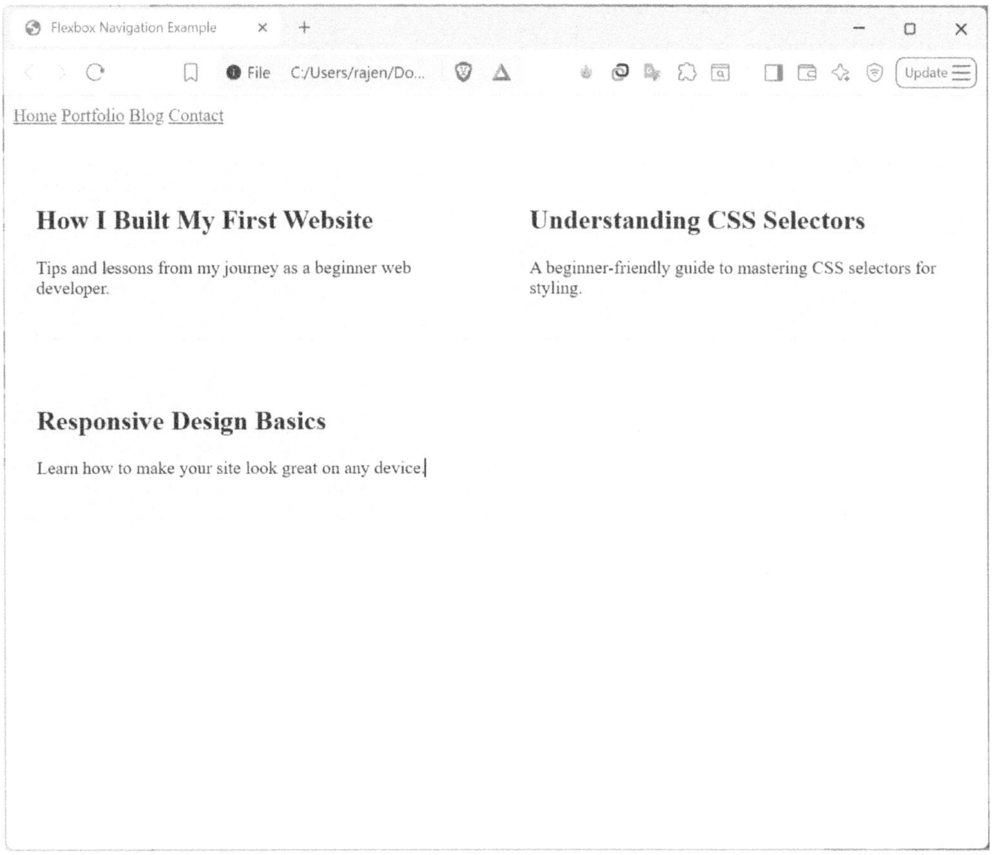

Resize your browser window. Notice how the blog posts automatically stack into new rows as the screen gets smaller. You've just created a responsive blog preview section using only a few lines of Flexbox CSS!

Experimenting with Flex Direction

Flexbox lets you control the direction in which items are laid out. By default, items are placed in a row (side by side), but you can easily switch to a column layout.

Try modifying the `.blog-list` style:

```
.blog-list {
  display: flex;
  flex-direction: column; /* Stack blog posts vertically */  gap: 20px;
```

```
}
```

Now, the blog posts appear in a single column. Change `flex-direction` back to `row` to restore the horizontal layout.

Building a Product Grid for an E-commerce Site

Let's try another practical scenario: displaying products in a grid, a common feature on e-commerce sites.

HTML Markup for Products

Add this to your HTML:

```
<section class= "product-grid">
  <div class= "product-card">T-shirt</div>
  <div class= "product-card">Sneakers</div>
  <div class= "product-card">Backpack</div>
  <div class= "product-card">Hat</div>
</section>
```

Flexbox Product Grid CSS

Add to your CSS:

```
.product-grid {
  display: flex;
  flex-wrap: wrap;
  gap: 16px;
}

.product-card {
  background: #fff;
  border: 1px solid #ddd;
  border-radius: 6px;
  flex: 1 1 150px;
  min-width: 120px;
  max-width: 220px;
  padding: 15px;
  text-align: center;
  font-weight: bold;
}
```

Just like the blog layout, the product cards will automatically fit as many items per row as possible, wrapping to the next line as needed. This makes your e-commerce product grid responsive out of the box.

In this section, you learned how to set up a Flexbox container, arrange navigation links, create responsive blog previews, and build a flexible product grid. Flexbox streamlines layout creation, helping you build modern, adaptable web pages with less code. Practice by experimenting with different numbers of items and adjusting properties like `flex-direction`, `gap`, and `flex-wrap` to see how your layouts respond.

Up next, you'll discover how to align and distribute space between Flexbox items, unlocking even more control over your layouts. Get ready to level up your designs with precise alignment techniques!

12.2. Aligning and Distributing Space

Flexbox gives you powerful tools to align and distribute space among elements with just a few lines of CSS. Whether you're working on a personal portfolio, a blog, or a modern e-commerce site, mastering alignment and spacing makes your layouts look professional and user-friendly. In this section, you'll learn how to control the position and spacing of Flexbox items using properties like **justify-content**, **align-items**, and **gap**. Let's dive in and see how these concepts work through real-world, hands-on examples.

Understanding Main Axis and Cross Axis

Before we start coding, it's important to understand two key concepts in Flexbox: the **main axis** and the **cross axis**.

The **main axis** is defined by the **flex-direction** property. By default, it runs horizontally (left to right) when **flex-direction: row** is set.

The **cross axis** is perpendicular to the main axis. If your main axis is horizontal, the cross axis is vertical.

Every alignment and distribution property in Flexbox works with one of these two axes. Let's see this in action.

Distributing Space with `justify-content`

The **justify-content** property aligns flex items along the main axis. This is especially useful for distributing space between items, centering them, or pushing them to the ends of your flex container.

Suppose you're building a simple navigation bar for a portfolio website. You want each navigation link to be spaced evenly across the bar.

HTML:

```
<nav class= "navbar">
  <a href= "#">Home</a>
```

```
  <a href= "#">Portfolio</a>
  <a href= "#">About</a>
  <a href= "#">Contact</a>
</nav>
```

CSS:

```
.navbar {
  display: flex;
  justify-content: space-between; /* Distributes space evenly */
background-color: #262a34;
  padding: 1rem;
}

.navbar a {
  color: #fff;
  text-decoration: none;
  font-weight: bold;
}
```

With **justify-content: space-between**, the links are pushed to the edges of the navbar, with equal space between them. Try changing the value to experiment:

- **flex-start**: All items are packed toward the start (left) of the main axis.
- **flex-end**: All items are packed toward the end (right).
- **center**: Items are centered along the main axis.
- **space-around**: Equal space is placed around every item.
- **space-evenly**: Equal space is distributed between and around items.

Change the **justify-content** value and refresh your browser to see the differences.

Aligning Items Vertically with `align-items`

While **justify-content** controls horizontal alignment (main axis), **align-items** manages vertical alignment (cross axis) of flex items within the container.

Imagine you have a pricing section on a blog or shop site where different pricing cards have varying heights. You want the cards to line up nicely along their tops or centers.

HTML:

```
<section class= "pricing">
  <div class= "card">Basic<br>$10</div>
  <div class= "card tall">Pro<br>$25<br>Popular</div>
  <div class= "card">Premium<br>$40</div>
</section>
```

CSS:

```
.pricing {
  display: flex;
  align-items: flex-start; /* Aligns items to the top */  gap: 1.5rem;
}

.card {
  background: #f5f5f5;
  padding: 1.2rem;
  border-radius: 8px;
  box-shadow: 0 2px 8px rgba(80, 80, 80, 0.1);
  min-width: 120px;
  text-align: center;
}

.card.tall {
  min-height: 130px;
}
```

Here, **align-items: flex-start** makes sure all cards align at the top of the section, regardless of content height.

Try these other values for **align-items**:

- **center**: Vertically centers all cards.
- **flex-end**: Aligns cards to the bottom.
- **stretch**: Stretches cards to fill the container's height (default if no height is set).

Experiment by switching the **align-items** value and observe the changes in your browser.

Combining Alignment Properties for Professional Layouts

Often, you'll use both **justify-content** and **align-items** together for precise control. Let's create a feature section for a modern business homepage, where you want features centered both horizontally and vertically.

HTML:

```
<section class= "features">
  <div class= "feature">Fast Delivery</div>
  <div class= "feature">Secure Payment</div>
  <div class= "feature">24/7 Support</div>
</section>
```

CSS:

```
.features {
  display: flex;
  justify-content: center; /* Center features horizontally */  align-
items: center; /* Center features vertically */  min-height: 200px;
  background: #edf0f3;
  gap: 2rem;
}

.feature {
  background: #fff;
  padding: 1.5rem 2rem;
  border-radius: 6px;
  box-shadow: 0 1px 4px rgba(0,0,0,0.05);
  font-size: 1.1rem;
}
```

This approach creates a visually balanced section, even as screen sizes change. The **gap** property also provides consistent spacing between items, making your layout clean and modern.

Using the `gap` Property for Spacing

The **gap** property (formerly **grid-gap**) lets you easily add space between flex items, without extra margins or complex CSS.

Let's revisit a simplified e-commerce product row where each product card should have equal spacing.

HTML:

```
<div class= "product-row">
  <div class= "product">Socks</div>
  <div class= "product">T-shirt</div>
  <div class= "product">Sneakers</div>
</div>
```

CSS:

```
.product-row {
  display: flex;
  gap: 2rem; /* Equal space between product items */  background:
#f7fafb;
  padding: 1rem;
}

.product {
  background: #d3eaf2;
  padding: 1rem 1.5rem;
  border-radius: 5px;
  font-weight: 500;
}
```

Unlike using margins, **gap** automatically spaces items evenly, making your code simpler and easier to maintain.

Practical Exercise: Create a Balanced Header

Let's put your new skills into action by building a responsive header where the logo is on the left, navigation links are centered, and a login button sits on the right.

Step-by-step instructions:

HTML Structure

```
<header class= "site-header">
  <div class= "logo">WebDev</div>
  <nav class= "main-nav">
    <a href= "#">Home</a>
    <a href= "#">Features</a>
    <a href= "#">Pricing</a>
  </nav>
  <button class= "login-btn">Login</button>
</header>
```

CSS Flexbox Layout

```
.site-header {
  display: flex;
  justify-content: space-between; /* Logo left, nav center, button
right */     align-items: center; /* Vertically aligned */
background: #1e2235;
  color: #fff;
  padding: 1rem 2rem;
}

.logo {
  font-size: 1.5rem;
  font-weight: bold;
  letter-spacing: 1px;
}

.main-nav {
  display: flex;
  gap: 2rem;
}

.main-nav a {
  color: #fff;
  text-decoration: none;
  font-size: 1.1rem;
}

.login-btn {
  background: #fff;
  color: #1e2235;
  border: none;
```

```
    padding: 0.5rem 1.2rem;
    border-radius: 20px;
    font-size: 1rem;
    cursor: pointer;
  }
```

This is a common layout you'll find on sites from tech portfolios to online stores. Notice how **justify-content: space-between** pushes the logo and login button to opposite ends, while the navigation stays centered. The **gap** property keeps navigation links evenly spaced.

You've just unlocked the essential skills of aligning and distributing space with Flexbox using **justify-content**, **align-items**, and **gap**. Experiment with these properties in your own projects to create balanced, visually appealing layouts. Try building a header, a feature section, or a product row using different alignment and gap values for practice.

Next, you'll take your Flexbox skills even further by building responsive navigation bars that adapt seamlessly to any device. Get ready to design menus that not only look great but are also mobile-friendly and interactive!

12.3. Responsive Navigation Bars with Flexbox

Navigation bars are a defining feature of every modern website, guiding visitors effortlessly through your content. Thanks to Flexbox, you can create navigation menus that are not only visually appealing but also automatically adapt to screens of all sizes. In this section, you'll learn how to build a responsive navigation bar step by step, empowering you to deliver a seamless browsing experience whether your site is viewed on a desktop, tablet, or smartphone.

Responsive navigation is especially important for real-world websites like blogs, portfolios, and e-commerce stores. For instance, consider a personal portfolio where you want visitors to easily access your About, Projects, and Contact pages, or an online shop where shoppers quickly jump between categories. With Flexbox, you can build a navigation menu that expands, shrinks, or rearranges itself based on the device. Let's dive into the hands-on process.

Understanding the Responsive Navigation Structure

Before coding, let's break down the typical structure of a navigation bar:

A **container** to hold the entire navigation bar.
A **logo** or site name for branding.
A **navigation menu** with links to different site sections.

(Optional) A **menu button** for mobile devices, sometimes called a "hamburger" icon.

Flexbox excels at arranging these elements in a row, spacing them evenly, and stacking or hiding them on smaller screens.

Step 1: Building the HTML Markup

Start by creating the basic navigation bar structure. Open your code editor and enter the following HTML:

```
<nav class= "navbar">
  <div class= "navbar-logo">MyPortfolio</div>
  <ul class= "navbar-menu">
    <li><a href= "#about">About</a></li>
    <li><a href= "#projects">Projects</a></li>
    <li><a href= "#blog">Blog</a></li>
    <li><a href= "#contact">Contact</a></li>
  </ul>
  <button class= "navbar-toggle"aria-label= "Toggle Menu">
    &#9776;
  </button>
</nav>
```

Let's quickly review what you've written:

The **nav** element is a semantic container for the navigation bar.
Inside, there's a logo, an unordered list for navigation links, and a button for toggling the menu on mobile.

Step 2: Styling the Navigation Bar with Flexbox

Next, you'll use CSS to transform the static HTML into a dynamic, responsive navigation bar.

First, set up some basic styles and enable Flexbox for the navigation container:

```
.navbar {
  display: flex;
  align-items: center;
  justify-content: space-between;
  background: #263859;
  padding: 1rem 2rem;
  color: #fff;
  position: relative;
}
```

display: flex; turns the navbar into a flex container.
align-items: center; vertically centers the logo and menu.
justify-content: space-between; places the logo on the left and menu on the right.

Now, style the logo and menu links:

```
.navbar-logo {
  font-size: 1.5rem;
  font-weight: bold;
  letter-spacing: 2px;
}

.navbar-menu {
  list-style: none;
  display: flex;
  gap: 2rem;
  margin: 0;
  padding: 0;
}

.navbar-menu li a {
  text-decoration: none;
  color: #fff;
  font-size: 1rem;
  transition: color 0.2s;
}

.navbar-menu li a:hover {
  color: #f7c873;
}
```

display: flex; on the menu ensures links sit in a row.
gap: adds space between each link.
Hover effects make the navigation interactive and professional.

Hide the mobile toggle button by default:

```
.navbar-toggle {
  display: none;
  background: none;
  border: none;
  font-size: 2rem;
  color: #fff;
  cursor: pointer;
}
```

Step 3: Making the Navbar Responsive

To adapt the navigation for smaller screens, use a **media query**. On mobile devices, it's common to hide the menu links and show the menu button instead. When the button is clicked, the menu appears in a column.

Add this CSS to handle responsiveness:

```
@media (max-width: 768px) {
  .navbar-menu {
```

```
    display: none;
    flex-direction: column;
    background: #263859;
    position: absolute;
    top: 100%;
    right: 0;
    width: 100%;
    text-align: center;
    padding: 1rem 0;
    box-shadow: 0 2px 8px rgba(0,0,0,0.1);
  }
  .navbar-menu.active {
    display: flex;
  }
  .navbar-toggle {
    display: block;
  }
}
```

display: none; hides the menu on small screens.
flex-direction: column; stacks the links vertically when shown.
The toggle button appears only on small screens.

Step 4: Adding Simple JavaScript for Interactivity

To show or hide the menu when the button is clicked, you'll need a small amount of JavaScript. Add this script just before the closing `</body>` tag:

```
<script>
  const toggleButton = document.querySelector( '.navbar-toggle');
  const menu = document.querySelector( '.navbar-menu');

  toggleButton.addEventListener( 'click', () => {
    menu.classList.toggle( 'active');
  });
</script>
```

This code listens for clicks on the menu button and toggles the **active** class, which reveals or hides the menu.

Real-World Application: Portfolio Site Navigation

Imagine you're building a personal portfolio. Visitors might arrive from desktops, tablets, or phones, and need to find your Projects or Contact page quickly. With the above setup, your navigation bar will look sharp on large screens and morph into a touch-friendly menu on mobile, just like top portfolio sites and tech blogs worldwide.

Real-World Application: E-Commerce Store Navigation

For an e-commerce shop, you could easily adapt this approach by adding links for Home, Shop, Cart, and Account. Flexbox ensures that even if you add or remove links, the menu remains perfectly spaced and responsive.

Coding Recap: Responsive Navigation with Flexbox

Let's summarize the key steps you just practiced:

HTML Structure: Build a semantic navigation bar with a logo, menu, and toggle button.
Flexbox Layout: Use Flexbox to arrange elements in a horizontal row and space them evenly.
Responsive Design: Apply media queries to hide the menu and show a toggle on small screens.
Interactivity: Add JavaScript to open and close the menu on mobile devices.

By mastering these techniques, you've created a foundation for stylish, responsive navigation bars suitable for any project, big or small.

Practice Makes Perfect

Try customizing your navigation bar for different scenarios. Change the color scheme to match your brand, add icons, or adjust the menu for a blog or online store. Experimenting with Flexbox properties like **justify-content**, **align-items**, and **gap** can help you fine-tune your design and become more comfortable with responsive layouts.

Up next, you'll apply your Flexbox skills to another essential component of modern web design: building a responsive portfolio grid. Get ready to showcase your projects in a flexible, visually engaging layout that adapts beautifully to any screen size.

12.4. Hands-On Exercise: Creating a Responsive Portfolio Grid

Let's dive straight into a hands-on project that will bring your new Flexbox skills to life: building a responsive portfolio grid. This is a practical layout seen on countless real-world websites, from creative portfolios to e-commerce galleries. By the end of this exercise, you'll have a professional-looking showcase that adapts smoothly to any device—exactly the kind of result that modern web designers deliver every day.

Understanding the Portfolio Grid Layout

Before we write any code, let's clarify what we're creating. A **portfolio grid** is a layout that displays multiple items (like project cards, product images, or blog post previews) in a neat, evenly spaced grid. The grid should:

Display several cards per row on large screens.
Automatically adjust to fewer cards per row on tablets and phones.
Ensure that every card is well-aligned, visually balanced, and accessible.

This approach is used on many sites, such as Behance (for creative portfolios) and online shops like Etsy. You'll use Flexbox, a powerful CSS layout model, to achieve this flexibility without complicated floats or manual sizing.

Step 1: Setting Up the HTML Structure

Let's start by creating a simple HTML skeleton for your portfolio. This will include a container that holds several project cards, each with a project image, title, and short description.

Open your code editor and create a new HTML file (for example, `portfolio.html`).
Add the following markup inside your `<body>` tag:

```
<!DOCTYPE html>
<html lang="en">
<head>
  <meta charset="UTF-8">
  <meta name="viewport" content="width=device-width, initial-
scale=1.0">
  <title>Responsive Portfolio Grid</title>
  <link rel="stylesheet" href="styles.css">
</head>
<body>
  <section class="portfolio-section">
    <h2>My Portfolio</h2>
    <div class="portfolio-grid">
      <div class="portfolio-card">
        <img src="images/project1.png" alt="Project 1 Screenshot">
        <h3>Project One</h3>
        <p>A web app for tracking daily habits.</p>
      </div>
      <div class="portfolio-card">
        <img src="images/project2.png" alt="Project 2 Screenshot">
        <h3>Project Two</h3>
        <p>Responsive homepage for a local bakery.</p>
      </div>
      <div class="portfolio-card">
        <img src="images/project3.png" alt="Project 3 Screenshot">
        <h3>Project Three</h3>
        <p>Personal blog with custom CMS.</p>
      </div>
      <!-- Add more portfolio-card elements as needed -->
    </div>
  </section>
</body>
</html>
```

This setup is simple and repeatable. Each `.portfolio-card` represents a project. You can add as many cards as you like.

Step 2: Basic Styling for the Portfolio

Now, let's style the grid using CSS. Open or create a file named `styles.css` in the same directory. We'll start with some basic resets and overall styling to ensure a clean, modern look.

```
body {
  font-family: Arial, sans-serif;
  background: #f5f7fa;
  margin: 0;
  padding: 0;
  color: #222;
}

.portfolio-section {
  max-width: 1100px;
  margin: 40px auto;
  padding: 20px;
}

.portfolio-section h2 {
  text-align: center;
  margin-bottom: 32px;
  font-size: 2.5rem;
  letter-spacing: 1px;
}
```

This provides a light background, centers the portfolio, and adds some breathing room. Now, let's move on to the grid layout.

Step 3: Building the Flexbox Grid

Flexbox makes it easy to create a responsive, evenly spaced grid. Here's how to turn `.portfolio-grid` into a flexible container:

```
.portfolio-grid {
  display: flex;
  flex-wrap: wrap;
  gap: 24px;
  justify-content: center;
}
```

display: flex turns the container into a flexbox, allowing its children to align automatically.

flex-wrap: wrap ensures that cards move to a new line if there isn't enough space.

gap: 24px adds spacing between cards.

justify-content: center centers the grid content.

Now, let's define how each card should look and behave:

```
.portfolio-card {
  background: #fff;
  border-radius: 10px;
  box-shadow: 0 2px 10px rgba(0,0,0,0.07);
  padding: 20px;
  width: 300px;
  box-sizing: border-box;
  transition: transform 0.2s, box-shadow 0.2s;
  display: flex;
  flex-direction: column;
  align-items: center;
}

.portfolio-card img {
  width: 100%;
  height: 170px;
  object-fit: cover;
  border-radius: 8px;
  margin-bottom: 16px;
}

.portfolio-card h3 {
  margin: 0 0 10px 0;
  font-size: 1.25rem;
}

.portfolio-card p {
  color: #666;
  font-size: 1rem;
  text-align: center;
}

.portfolio-card:hover {
  transform: translateY(-6px) scale(1.03);
  box-shadow: 0 8px 24px rgba(0,0,0,0.12);
}
```

Each card is visually separated, with rounded corners and a drop shadow.
The image is responsive and covers the top portion of the card.
On hover, cards lift slightly and cast a deeper shadow for interactivity.

Step 4: Making the Grid Responsive

A crucial feature of modern layouts is responsiveness. With a few lines of CSS, Flexbox lets your grid adapt to any screen size. We'll use media queries to adjust card width at different breakpoints:

```
@media (max-width: 900px) {
  .portfolio-card {
    width: 45%;
  }
}

@media (max-width: 600px) {
  .portfolio-card {
    width: 100%;
```

```
  }
  .portfolio-grid {
    gap: 16px;
  }
}
```

On screens up to 900px wide (like tablets), cards take up about half the width, creating two cards per row.

On small screens (phones), each card takes the full width, stacking vertically for easy scrolling.

The grid gap reduces slightly on the smallest screens for a snug fit.

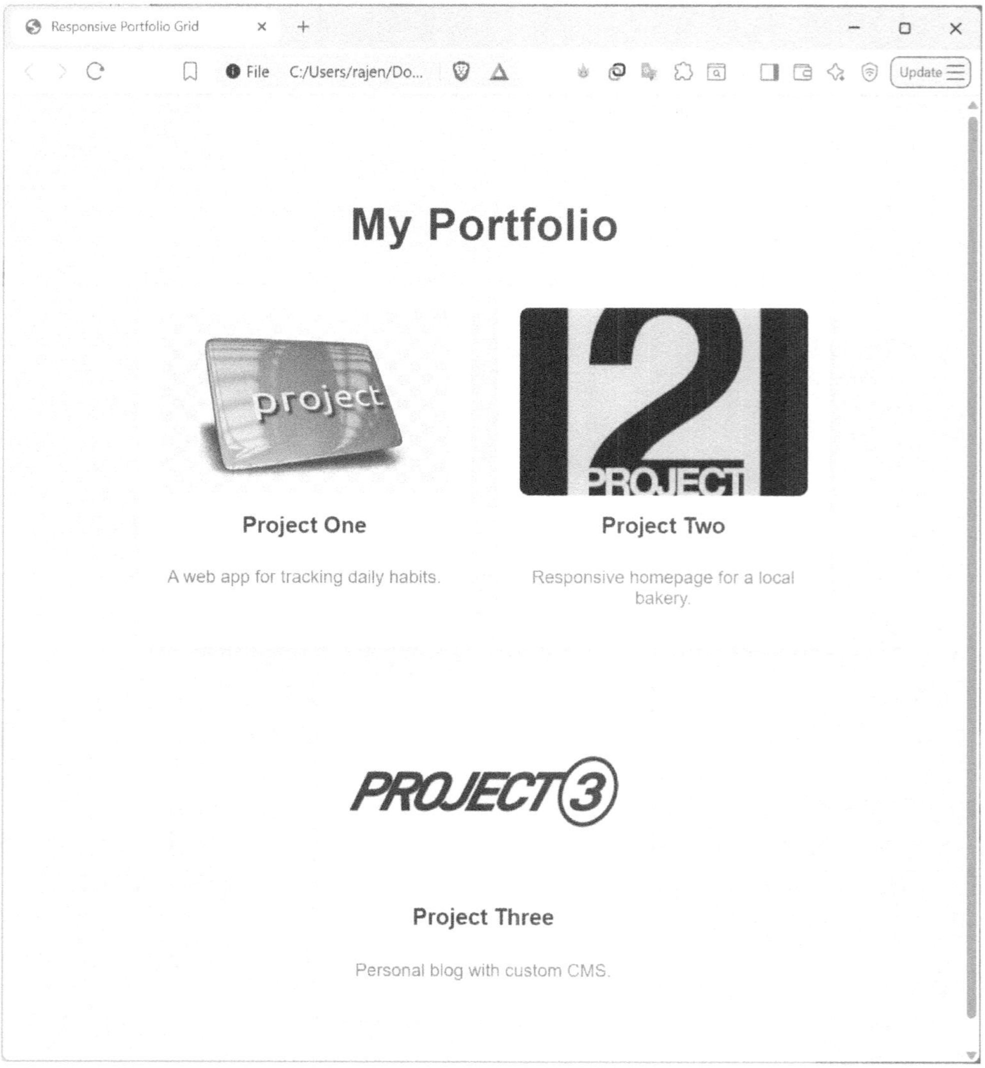

Step 5: Testing and Customizing Your Portfolio Grid

Open `portfolio.html` in your browser and resize the window. You'll see the grid shift from three (or more) cards per row on desktop, to two on tablets, and one per row on mobile devices. This mirrors the adaptability seen on professional portfolio sites like Dribbble or agency landing pages.

Feel free to personalize your project by:

Adding more `.portfolio-card` items to showcase additional work.
Swapping placeholder images and descriptions for your real projects.
Adjusting colors, font sizes, or gaps to match your brand or style inspiration.

Step 6: Real-World Scenario—Freelance Portfolio Example

Imagine you're a freelance web designer building your portfolio. Clients need to see your best work at a glance, whether they're on a laptop or their phone. By using this responsive grid, your site stands out as professional and user-friendly, increasing your chances of impressing potential employers or clients.

In this exercise, you created a fully responsive portfolio grid using Flexbox, modern CSS, and semantic HTML. You learned to structure HTML for repeated content, style cards with Flexbox, and make your layout adapt smoothly to all devices using media queries. Practicing these techniques will solidify your understanding of responsive design—a must-have skill for every aspiring web developer.

13. ADVANCED LAYOUTS WITH CSS GRID

If you've ever wondered how today's most visually striking websites achieve their crisp, organized layouts, the answer often lies in the power of CSS Grid. This chapter unlocks the secrets behind building sophisticated page structures that look professional and adapt beautifully to any device. With CSS Grid, you'll discover how to move beyond basic one-column layouts and unlock true creative control over how your content is arranged.

As you work through this chapter, you'll master skills such as placing and spanning grid items, designing multi-column and multi-row sections, and creating magazine-style homepages that mirror real-world digital publications. You'll see firsthand how CSS Grid can dramatically simplify the process of building complex layouts—no more wrestling with floats or stacking endless containers. We'll also explore how to combine the strengths of CSS Grid and Flexbox, giving you a modern toolkit for any design challenge.

You won't just read about these techniques—you'll put them into practice by building a hands-on project that brings your new skills to life. By designing a magazine-style homepage, you'll gain practical experience that applies to portfolio sites, blogs, and even e-commerce platforms. These are the same layout strategies used by professional web designers around the globe.

Whether your goal is to launch your own website or step confidently into the world of web development, mastering CSS Grid will set you apart. Get ready to level up your layout skills and transform your ideas into polished, responsive designs. In the next sections, you'll see just how intuitive and powerful modern web layouts can be.

13.1. What is CSS Grid?

CSS Grid is a breakthrough in web design, making it easier than ever to create complex, flexible layouts that adapt beautifully to any screen size. Unlike older layout methods, CSS Grid provides a two-dimensional system, allowing you to control rows and columns at the same time. This means you can place content exactly where you want it, without resorting to messy workarounds or extra markup. In this section, you'll discover how to harness CSS Grid for real-world projects, building practical skills step by step.

Why Use CSS Grid?

Before you dive into code, it's important to understand why CSS Grid is such a game changer for modern web development:

Two-dimensional layouts: Arrange elements in rows and columns simultaneously.
Responsive design made simple: Easily adjust layouts for different devices and screen sizes.
Cleaner code: Fewer `div` wrappers and less reliance on floats or positioning hacks.
Intuitive placement: Place items exactly where you want them within your layout.

Let's see how CSS Grid works in practice by creating the foundation for a responsive portfolio gallery—a layout that's common on personal and professional websites around the world.

Setting Up a CSS Grid Container

To start using CSS Grid, you first need a grid container. This is any element to which you apply the `display: grid;` property. The container then controls the layout of its direct child elements (the grid items).

Suppose you want to display a series of portfolio pieces in a neat grid. Here's how you can set up the HTML and CSS:

HTML:

```
<section class= "portfolio-grid">
  <div class= "project">Project 1</div>
  <div class= "project">Project 2</div>
  <div class= "project">Project 3</div>
  <div class= "project">Project 4</div>
  <div class= "project">Project 5</div>
  <div class= "project">Project 6</div>
</section>
```

This section represents six portfolio projects you want to showcase.

CSS:

```
.portfolio-grid {
  display: grid;
  grid-template-columns: 1fr 1fr 1fr;
  grid-gap: 20px;
}

.project {
  background: #f0f0f0;
  padding: 30px;
  text-align: center;
```

```
  font-size: 1.2rem;
  border-radius: 8px;
}
```

The `display: grid;` property turns `.portfolio-grid` into a grid container.
`grid-template-columns: 1fr 1fr 1fr;` creates three columns of equal width. The `fr` unit stands for "fraction," making each column take up an equal share of available space.
`grid-gap: 20px;` adds space between the grid items, keeping your design clean and readable.

Try adding or removing `.project` divs in the HTML to see how the grid automatically adjusts!

Responsive Grids: Adapting to Different Devices

One of the greatest strengths of CSS Grid is its responsiveness. You can use media queries or newer CSS functions to make layouts that work well on any screen size.

Let's upgrade the portfolio grid so it looks great on mobile devices:

CSS:

```
@media (max-width: 800px) {
  .portfolio-grid {
    grid-template-columns: 1fr 1fr;
  }
}

@media (max-width: 500px) {
  .portfolio-grid {
    grid-template-columns: 1fr;
  }
}
```

When the screen width is 800 pixels or less, the grid switches to two columns.
At 500 pixels or less, it shifts to a single column, perfect for mobile phones.

This approach ensures your layout is user-friendly and visually appealing on desktops, tablets, and smartphones alike—a key skill for every web developer.

Real-World Example: E-commerce Product Grid

Let's look at another scenario: an e-commerce homepage featuring a grid of product cards. CSS Grid makes this simple and efficient.

HTML:

```
<section class= "product-grid">
  <div class= "product-card">Product A</div>
  <div class= "product-card">Product B</div>
  <div class= "product-card">Product C</div>
  <div class= "product-card">Product D</div>
</section>
```

CSS:

```
.product-grid {
  display: grid;
  grid-template-columns: repeat(auto-fit, minmax(220px, 1fr));
  gap: 24px;
}

.product-card {
  background: #fff;
  padding: 24px;
  box-shadow: 0 2px 8px rgba(0,0,0,0.07);
  border-radius: 6px;
  text-align: center;
}
```

Here, `grid-template-columns: repeat(auto-fit, minmax(220px, 1fr));` creates as many columns as will fit in the container, with each column at least 220 pixels wide but able to grow to fill available space. This technique is especially handy for e-commerce product grids and is used by modern online stores worldwide.

Hands-On Coding Exercise: Build Your Own Image Gallery Grid

Now it's your turn! Create a simple image gallery using CSS Grid. Follow these steps:

Add a section with the class `gallery` and at least four image elements inside your HTML.
In your CSS, make `.gallery` a grid container with four columns on desktop.
Add a gap between images for better visual separation.
Use media queries so the gallery switches to two columns on tablets and one column on phones.

Sample HTML:

```
<section class= "gallery">
  <img src= "img1.jpg"alt= "Gallery Image 1">
  <img src= "img2.jpg"alt= "Gallery Image 2">
  <img src= "img3.jpg"alt= "Gallery Image 3">
  <img src= "img4.jpg"alt= "Gallery Image 4">
</section>
```

Sample CSS:

```
.gallery {
  display: grid;
  grid-template-columns: repeat(4, 1fr);
  gap: 16px;
}
@media (max-width: 900px) {
  .gallery {
    grid-template-columns: repeat(2, 1fr);
  }
}
@media (max-width: 500px) {
  .gallery {
    grid-template-columns: 1fr;
  }
}
.gallery img {
  width: 100%;
  border-radius: 4px;
}
```

This simple gallery is a perfect way to get comfortable with CSS Grid's core features.

You've learned how to create flexible, responsive layouts using CSS Grid, from simple three-column portfolios to advanced, auto-adjusting product grids and image galleries. CSS Grid's intuitive syntax and powerful features make it an essential tool for every modern web developer. Practice by building grids for different types of websites and experimenting with column numbers, gaps, and responsive behaviors.

As you become familiar with grid containers, you'll want to take even more control over how individual items are placed and how they can span multiple rows or columns. In the next section, "Placing and Spanning Grid Items," you'll discover how to position content exactly where you want it within your grid—unlocking even more creative web design possibilities!

13.2. Placing and Spanning Grid Items

When you start working with CSS Grid, you quickly realize its real superpower: giving you precise control over where each item sits within your layout. This level of control is essential for creating beautiful, responsive web pages that look professional and function perfectly on any device. By learning how to place and span grid items, you'll be able to craft layouts found in modern portfolios, blogs, and even e-commerce sites. Let's dive into hands-on coding to see these skills in action.

Understanding Grid Item Placement

By default, grid items are placed in the grid in the order they appear in your HTML. However, CSS Grid allows you to override this automatic placement. Using the **grid-column** and **grid-row** properties, you can tell any item exactly where it should appear.

Here's the basic idea:

grid-column determines the starting and ending vertical lines for an item (left/right).
grid-row determines the starting and ending horizontal lines for an item (top/bottom).

Let's see how this works with a simple example.

Practice: Building a Simple Portfolio Grid

Imagine you're creating a personal portfolio website. You want a section where your featured project takes up more space, while other projects are smaller. We'll use CSS Grid to achieve this.

Set Up the HTML Structure

```
<section class= "portfolio-grid">
  <div class= "featured">Featured Project</div>
  <div class= "project">Project 1</div>
  <div class= "project">Project 2</div>
  <div class= "project">Project 3</div>
  <div class= "project">Project 4</div>
</section>
```

Create the Grid and Place Items

Let's style the grid container and place the featured project so it spans multiple columns and rows.

```
.portfolio-grid {
  display: grid;
  grid-template-columns: repeat(3, 1fr);
  grid-template-rows: repeat(2, 200px);
  gap: 16px;
  max-width: 900px;
  margin: 0 auto;
}

.featured {
  background: #4fc3f7;
  color: #fff;
  display: flex;
  align-items: center;
  justify-content: center;
```

```
  font-size: 1.5em;
  /* Place and span the featured item */  grid-column: 1 / span 2; /* Start at column 1, span 2 columns */  grid-row: 1 / span 2; /* Start at row 1, span 2 rows */  border-radius: 8px;
}

.project {
  background: #ececec;
  display: flex;
  align-items: center;
  justify-content: center;
  font-size: 1em;
  border-radius: 8px;
}
```

What's happening here?

The `.portfolio-grid` uses three columns and two rows.
The `.featured` item is told to start on the first column line and span two columns, and to start on the first row line and span two rows.
The other `.project` items fill in the remaining grid spaces.

This technique is at the heart of many modern portfolio and gallery layouts.

Spanning Multiple Columns and Rows

What if you want an item to cover just a single row but multiple columns, or vice versa? CSS Grid makes this easy with the **span** keyword.

Suppose you're designing a blog layout where you want a "Latest Post" to stretch across the grid's full width at the top, with smaller posts below.

HTML for a Blog Section

```
<section class= "blog-grid">
  <div class= "latest-post">Latest Post</div>
  <div class= "post">Post 1</div>
  <div class= "post">Post 2</div>
  <div class= "post">Post 3</div>
</section>
```

CSS for Spanning Columns

```
.blog-grid {
  display: grid;
  grid-template-columns: repeat(3, 1fr);
  grid-auto-rows: 180px;
  gap: 20px;
  max-width: 900px;
  margin: 32px auto;
}
```

```
.latest-post {
  background: #ffa726;
  color: #fff;
  font-weight: bold;
  display: flex;
  align-items: center;
  justify-content: center;
  font-size: 1.3em;
  grid-column: 1 / span 3; /* Span across all 3 columns */  grid-row:
1; /* Stay on the first row */  border-radius: 8px;
}

.post {
  background: #fffde7;
  display: flex;
  align-items: center;
  justify-content: center;
  border-radius: 8px;
  box-shadow: 0 0 6px #fbc02d22;
}
```

Here, the `.latest-post` stretches across all three columns at the top of the grid, while the other posts automatically fill in the next available spaces.

Explicit Grid Line Placement

Sometimes you need even more control. You can use specific grid lines by referencing their numbers, giving you pixel-perfect layouts. For example, in an e-commerce product grid, you might want to highlight a featured product in a specific grid spot.

```
.product-highlight {
  grid-column: 2 / 4; /* Start at line 2, end at line 4 */  grid-row: 1
/ 2; /* Start at line 1, end at line 2 */  background: #81c784;
  color: #fff;
  font-size: 1.2em;
  display: flex;
  align-items: center;
  justify-content: center;
  border-radius: 8px;
}
```

This rule places `.product-highlight` precisely in the second and third columns of the first row, perfect for drawing attention to special products.

Hands-On Exercise: Build Your Own Custom Grid

Let's put these skills into practice. Try creating a grid with the following layout:

The first item spans two columns.
The second item spans two rows.

The remaining items fill the grid normally.

HTML Example

```
<section class= "custom-grid">
  <div class= "item item1">Item 1</div>
  <div class= "item item2">Item 2</div>
  <div class= "item item3">Item 3</div>
  <div class= "item item4">Item 4</div>
</section>
```

CSS Example

```
.custom-grid {
  display: grid;
  grid-template-columns: repeat(3, 1fr);
  grid-template-rows: repeat(2, 120px);
  gap: 12px;
  max-width: 600px;
  margin: 40px auto;
}

.item1 {
  background: #ba68c8;
  color: #fff;
  grid-column: 1 / span 2; /* Spans first two columns */}

.item2 {
  background: #81d4fa;
  color: #333;
  grid-row: 1 / span 2; /* Spans first two rows */  grid-column: 3; /*
Sits in the third column */}

.item3, .item4 {
  background: #fff;
  color: #333;
}
```

Experiment by changing the span values and observe how the layout shifts in your browser. This is the best way to learn how grid placement affects your design.

You've learned how to place and span grid items using **grid-column**, **grid-row**, and the **span** keyword. These techniques give you the flexibility to create dynamic layouts for portfolios, blogs, product grids, and more. Practice by adjusting spans and placements to get a feel for different layouts. The more you experiment, the more intuitive CSS Grid will become.

Next, you'll discover how to combine these skills to create complex, magazine-style layouts effortlessly with CSS Grid. Get ready to take your web design to the next level in "Creating Complex Layouts Simply"!

13.3. Creating Complex Layouts Simply

When you start exploring advanced layouts, CSS Grid becomes your most powerful tool. It allows you to organize content visually in a way that's both flexible and clean, making seemingly complex website structures surprisingly simple. With CSS Grid, you can create magazine-style columns, overlapping elements, and intricate gallery layouts—all with just a few lines of code. Let's dive in and see how you can build professional-quality, responsive layouts step by step.

Understanding the Power of CSS Grid

CSS Grid is a layout system designed to handle both rows and columns at the same time. Unlike older layout methods (like floats or even Flexbox, which is mainly one-dimensional), CSS Grid enables you to design two-dimensional layouts with ease. Imagine building a photo gallery, a blog homepage, or a portfolio showcase—Grid can handle it all.

Step-by-Step: Building a Responsive Blog Layout

Suppose you want to create a modern blog homepage with a featured article and a dynamic grid of smaller articles below. This structure is common on news sites and personal blogs.

Start with the HTML Structure

Let's define the basic layout elements in HTML. We'll use semantic tags for clarity, such as `` `<section>` ``, `` `<article>` ``, and `` `<aside>` ``.

```
<section class= "blog-layout">
  <article class= "featured">
    <h2>Featured Article</h2>
    <p>This is the main highlight of the blog.</p>
  </article>
  <article class= "post">Blog Post 1</article>
  <article class= "post">Blog Post 2</article>
  <article class= "post">Blog Post 3</article>
  <article class= "post">Blog Post 4</article>
  <aside class= "sidebar">
    <h3>About Me</h3>
    <p>Short bio or links.</p>
  </aside>
</section>
```

Define the CSS Grid Container

Now, let's turn the `` `.blog-layout` `` section into a grid container. We'll create a layout with two columns: the main content and a sidebar. The featured article will span both columns at the top.

```
.blog-layout {
  display: grid;
  grid-template-columns: 2fr 1fr; /* Main area and sidebar */  grid-
template-rows: auto 1fr;
  grid-gap: 20px;
  grid-template-areas:
    "featured featured"
    "posts sidebar";
  max-width: 1200px;
  margin: 40px auto;
  padding: 20px;
}
```

Position Grid Items with Grid Areas

Assign each section to its place using the **grid-area** property.

```
.featured {
  grid-area: featured;
  background: #f9fafc;
  padding: 24px;
  border-radius: 8px;
}

.sidebar {
  grid-area: sidebar;
  background: #eaf4fa;
  padding: 16px;
  border-radius: 8px;
}

.post {
  background: #fff;
  padding: 16px;
  border: 1px solid #e2e2e2;
  border-radius: 6px;
}
```

To have the blog posts automatically flow into the posts area, let's wrap them in a container:

```
<div class= "posts">
  <article class= "post">Blog Post 1</article>
  <article class= "post">Blog Post 2</article>
  <article class= "post">Blog Post 3</article>
  <article class= "post">Blog Post 4</article>
</div>
```

And update the grid areas:

```
.blog-layout {
  display: grid;
  grid-template-columns: 2fr 1fr;
```

```
  grid-template-rows: auto 1fr;
  grid-gap: 20px;
  grid-template-areas:
    "featured featured"
    "posts sidebar";
}

.featured { grid-area: featured; }
.posts { grid-area: posts; display: grid; grid-template-columns: 1fr
1fr; grid-gap: 16px; }
.sidebar { grid-area: sidebar; }
```

Now, the `.posts` container itself becomes a grid for the individual posts, easily creating a two-column mini-grid within the main grid.

Responsive Design: Making the Layout Adapt

A key advantage of CSS Grid is how easily you can make layouts responsive. Let's add a media query so the layout stacks vertically on smaller screens.

```
@media (max-width: 800px) {
  .blog-layout {
    grid-template-columns: 1fr;
    grid-template-areas:
      "featured"
      "posts"
      "sidebar";
  }
  .posts {
    grid-template-columns: 1fr;
  }
}
```

With just a few lines, your grid transforms from a side-by-side desktop layout to a stacked mobile-friendly view.

Real-World Example: E-Commerce Product Grid

Another common scenario is an e-commerce product listing page, where products are displayed in a grid that automatically adjusts to any screen width.

HTML Markup for the Product Grid

```
<section class= "product-grid">
  <div class= "product">Product 1</div>
  <div class= "product">Product 2</div>
  <div class= "product">Product 3</div>
  <div class= "product">Product 4</div>
  <div class= "product">Product 5</div>
  <div class= "product">Product 6</div>
</section>
```

CSS for a Flexible Product Grid

Let's make the grid automatically fit as many columns as possible, shrinking to one column on small devices.

```
.product-grid {
  display: grid;
  grid-template-columns: repeat(auto-fit, minmax(220px, 1fr));
  gap: 24px;
  max-width: 1000px;
  margin: 32px auto;
  padding: 16px;
}

.product {
  background: #fff;
  border: 1px solid #e0e0e0;
  border-radius: 8px;
  padding: 24px;
  text-align: center;
  box-shadow: 0 2px 8px rgba(0,0,0,0.04);
}
```

This technique lets your product grid expand or contract seamlessly, making it ideal for a modern, responsive online store.

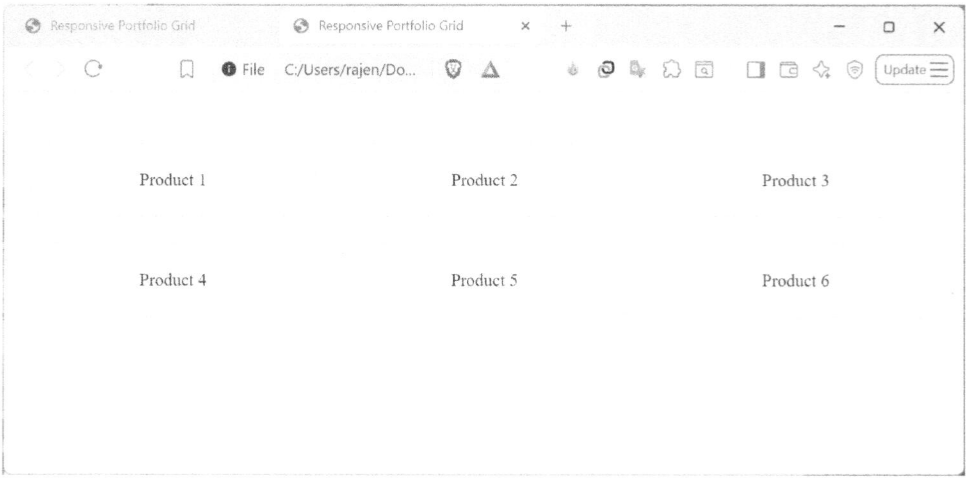

In this section, you've learned how to create advanced, responsive layouts using CSS Grid, from blog homepages to e-commerce product grids. CSS Grid allows you to design complex structures quickly, keeping your code clean and maintainable. Try customizing the layouts above with your own content or colors to reinforce your skills.

Get ready for the next step in modern web design! In the following section, you'll discover how to combine the strengths of CSS Grid and Flexbox, unlocking even more creative possibilities for layouts that adapt beautifully to any device or screen size.

13.4. Combining Grid and Flexbox for Modern Designs

Modern web design often calls for layouts that are both powerful and flexible. CSS Grid and Flexbox are two layout systems that, when combined, give you the ultimate toolkit for crafting responsive, dynamic designs. Each has its strengths: **Grid** excels at creating two-dimensional layouts (rows and columns), while **Flexbox** is perfect for arranging content in a single direction and distributing space within smaller components. By mixing these techniques, you can solve complex layout challenges and create modern, user-friendly websites.

Let's explore how you can combine Grid and Flexbox to build a real-world responsive layout, step by step.

Why Combine Grid and Flexbox?

Before jumping into code, it's important to understand why you'd mix these systems:

Grid is great for overall page or section structure, defining large-scale areas like headers, footers, and main content zones.

Flexbox shines for arranging the elements inside those areas, such as navigation links, cards, or buttons.

By combining them, you get precise control over both macro and micro layouts, making your designs more adaptable and visually appealing.

Real-World Scenario: Portfolio Website Layout

Imagine you're creating a portfolio homepage. You want a header with navigation, a main content area with a grid of featured projects, and a footer. Inside the header and footer, the content should be spaced out horizontally. The main area should showcase projects in a responsive grid, and each project card will use Flexbox to align its contents neatly.

Let's build this step by step.

Step 1: Setting Up the HTML Structure

Start by creating the basic HTML structure for your layout.

```
<body>
  <div class= "container">
    <header class= "header">
```

```
      <nav class= "nav">
        <a href= "#">Home</a>
        <a href= "#">Portfolio</a>
        <a href= "#">About</a>
        <a href= "#">Contact</a>
      </nav>
    </header>
    <main class= "main">
      <section class= "projects">
        <div class= "project-card">
          <img src= "project1.jpg"alt= "Project 1">
          <div class= "card-content">
            <h3>Project One</h3>
            <p>Creative branding for global clients.</p>
            <button>View Details</button>
          </div>
        </div>
        <div class= "project-card">
          <img src= "project2.jpg"alt= "Project 2">
          <div class= "card-content">
            <h3>Project Two</h3>
            <p>Responsive web development project.</p>
            <button>View Details</button>
          </div>
        </div>
        <!-- Add more project cards as needed -->        </section>
    </main>
    <footer class= "footer">
      <div class= "footer-content">
        <span>© 2024 Your Name</span>
        <span>Follow me: <a href= "#">Twitter</a> | <a href=
"#">GitHub</a></span>
      </div>
    </footer>
  </div>
</body>
```

This structure uses semantic tags and clearly separates the header, main content, and footer.

Step 2: Using CSS Grid for Page Structure

Let's use **CSS Grid** to define the overall layout for the `.container`, placing the header, main, and footer in specific rows.

```
.container {
  display: grid;
  grid-template-rows: auto 1fr auto;
  min-height: 100vh;
  grid-template-areas:
    "header"
    "main"
    "footer";
}

.header {
```

```
  grid-area: header;
  background: #222;
  color: #fff;
  padding: 1rem;
}

.main {
  grid-area: main;
  padding: 2rem;
  background: #f8f8f8;
}

.footer {
  grid-area: footer;
  background: #222;
  color: #fff;
  padding: 1rem 0;
}
```

Here, the **container** uses Grid to create three rows. The **header**, **main**, and **footer** are assigned to their respective grid areas, keeping the layout organized and responsive.

Step 3: Flexbox for Navigation and Footer Alignment

Now, use **Flexbox** inside the header's navigation and the footer's content to space items evenly.

```
.nav {
  display: flex;
  justify-content: space-between;
  align-items: center;
  gap: 2rem;
}

.nav a {
  color: #fff;
  text-decoration: none;
  font-weight: bold;
}

.footer-content {
  display: flex;
  justify-content: space-between;
  align-items: center;
  max-width: 900px;
  margin: 0 auto;
}
```

display: flex makes the `.nav` and `.footer-content` flex containers.

justify-content: space-between pushes the links or text to the opposite sides, creating a clean, balanced look.

gap adds spacing between navigation links.

Step 4: Grid for Project Cards Layout

Next, use **Grid** again to lay out the projects in the main area, adapting the number of columns as the screen size changes.

```
.projects {
  display: grid;
  grid-template-columns: repeat(auto-fit, minmax(260px, 1fr));
  gap: 2rem;
}
```

repeat(auto-fit, minmax(260px, 1fr)) automatically adjusts the number of columns to fit the screen, making the grid responsive.
gap creates spacing between the cards.

Step 5: Flexbox Inside Each Project Card

For each project card, use Flexbox to align the image and content vertically, and to center the button inside the card content.

```
.project-card {
  background: #fff;
  border-radius: 8px;
  overflow: hidden;
  box-shadow: 0 2px 8px rgba(0,0,0,0.05);
  display: flex;
  flex-direction: column;
  height: 100%;
  transition: box-shadow 0.2s;
}

.project-card img {
  width: 100%;
  height: 180px;
  object-fit: cover;
}

.card-content {
  display: flex;
  flex-direction: column;
  justify-content: space-between;
  padding: 1rem;
  flex: 1;
}

.card-content h3 {
  margin: 0 0 0.5rem 0;
}

.card-content button {
  align-self: flex-start;
  margin-top: 1rem;
```

```
  background: #0077ff;
  color: #fff;
  border: none;
  border-radius: 4px;
  padding: 0.5rem 1rem;
  cursor: pointer;
}
```

display: flex and **flex-direction: column** stack elements vertically.
justify-content: space-between in `.card-content` ensures the heading, description, and button have even spacing.
align-self: flex-start aligns the button to the left.

Responsive Design in Action

The combination of Grid and Flexbox adapts beautifully to any screen size. On wider screens, you'll see multiple project cards per row. On smaller screens, the grid stacks cards vertically, while Flexbox keeps navigation and content spaced out and easy to read.

Real-World Example: E-Commerce Product Gallery

This approach is just as powerful for an e-commerce homepage, where you might use Grid to display products and Flexbox to arrange price, title, and "Add to Cart" buttons within each product card. The principles are the same: Grid for the big picture, Flexbox for fine-tuned alignment.

Summary: Key Takeaways and Practice

You've just combined the strengths of **CSS Grid** and **Flexbox** to build a modern, adaptable portfolio layout. Grid handled the overall page and project gallery structure, while Flexbox made navigation, footers, and card content visually balanced and responsive. Practice combining these techniques in your own projects, experimenting with different layouts and content types to build confidence.

Next, you'll put these skills to the test by designing a magazine-style homepage. Get ready to apply everything you've learned about advanced layouts and responsive design in a hands-on coding challenge that brings your web development skills to life!

13.5. Hands-On Exercise: Designing a Magazine-Style Homepage

Designing a magazine-style homepage is a fantastic way to practice your new CSS Grid skills and experience the power of advanced layout techniques in a real-world context. Many popular news sites, digital magazines, and lifestyle blogs use magazine-style layouts to organize their content visually, making articles, images, and featured stories stand out. In this hands-on exercise, you'll build a stylish, responsive homepage that could serve as the foundation for a blog, news portal, or creative portfolio.

Let's break down the process step by step, focusing on structuring your HTML and applying CSS Grid to achieve an elegant, modern layout. By the end, you'll have a multi-section homepage that adapts beautifully to different screen sizes—just like the websites you admire.

Planning the Magazine-Style Layout

Before diving into code, it's helpful to visualize the layout. Imagine the homepage of a design magazine:

A bold, eye-catching **header** with the site title and navigation links.
A prominent **featured article** section.
A **grid of secondary stories** (usually two or three columns).
A **sidebar** for announcements or popular posts.
A **footer** with copyright or contact links.

This structure is popular on sites like Vogue, Wired, or Medium. It balances visual interest and readability, guiding visitors to key content.

Step 1: Building the HTML Structure

Start by creating a simple HTML file. Use meaningful tags and class names to make your layout easy to style and understand.

```
<!DOCTYPE html>
<html lang= "en">
<head>
  <meta charset= "UTF-8">
  <title>Magazine Homepage Example</title>
  <link rel= "stylesheet"href= "styles.css">
</head>
<body>
  <header class= "site-header">
    <h1>Modern Magazine</h1>
    <nav>
      <ul>
        <li><a href= "#">Home</a></li>
        <li><a href= "#">Features</a></li>
        <li><a href= "#">Culture</a></li>
        <li><a href= "#">Contact</a></li>
      </ul>
    </nav>
  </header>

  <main class= "grid-container">
    <section class= "featured-article">
      <img src= "featured.jpg"alt= "Featured Story">
```

```
      <h2>Reimagining Creativity in the Digital Age</h2>
      <p>Explore how emerging artists are transforming the web’s visual
landscape.</p>
    </section>

    <section class= "secondary-stories">
      <article>
        <img src= "images/story1.jpg"alt= "Story 1">
        <h3>Design Trends 2024</h3>
        <p>This year’s biggest web design inspirations from around the
globe.</p>
      </article>
      <article>
        <img src= "images/story2.jpg"alt= "Story 2">
        <h3>Interview: Rising Star</h3>
        <p>Meet the young coder making waves in the creative tech
scene.</p>
      </article>
      <article>
        <img src= "images/story3.jpg"alt= "Story 3">
        <h3>How to Build Your Portfolio</h3>
        <p>Tips and tricks for showcasing your projects online.</p>
      </article>
    </section>

    <aside class= "sidebar">
      <h4>Editor’s Picks</h4>
      <ul>
        <li><a href= "#">Best of 2024</a></li>
        <li><a href= "#">Web Typography Essentials</a></li>
        <li><a href= "#">Responsive Design Tricks</a></li>
      </ul>
    </aside>
  </main>

  <footer class= "site-footer">
    <p>&copy; 2024 Modern Magazine | Designed by You</p>
  </footer>
</body>
</html>
```

Notice how each section is wrapped in semantic HTML tags, making your code more readable and accessible.

Step 2: Creating the CSS Grid Layout

Now, let’s bring your homepage to life with CSS Grid. Open `styles.css` and start by setting up the overall grid for the main content area.

```
body {
  margin: 0;
  font-family: 'Segoe UI', Arial, sans-serif;
  background: #f6f6f6;
  color: #222;
}
```

```
.site-header, .site-footer {
  background: #212121;
  color: #fff;
  padding: 1em 2em;
}

.site-header h1 {
  margin: 0;
  font-size: 2em;
  letter-spacing: 2px;
}

.site-header nav ul {
  list-style: none;
  padding: 0;
  display: flex;
  gap: 1.5em;
}

.site-header nav a {
  color: #fff;
  text-decoration: none;
  font-weight: bold;
}

/* The main grid container */.grid-container {
  display: grid;
  grid-template-columns: 2fr 1fr;
  grid-template-rows: auto 1fr;
  gap: 2em;
  padding: 2em;
  max-width: 1200px;
  margin: auto;
}

/* Featured article spans full width */.featured-article {
  grid-column: 1 / 3;
  background: #fff;
  padding: 1.5em;
  border-radius: 8px;
  box-shadow: 0 2px 8px rgba(0,0,0,0.08);
  display: flex;
  flex-direction: column;
  align-items: flex-start;
}

.featured-article img {
  width: 100%;
  border-radius: 8px;
  margin-bottom: 1em;
}

/* Secondary stories in a grid */.secondary-stories {
  display: grid;
  grid-template-columns: repeat(2, 1fr);
  gap: 1.5em;
}
```

```
.secondary-stories article {
  background: #fff;
  padding: 1em;
  border-radius: 6px;
  box-shadow: 0 1px 4px rgba(0,0,0,0.05);
}

.secondary-stories img {
  width: 100%;
  border-radius: 6px;
  margin-bottom: 0.5em;
}

/* Sidebar styles */.sidebar {
  background: #fafafa;
  padding: 1em;
  border-radius: 6px;
  box-shadow: 0 1px 4px rgba(0,0,0,0.04);
}

.sidebar h4 {
  margin-top: 0;
}

/* Footer styles */.site-footer {
  text-align: center;
  font-size: 0.9em;
  margin-top: 3em;
}
```

This CSS creates a two-column layout, with the **featured article** stretching across the top, **secondary stories** on the left, and a **sidebar** on the right. The use of `grid-column` and `grid-template-columns` enables flexible arrangements without complicated floats or positioning.

Step 3: Making It Responsive

Modern magazine sites must look great on every device. Let's add a media query to make your layout adapt to smaller screens, such as tablets and smartphones.

```
@media (max-width: 900px) {
  .grid-container {
    grid-template-columns: 1fr;
  }
  .featured-article {
    grid-column: 1;
  }
  .sidebar {
    order: 2;
    margin-top: 2em;
  }
  .secondary-stories {
    grid-template-columns: 1fr;
  }
```

}

With this code, the layout gracefully stacks on small screens: the featured article appears first, followed by secondary stories and the sidebar. This responsive design approach is used by major digital publications and ensures accessibility for all visitors.

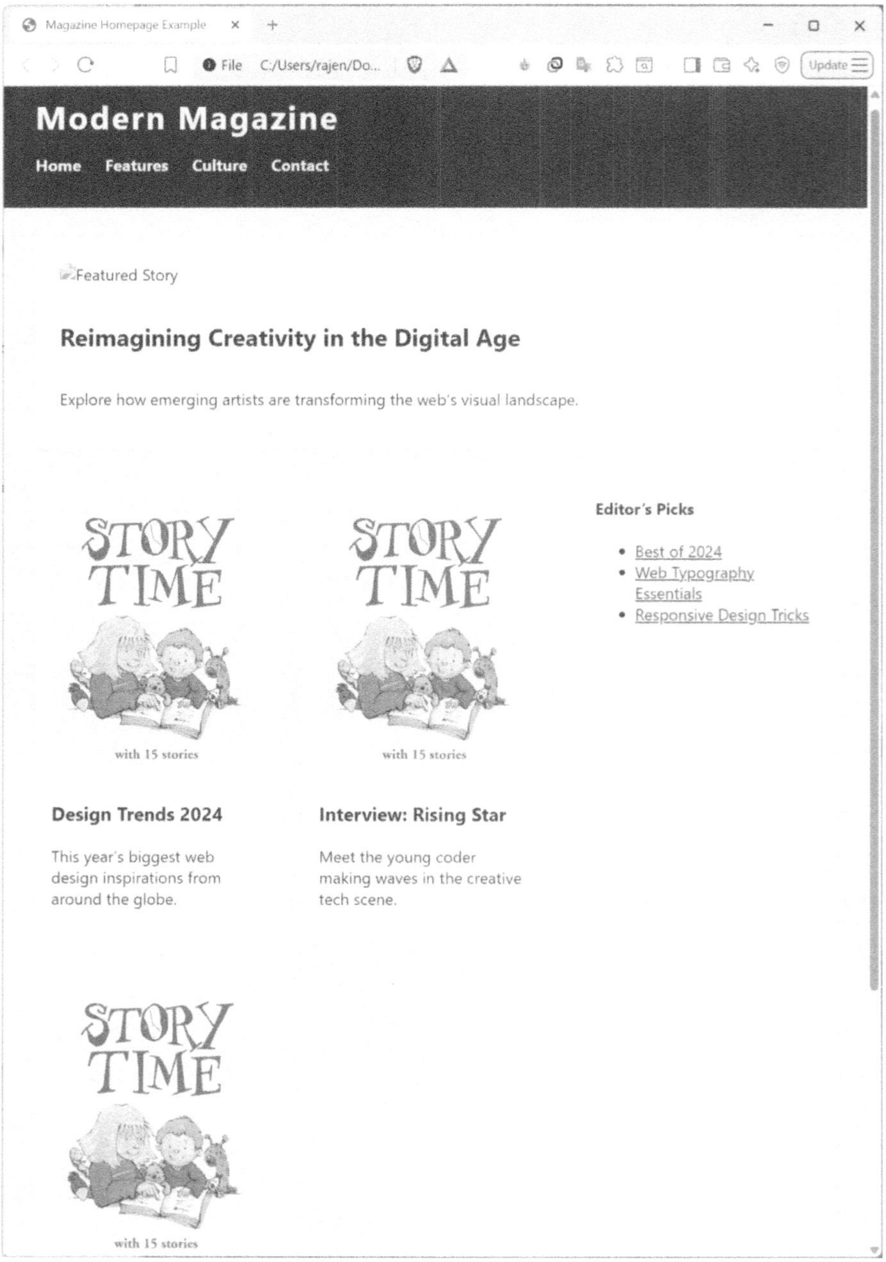

Step 4: Customizing and Experimenting

Try swapping in your own images, colors, and text to personalize the homepage. For inspiration, browse sites like [Vogue](https://www.vogue.com/) or [Medium](https://medium.com/) and notice how they use grid layouts to highlight stories.

You can also experiment with more advanced CSS Grid features, such as `grid-area` or fractional units, to further refine your design. Adding hover effects or subtle transitions will enhance the user experience and make your homepage even more engaging.

You've just built a responsive, magazine-style homepage using HTML and CSS Grid. This exercise brought together semantic markup, advanced grid layouts, and mobile-friendly design—all crucial skills for modern web developers. Keep experimenting with different grid arrangements and content types to deepen your understanding and creativity.

14. RESPONSIVE WEB DESIGN PRINCIPLES

Picture this: you've crafted a beautiful website, but when you check it on your friend's phone or a tablet, the layout falls apart—images overflow, text shrinks, and navigation becomes nearly impossible. In today's highly connected world, websites are accessed from an incredible variety of devices and screen sizes. That's why understanding how to build responsive websites isn't just a bonus skill—it's essential for every modern web developer.

In this chapter, you'll discover how to make your websites look and function perfectly whether viewed on a large desktop monitor, a slim laptop, or a compact mobile device. We'll introduce you to the fundamentals of **responsive design**, a powerful approach that ensures your web projects automatically adapt and shine on any screen. You'll learn how to use **CSS media queries** to apply different styles based on device characteristics, making your layouts flexible and user-friendly.

We'll also explore how to create **fluid images** and **responsive typography**, so your visuals and text remain clear and attractive no matter the context. Practical navigation solutions—like collapsible menus and mobile-friendly navigation patterns—will be covered to help you build sites that are easy to explore on any device.

To put your new skills into action, you'll work through a hands-on exercise: transforming a blog layout into a fully responsive experience. By the end of this chapter, you'll be able to confidently create websites that stand out in today's multi-device world, boosting your portfolio and opening doors to exciting web development opportunities. Let's dive in and unlock the secrets of professional, responsive web design together.

14.1. Introduction to Responsive Design

Responsive web design is the secret to creating websites that look great and function smoothly on any device, whether it's a smartphone, tablet, laptop, or desktop. In today's world, people access the web on screens of all shapes and sizes, so mastering responsive design is essential for building professional-quality sites that deliver a seamless experience to everyone. This section will guide you through the core concepts and practical steps to make your web pages adapt beautifully to any screen.

What is Responsive Web Design?

Responsive web design means your site automatically adjusts its layout, images, and text to fit the device it's viewed on. Instead of building separate sites for mobile and desktop users, you create one flexible page that works everywhere. This is achieved with a mix of flexible layouts, scalable images, and clever CSS.

Why Responsive Design Matters

It improves user experience, making your site accessible and enjoyable on all devices.

It helps your site rank better in search engines, as mobile-friendliness is a key ranking factor.

It increases engagement, as visitors are less likely to leave a site that works well on their device.

Let's jump in and start building a responsive foundation!

Building a Simple Responsive Page: Hands-On Practice

To see responsive design in action, we'll create a basic personal portfolio homepage. This example will introduce you to the core techniques: using a flexible layout, setting responsive images, and applying fluid typography.

Start by creating a new HTML file. Here's the basic structure:

```
<!DOCTYPE html>
<html lang= "en">
<head>
  <meta charset= "UTF-8">
  <meta name= "viewport"content= "width=device-width, initial-
scale=1.0">
  <title>My Responsive Portfolio</title>
  <link rel= "stylesheet"href= "styles.css">
</head>
<body>
  <header>
    <h1>Jane Doe</h1>
    <p>Web Developer & Designer</p>
  </header>
  <main>
    <section class= "about">
      <img src= "profile.jpg"alt= "Jane's portrait"class= "profile-
pic">
      <p>Hello! I build modern, responsive websites. Let's work
together!</p>
    </section>
  </main>
  <footer>
    <p>&copy; 2024 Jane Doe</p>
  </footer>
</body>
</html>
```

Let's break down the essential steps to make this page responsive.

Setting the Viewport

The **meta viewport tag** in the `<head>` is crucial. It tells browsers how to control the page's dimensions and scaling. Without it, your site won't display properly on mobile devices.

```
<meta name= "viewport"content= "width=device-width, initial-scale=1.0">
```

Add this tag to every HTML page you want to be responsive.

Creating Flexible Layouts with CSS

Now, let's create the CSS file (`styles.css`) that gives your site its responsive magic.

Start by making the layout flexible:

```
body {
  font-family: Arial, sans-serif;
  margin: 0;
  padding: 0;
  line-height: 1.6;
}

header, footer {
  background-color: #374151;
  color: #fff;
  text-align: center;
  padding: 1.5rem 0;
}

main {
  padding: 1rem;
}

.about {
  display: flex;
  flex-direction: column;
  align-items: center;
  max-width: 600px;
  margin: 2rem auto 0 auto;
  background: #f3f4f6;
  border-radius: 8px;
  box-shadow: 0 2px 8px rgba(0,0,0,0.05);
  padding: 2rem 1rem;
}
```

Here, the `.about` section uses **flexbox** to align its content in a column that centers everything. The layout adapts to the available space.

Making Images Responsive

Images need to scale with the layout so they don't overflow on smaller screens. Use **max-width** and **height** in your CSS:

```
.profile-pic {
  width: 160px;
  max-width: 100%;
  border-radius: 50%;
  margin-bottom: 1rem;
  display: block;
}
```

This ensures your profile image looks good on all devices and never stretches outside its container.

Fluid Typography

To keep text readable, use relative units like **rem** and **em** instead of fixed pixel sizes. This allows fonts to scale with the user's device settings:

```
h1 {
  font-size: 2.5rem;
  margin-bottom: 0.3rem;
}

p {
  font-size: 1.1rem;
}
```

Real-World Scenario: E-Commerce Product Card

Imagine you're building a product card for a small online shop. You want each card to look great on mobile and grow gracefully on larger screens.

Here's the HTML:

```
<section class= "product-card">
  <img src= "shoes.jpg"alt= "Running Shoes"class= "product-img">
  <h2>Running Shoes</h2>
  <p>Lightweight, comfortable, and stylish. Available in all sizes.</p>
  <button>Add to Cart</button>
</section>
```

Add the following to your CSS for a responsive product card:

```
.product-card {
  max-width: 320px;
  margin: 2rem auto;
  background: #fff;
  border-radius: 12px;
  box-shadow: 0 4px 12px rgba(0,0,0,0.08);
  padding: 1.5rem;
  text-align: center;
}

.product-img {
```

```
  width: 100%;
  height: auto;
  border-radius: 8px;
  margin-bottom: 1rem;
}

button {
  background: #2563eb;
  color: #fff;
  border: none;
  border-radius: 6px;
  padding: 0.75rem 1.5rem;
  font-size: 1rem;
  cursor: pointer;
}
```

The product card layout and image automatically scale to the screen width, ensuring a polished look everywhere.

Hands-On Exercise: Test Your Responsive Site

Open your new portfolio page in your browser.

Resize the browser window or open the page on different devices.

Observe how the layout, images, and text adapt smoothly.

Try swapping out the `.about` section for the `.product-card` to see both examples in action.

You've just taken your first steps into responsive web design, learning how to make layouts and images adapt to different screens using flexible containers, fluid images, and scalable typography. By practicing with real-world examples like a personal portfolio and product card, you're building the practical skills to create user-friendly sites.

Next, you'll unlock even more responsive power by learning to use **media queries** in CSS. Media queries allow you to apply styles based on device characteristics, making your sites look flawless on any screen. Get ready to dive deeper and make your web pages truly dynamic!

14.2. Using Media Queries in CSS

Media queries are the backbone of responsive web design, empowering you to tailor your website's appearance to different devices and screen sizes. With this technique, you'll move beyond static layouts and create web pages that automatically adapt, whether viewed on a smartphone, tablet, or desktop.

In this section, you'll gain hands-on experience writing your own media queries, explore practical CSS rules, and apply them to real-world scenarios. By the end, you'll be able to make your layouts flexible and visually appealing on any device.

What Are Media Queries?

Media queries are special CSS rules that let you apply styles only when certain conditions are met. These conditions typically relate to the dimensions of the browser window or the device itself, such as width, height, orientation, or even resolution.

A basic media query uses the `@media` keyword, followed by a set of conditions, and a block of CSS rules inside curly braces. Here's the structure:

```
@media (condition) {
  /* CSS rules here */
}
```

For example, to change the background color when the screen width is less than 600 pixels:

```
@media (max-width: 600px) {
  body {
    background-color: lightblue;
  }
}
```

Step-by-Step: Making a Simple Web Page Responsive

Let's walk through a practical scenario: you're building a personal blog homepage. You want the layout to look great on both desktops and smartphones.

Start with Your HTML Structure

Create a simple HTML file for your blog homepage:

```
<!DOCTYPE html>
<html lang="en">
<head>
  <meta charset="UTF-8">
  <title>My Responsive Blog</title>
  <link rel="stylesheet" href="styles.css">
</head>
<body>
  <header>
    <h1>Welcome to My Blog</h1>
    <nav>
      <a href="#">Home</a>
      <a href="#">About</a>
      <a href="#">Contact</a>
    </nav>
  </header>
  <main>
```

```
    <article>
      <h2>First Post</h2>
      <p>This is an introduction to my blog. Stay tuned for
updates!</p>
    </article>
  </main>
</body>
</html>
```

Basic CSS for Default (Desktop) View

Add some desktop-friendly styles in `styles.css`:

```
body {
  font-family: Arial, sans-serif;
  margin: 0;
  padding: 0;
}

header {
  background: #333;
  color: white;
  padding: 20px;
  text-align: center;
}

nav a {
  color: white;
  text-decoration: none;
  margin: 0 15px;
}

main {
  max-width: 800px;
  margin: 40px auto;
  padding: 20px;
}
```

Add a Media Query for Small Screens

Now, use a media query to adjust the layout when viewed on screens 600px wide or smaller:

```
@media (max-width: 600px) {
  header {
    padding: 10px;
  }
  nav {
    display: flex;
    flex-direction: column;
    gap: 10px;
  }
  main {
    padding: 10px;
    margin: 10px;
```

```
  }
}
```

Let's break down what's happening:

When the screen width is 600px or less, the navigation links stack vertically instead of horizontally.
The header and main content get smaller paddings for a more compact appearance on mobile devices.
The overall design feels more comfortable for touch and smaller screens.

Practical Example: Responsive E-Commerce Product Grid

E-commerce sites often need to display products in a grid that adapts to various screen sizes. Let's see how media queries help you achieve this.

HTML for a Simple Product Grid

```
<section class="products">
  <div class="product-card">
    <img src="shirt.jpg" alt="T-shirt">
    <h3>Classic T-shirt</h3>
    <p>$19.99</p>
  </div>
  <div class="product-card">
    <img src="jeans.jpg" alt="Jeans">
    <h3>Denim Jeans</h3>
    <p>$39.99</p>
  </div>
  <div class="product-card">
    <img src="sneakers.jpg" alt="Sneakers">
    <h3>Sport Sneakers</h3>
    <p>$49.99</p>
  </div>
</section>
```

Desktop Styles for the Product Grid

```
.products {
  display: grid;
  grid-template-columns: repeat(3, 1fr);
  gap: 24px;
  max-width: 1000px;
  margin: 40px auto;
}
.product-card {
  background: #f8f8f8;
  padding: 20px;
  text-align: center;
  border-radius: 8px;
}
.product-card img {
  max-width: 100%;
```

```
  height: auto;
}
```

Media Query for Tablets and Smaller Devices

```
@media (max-width: 900px) {
  .products {
    grid-template-columns: repeat(2, 1fr);
  }
}

@media (max-width: 600px) {
  .products {
    grid-template-columns: 1fr;
  }
  .product-card {
    padding: 10px;
  }
}
```

Here's how this works:

On large screens, products display in a three-column grid.

On tablets (900px or less), the grid switches to two columns.

On small screens (600px or less), products stack in a single column, making them easy to view and interact with on mobile devices.

Best Practices for Using Media Queries

To ensure your websites are flexible and future-proof, keep these tips in mind:

Place your media queries after your base styles in your CSS file, so they override default rules when conditions are met.

Use relative units (like percentages or `em`) whenever possible for widths, padding, and font sizes.

Test your site on multiple devices or resize your browser window to preview how your media queries work in real time.

Combine multiple conditions with the `and` keyword for more precise targeting. For example:

```
@media (min-width: 600px) and (max-width: 900px) {
  /* Styles for devices between 600 and 900 pixels wide */
}
```

Hands-On Exercise: Make Your Navigation Responsive

Try this challenge to reinforce your understanding:

Create a simple navigation bar with links.

Use media queries to switch the navigation from a horizontal layout to a vertical stack when the screen is 500px wide or smaller.
Adjust padding and font size for touch-friendly navigation on mobile.

Use the HTML and CSS examples above as a guide, but add your own creative touches.

You've taken a major step forward in responsive design by learning how to use media queries in CSS. Now, you can write CSS that adapts to different screen sizes, ensuring your web pages look polished on any device. Keep practicing by tweaking and experimenting with your own layouts and observing how they respond at different breakpoints.

Up next, you'll discover how to make your images flexible and your text readable across all devices. Get ready to explore flexible images and responsive typography, the next essential skills in your responsive web design toolkit!

14.3. Flexible Images and Responsive Typography

Images and text are two of the most visible elements on any website. In the era of smartphones, tablets, laptops, and ultra-wide monitors, your site must deliver beautiful visuals and readable content on every screen. This section will guide you through practical strategies for making images flexible and typography responsive, ensuring your designs look sharp and feel comfortable to read, whether on a mobile device or a desktop.

Why Flexible Images Matter

When an image is too large for a small screen, it can break layouts or force users to scroll awkwardly. On the other hand, tiny images on big screens can seem unimpressive and pixelated. Flexible images scale up or down with their container, always fitting perfectly within their parent element.

Let's see how to code flexible images in a simple, real-world scenario: a personal portfolio site showcasing your artwork.

Step-by-Step: Making Images Responsive

Suppose you have an HTML structure for a portfolio site with images in a gallery section. Here's a basic example:

```
<section class= "gallery">
  <img src= "images/artwork1.jpg"alt= "Abstract painting in blue and gold">
  <img src= "images/artwork2.jpg"alt= "Minimalist landscape painting">
```

```
</section>
```

To make these images flexible, you'll use CSS to set their `max-width` to 100% and their `height` to auto. This approach ensures the images scale down for smaller screens but never stretch beyond their natural width.

Add this CSS:

```
.gallery img {
  max-width: 100%;
  height: auto;
  display: block;
  margin-bottom: 1rem;
}
```

How it works:

max-width: 100% means the image can never be wider than its container (such as a column or the viewport).
height: auto keeps the original aspect ratio, so images don't look squished.
display: block ensures images break onto a new line, preventing small spacing bugs.
margin-bottom: 1rem adds comfortable spacing between stacked images.

Try this: Resize your browser window or preview on a phone. Notice how the artwork images always fit nicely, no matter the screen size.

Advanced: Responsive Images with srcset

Sometimes, you want to serve different image sizes for different devices. The HTML `srcset` attribute helps browsers pick the best image for the screen's resolution or width.

Update your HTML like this:

```
<img
  src= "images/artwork1-small.jpg"
  srcset= "images/artwork1-small.jpg 600w, images/artwork1-large.jpg
1200w"
  sizes= "(max-width: 600px) 100vw, 600px"
  alt= "Abstract painting in blue and gold">
```

srcset lists different versions of your image with their widths.
sizes tells the browser how wide the image will be in different situations.

This technique is especially useful for blogs, news sites, or e-commerce platforms where images are vital but bandwidth is limited.

Responsive Typography: Making Text Work Everywhere

Typography plays a crucial role in user experience. If your text is too small on mobile or uncomfortably huge on desktop, users will quickly lose interest. Responsive typography adapts font size and spacing to every device, ensuring readability and a clean look.

Let's practice with a blog site example. Imagine you're building a simple blog post:

```
<article class= "blog-post">
  <h1>Learning to Code: My Journey</h1>
  <p>When I started coding, I had no idea where to begin. Now, I build
websites that work on every device!</p>
</article>
```

Traditionally, you might set font sizes in pixels:

```
.blog-post h1 {
  font-size: 32px;
}
.blog-post p {
  font-size: 16px;
}
```

However, this doesn't scale on different devices. Instead, use **relative units** like `em`, `rem`, or `vw` for flexible sizing.

Using rems for Scalable Text

The `rem` unit is based on the root (html) font size, which is usually 16px by default.

```
.blog-post h1 {
  font-size: 2.25rem; /* 36px if root is 16px */  line-height: 1.2;
}
.blog-post p {
  font-size: 1rem; /* 16px if root is 16px */  line-height: 1.6;
}
```

This approach scales up or down if the root font size changes, making your typography more flexible.

Fluid Typography with clamp()

Modern CSS introduces the `clamp()` function, allowing font sizes to scale between a minimum and maximum value, based on the width of the viewport. This is called **fluid typography**.

Try this for your blog post title:

```
.blog-post h1 {
  font-size: clamp(1.5rem, 5vw, 3rem);
}
```

The font size will never be smaller than **1.5rem** or larger than **3rem**, but in between, it scales smoothly with the viewport width (`5vw` is 5% of the viewport width).

Practice Tip: Adjust the browser width. Watch the heading size grow or shrink smoothly, always staying within the limits you set.

Responsive Typography with Media Queries

For even more control, use **media queries** to change font sizes at specific breakpoints:

```
.blog-post p {
  font-size: 1rem;
}

@media (min-width: 600px) {
  .blog-post p {
    font-size: 1.125rem;
  }
}
@media (min-width: 900px) {
  .blog-post p {
    font-size: 1.25rem;
  }
}
```

This method ensures your text is always legible, whether on a small phone, tablet, or wide desktop monitor.

Real-World Scenario: E-Commerce Product Cards

E-commerce sites rely on both images and text to sell products. Imagine you're building a product card for an online store:

```
<div class= "product-card">
  <img src= "images/shoe.jpg"alt= "Classic running shoe">
  <h2>Classic Running Shoe</h2>
  <p class= "price">$59.99</p>
</div>
```

Apply flexible image and responsive typography styles:

```
.product-card img {
  width: 100%;
  height: auto;
}
```

```
.product-card h2 {
  font-size: clamp(1.25rem, 4vw, 2rem);
}
.product-card .price {
  font-size: 1.125rem;
  color: #2a8f45;
  font-weight: bold;
}
```

Now your product card looks sharp and readable on every device, giving customers a great shopping experience.

You've just practiced making images responsive with `max-width`, `height: auto`, and the `srcset` attribute for performance. You also learned to use rem, clamp, and media queries for responsive typography, ensuring your text always looks great and reads easily. Try combining these techniques in your next project, such as a personal portfolio, blog, or online shop, to deliver a polished, professional experience for every user.

Next up: Get ready to tackle **responsive navigation patterns**. You'll learn how to make menus and navigation bars adapt seamlessly to all devices, creating intuitive browsing experiences for your visitors. Let's unlock the secrets behind mobile-friendly menus and modern navigation design!

14.4. Responsive Navigation Patterns

Navigating a website should feel effortless, whether someone is on a desktop, tablet, or smartphone. Responsive navigation patterns are essential to achieving this seamless experience, allowing menus to adapt their appearance and behavior to fit different screen sizes and devices. In this section, you'll explore practical techniques to create adaptable, user-friendly navigation menus using HTML and CSS.

Understanding Responsive Navigation

Modern web users access sites from a variety of devices. A navigation menu that works well on a wide desktop screen might become cluttered or unreadable on a small phone. Responsive navigation solves this by adjusting the menu's layout and interaction based on the device's screen size. The most common approach is to show a horizontal navigation bar on larger screens and a compact, expandable "hamburger" menu on smaller devices.

Let's see how this works in practice by building a responsive navigation bar, step by step.

Setting Up the HTML Structure

A simple yet effective navigation menu typically uses the **`<nav>`** element to define the navigation area. Inside, you might use an unordered list for your links. For a responsive menu, you'll also add a button that toggles the mobile navigation.

Below is an HTML structure commonly used on portfolio and blog websites:

```
<nav class= "main-nav">
  <div class= "logo">MySite</div>
  <button class= "nav-toggle"aria-label= "Open navigation">
    <span class= "hamburger"></span>
  </button>
  <ul class= "nav-links">
    <li><a href= "#home">Home</a></li>
    <li><a href= "#about">About</a></li>
    <li><a href= "#projects">Projects</a></li>
    <li><a href= "#contact">Contact</a></li>
  </ul>
</nav>
```

`<nav>`: Defines the navigation section.
`.logo`: Displays your site or brand name.
`.nav-toggle`: A button with a hamburger icon for mobile.
`.nav-links`: Contains the navigation links.

Styling the Navigation for Desktop

First, style your navigation bar for larger screens. You'll create a horizontal layout and hide the toggle button.

```
.main-nav {
  display: flex;
  align-items: center;
  justify-content: space-between;
  background: #222;
  padding: 0.5rem 2rem;
}

.logo {
  color: #fff;
  font-size: 1.7rem;
  letter-spacing: 2px;
  font-weight: bold;
}

.nav-links {
  display: flex;
  list-style: none;
  margin: 0;
}

.nav-links li {
```

```
  margin-left: 2rem;
}

.nav-links a {
  color: #fff;
  text-decoration: none;
  font-size: 1.1rem;
  transition: color 0.2s;
}

.nav-links a:hover {
  color: #ff9800;
}

.nav-toggle {
  display: none;
  background: none;
  border: none;
  cursor: pointer;
}

.hamburger {
  width: 25px;
  height: 3px;
  background: #fff;
  display: block;
  position: relative;
}

.hamburger::before,
.hamburger::after {
  content: "";
  width: 25px;
  height: 3px;
  background: #fff;
  position: absolute;
  left: 0;
  transition: 0.3s;
}

.hamburger::before {
  top: -8px;
}

.hamburger::after {
  top: 8px;
}
```

This code creates a clean, horizontal navigation bar with a simple brand logo and links. The hamburger menu is hidden on large screens.

Making the Navigation Responsive with Media Queries

To adapt your menu for tablets and phones, use a **media query** to change the layout when the screen width is below a certain size (commonly 768px).

```
@media (max-width: 768px) {
  .nav-links {
    flex-direction: column;
    position: absolute;
    top: 60px;
    left: 0;
    width: 100%;
    background: #222;
    display: none;
  }

  .nav-links li {
    margin: 1.5rem 0;
    text-align: center;
  }

  .nav-toggle {
    display: block;
  }
}
```

The **`.nav-links`** list stacks vertically, takes the full width, and is initially hidden.

The **`.nav-toggle`** (hamburger button) becomes visible, allowing users to open the menu.

Adding Interactivity with Simple JavaScript

To allow users to open and close the menu on mobile, you'll need a bit of JavaScript. This toggles the visibility of the navigation links when the hamburger button is clicked.

```
<script>
  const navToggle = document.querySelector( '.nav-toggle');
  const navLinks = document.querySelector( '.nav-links');

  navToggle.addEventListener( 'click', () => {
    navLinks.style.display = navLinks.style.display === 'flex'? 'none':
'flex';
  });
</script>
```

This script checks the menu's current state and toggles it between `display: flex` (shown) and `display: none` (hidden).

Real-World Scenarios: Blog and E-Commerce Navigation

Example 1: Blog Navigation

A personal blog, such as "Wanderlust Writings," might use a responsive menu to keep navigation easy for readers on the go. On a phone, the navigation collapses into a hamburger menu, ensuring the blog content remains the focus.

Example 2: E-Commerce Site Navigation

An online store like "GlobalThreads" could use a similar pattern but add icons for a shopping cart or user profile. The responsive menu ensures that customers can easily browse categories and products, regardless of device.

Enhancing Accessibility

It's important to ensure your navigation is accessible. The **`aria-label`** attribute on the toggle button helps screen readers understand its purpose. You can further improve accessibility by managing focus and keyboard navigation, though this can be explored in more advanced sections.

You've just built a responsive navigation menu that adapts to any screen size, using HTML for structure, CSS for layout, and a touch of JavaScript for interactivity. Responsive navigation ensures users can explore your site easily, whether browsing a portfolio, reading a blog, or shopping on an e-commerce store.

Continue practicing by customizing the navigation colors, adding dropdown menus, or refining the hamburger animation to make your navigation unique and user-friendly.

Next, get ready to put your new skills to the test! In the upcoming section, you'll learn how to make an entire blog layout responsive across devices, combining navigation with flexible grids and adaptable content. Let's dive into building a fully responsive web page step by step!

14.5. Hands-On Exercise: Making a Blog Layout Responsive Across Devices

Responsive web design is all about ensuring your website looks great and works smoothly on any device, from desktop monitors to tablets and smartphones. As more people browse the web on mobile devices, building layouts that adapt seamlessly is an essential skill for every web developer. In this exercise, we'll apply core responsive design principles to a simple blog layout, focusing on flexible grids, media queries, and user-friendly navigation. By the end, you'll be able to create a blog page that automatically adjusts its structure and style to fit any screen size—a skill used on real-world sites like Medium and WordPress blogs.

Planning the Responsive Blog Layout

Before you start coding, take a moment to visualize the basic structure of a typical blog layout:

A **header** section for the blog title and navigation links

A **main content** area for blog posts
A **sidebar** for widgets, categories, or recent posts
A **footer** for copyright or extra links

On wide screens, the sidebar sits next to the main content; on smaller screens, the sidebar stacks below or above the content for easier reading.

Step 1: Setting Up Basic HTML Structure

Start by writing the HTML markup for your responsive blog layout. This structure will serve as the foundation for applying responsive CSS.

```
<!DOCTYPE html>
<html lang="en">
<head>
  <meta charset="UTF-8">
  <title>Responsive Blog Layout</title>
  <link rel="stylesheet" href="style.css">
</head>
<body>
  <header class="site-header">
    <h1>My Awesome Blog</h1>
    <nav class="main-nav">
      <ul>
        <li><a href="#">Home</a></li>
        <li><a href="#">Posts</a></li>
        <li><a href="#">About</a></li>
      </ul>
    </nav>
  </header>
  <div class="container">
    <main class="content">
      <article>
        <h2>First Blog Post</h2>
        <p>Welcome to my responsive blog! This is the first post.</p>
      </article>
      <article>
        <h2>Another Article</h2>
        <p>Responsive design makes websites more user-friendly on any
device.</p>
      </article>
    </main>
    <aside class="sidebar">
      <h3>About Me</h3>
      <p>Hi! I'm a web developer learning HTML & CSS.</p>
      <h3>Categories</h3>
      <ul>
        <li>Coding</li>
        <li>Design</li>
        <li>Tips</li>
      </ul>
    </aside>
  </div>
  <footer class="site-footer">
    <p>&copy; 2024 My Awesome Blog</p>
```

```
  </footer>
</body>
</html>
```

This markup separates the site into clear sections, making it easier to apply styles and responsive rules.

Step 2: Creating the Desktop Layout with CSS Grid

Let's style your blog for desktop screens first using CSS Grid for a modern, flexible layout.

```
/* style.css */

/* Reset some default styles */
* {
  box-sizing: border-box;
  margin: 0;
  padding: 0;
}

body {
  font-family: Arial, sans-serif;
  background: #f8f9fa;
  color: #333;
}

.site-header {
  background: #0077cc;
  color: #fff;
  padding: 1.5rem 0;
  text-align: center;
}

.main-nav ul {
  list-style: none;
  display: flex;
  justify-content: center;
  gap: 2rem;
  margin-top: 1rem;
}

.main-nav a {
  color: #fff;
  text-decoration: none;
  font-weight: bold;
}

.container {
  display: grid;
  grid-template-columns: 3fr 1fr;
  gap: 2rem;
  max-width: 1100px;
  margin: 2rem auto;
  padding: 0 1rem;
}
```

```
.content {
  background: #fff;
  padding: 2rem;
  border-radius: 8px;
  box-shadow: 0 2px 8px rgba(0,0,0,0.05);
}

.sidebar {
  background: #e9ecef;
  padding: 2rem;
  border-radius: 8px;
}

.site-footer {
  background: #212529;
  color: #fff;
  text-align: center;
  padding: 1rem 0;
  margin-top: 2rem;
}
```

With this styling, your blog looks structured and clean on desktop screens, with main content and sidebar side-by-side.

Step 3: Making the Layout Responsive with Media Queries

To adjust the layout for tablets and smartphones, you'll use **CSS media queries**. These allow you to apply different styles based on the device's screen width.

```
/* Responsive adjustments */
@media (max-width: 900px) {
  .container {
    grid-template-columns: 1fr;
    gap: 1rem;
  }
  .sidebar {
    margin-top: 2rem;
  }
}

@media (max-width: 600px) {
  .site-header, .content, .sidebar, .site-footer {
    padding-left: 0.5rem;
    padding-right: 0.5rem;
  }
  .main-nav ul {
    flex-direction: column;
    gap: 0.5rem;
    align-items: center;
  }
  .content, .sidebar {
    padding: 1rem;
  }
}
```

Here's what these media queries accomplish:

When the screen is **900px or less**, the grid layout collapses into a single column so the sidebar stacks below the main content.

For screens **600px or less**, padding is reduced for smaller displays, and navigation links stack vertically for easier tapping on smartphones.

Step 4: Testing and Refining Responsiveness

Open your blog layout in a web browser. Resize the window to observe how your content adjusts smoothly at different widths. Try viewing the page on a mobile device or using browser developer tools to simulate various screens.

You'll notice how navigation, content, and sidebar all reorganize automatically. This is the essence of responsive design, providing a great reading experience for all users.

Step 5: Optional Enhancements

To make your blog even more engaging, try adding:

A subtle hover effect for navigation links
Responsive images that scale with screen size
A "hamburger" menu for mobile navigation

These enhancements are common on popular blogging platforms and help create a polished, professional look.

Real-World Inspiration

Websites like **Medium** and **WordPress.com** use responsive layouts to ensure their content is accessible and beautiful on any device. Explore these sites and observe how their navigation, articles, and sidebars adapt as you resize your browser window. Try recreating some of their design elements to practice your new skills.

Quick Coding Summary

You've created a fully responsive blog layout by combining semantic HTML structure, **CSS Grid** for desktop layouts, and **media queries** for handling different device widths. These techniques empower you to design flexible, modern websites that look fantastic everywhere.

Take a few minutes to experiment by adding more posts, tweaking colors, or customizing the navigation. Practice is the key to mastering responsive design.

15. REAL-WORLD PROJECTS AND NEXT STEPS

You've built a strong foundation in HTML and CSS, and now it's time to put your skills to the test by creating real websites that people can see and use. This chapter marks a pivotal moment in your journey from learning the basics to building professional, fully functional web projects that showcase your creativity and technical ability. Here, you'll discover how to plan and structure entire websites, moving beyond single pages to develop complete projects that reflect your personality, interests, or business goals.

Through step-by-step guidance, you'll design a personal portfolio to highlight your work, craft a simple blog or news site for sharing ideas, and create a small e-commerce product page that introduces essential features like product listings and calls to action. Along the way, you'll explore how to organize content, select effective layout strategies, and apply modern CSS styling for professional, visually appealing results. You'll also examine global web design trends, learning how to blend your unique vision with industry best practices to build sites that stand out and connect with users worldwide.

As you bring your projects to life, you'll gain hands-on experience with crucial techniques such as navigation menus, responsive grids, and consistent site-wide styling. The chapter also guides you through the process of publishing your site online, giving you the tools to share your creations with the world. By the end, you'll not only have a collection of real-world projects to add to your portfolio but also the confidence and know-how to continue growing as a web developer. Get ready to turn your coding knowledge into practical, eye-catching websites that open doors to new opportunities and creative possibilities.

15.1. Planning Your First Complete Website

Before you dive into building complex, eye-catching websites, it's important to understand how to thoughtfully plan your first complete project. Effective planning helps you avoid common mistakes, saves time, and leads to a smoother, more enjoyable development process. In this section, you'll learn how to outline, structure, and start coding a real website using modern HTML and CSS techniques. Let's walk through a practical scenario: planning and beginning the creation of a responsive blog homepage.

Setting Your Project Goals

Every great website starts with clear goals. Ask yourself:

What is the main purpose of my site?

Who is my target audience?
What content and features should appear on the homepage?

For our example, let's create a simple, responsive blog homepage that includes a header, featured post, list of recent articles, and a clean navigation menu.

Sketching a Basic Layout

Before writing any code, sketch a rough wireframe on paper or use a digital tool. Map out these sections:

Header: Site title and navigation links
Main content: Featured blog post and recent articles
Sidebar or footer (optional): About section or social links

This visual plan guides your HTML structure and CSS styling.

Structuring the HTML

Now, let's turn the sketch into a basic HTML skeleton. Start with semantic elements for clarity and accessibility.

```
<!DOCTYPE html>
<html lang= "en">
<head>
  <meta charset= "UTF-8">
  <meta name= "viewport"content= "width=device-width, initial-
scale=1.0">
  <title>My First Blog</title>
  <link rel= "stylesheet"href= "style.css">
</head>
<body>
  <header>
    <h1>My First Blog</h1>
    <nav>
      <ul>
        <li><a href= "#">Home</a></li>
        <li><a href= "#">About</a></li>
        <li><a href= "#">Contact</a></li>
      </ul>
    </nav>
  </header>
  <main>
    <section class= "featured-post">
      <h2>Featured Post</h2>
      <p>This is the summary of the most recent or most important blog
post.</p>
      <a href= "#">Read More</a>
    </section>
    <section class= "recent-posts">
      <h2>Recent Posts</h2>
      <article>
        <h3>Another Great Article</h3>
```

```
        <p>Short introduction to the article content.</p>
        <a href= "#">Continue Reading</a>
      </article>
      <article>
        <h3>Learning HTML & CSS</h3>
        <p>Tips and tricks for beginners starting their coding
journey.</p>
        <a href= "#">Continue Reading</a>
      </article>
    </section>
  </main>
  <footer>
    <p>&copy; 2024 My First Blog</p>
  </footer>
</body>
</html>
```

This markup sets up a clear, accessible layout. Notice the use of `<header>`, `<nav>`, `<main>`, `<section>`, `<article>`, and `<footer>` for logical structure.

Adding Responsive CSS Styling

Let's make the blog look modern and ensure it works well on any device. You'll use media queries and flexible units in your `style.css` file.

```
body {
  font-family: Arial, sans-serif;
  margin: 0;
  background: #f9f9f9;
  color: #222;
}

header {
  background: #2c3e50;
  color: #fff;
  padding: 1.5rem;
  text-align: center;
}

nav ul {
  list-style: none;
  padding: 0;
  display: flex;
  justify-content: center;
  gap: 1.5rem;
}

nav a {
  color: #fff;
  text-decoration: none;
  font-weight: bold;
}

main {
  max-width: 900px;
  margin: 2rem auto;
```

```
  padding: 1rem;
  background: #fff;
  border-radius: 8px;
  box-shadow: 0 2px 8px rgba(0,0,0,0.07);
}

.featured-post {
  background: #eaf6ff;
  padding: 1.5rem;
  border-radius: 6px;
  margin-bottom: 2rem;
}

.recent-posts article {
  margin-bottom: 1.5rem;
  border-bottom: 1px solid #eee;
  padding-bottom: 1rem;
}

footer {
  text-align: center;
  padding: 1rem 0;
  background: #2c3e50;
  color: #fff;
  margin-top: 2rem;
}

/* Responsive Design */@media (max-width: 600px) {
  main {
    padding: 0.5rem;
  }
  nav ul {
    flex-direction: column;
    gap: 0.75rem;
  }
}
```

This styling ensures your blog is visually appealing and adapts to smaller screens, a crucial part of modern web development.

Real-World Example: Simple E-Commerce Landing Page

Let's say you want to build a basic landing page for a store. The planning steps are similar: define your product focus, sketch the layout (header, product grid, call-to-action button), and then structure your HTML and CSS accordingly. This reinforces the importance of planning before coding and helps you adapt your skills to various project types.

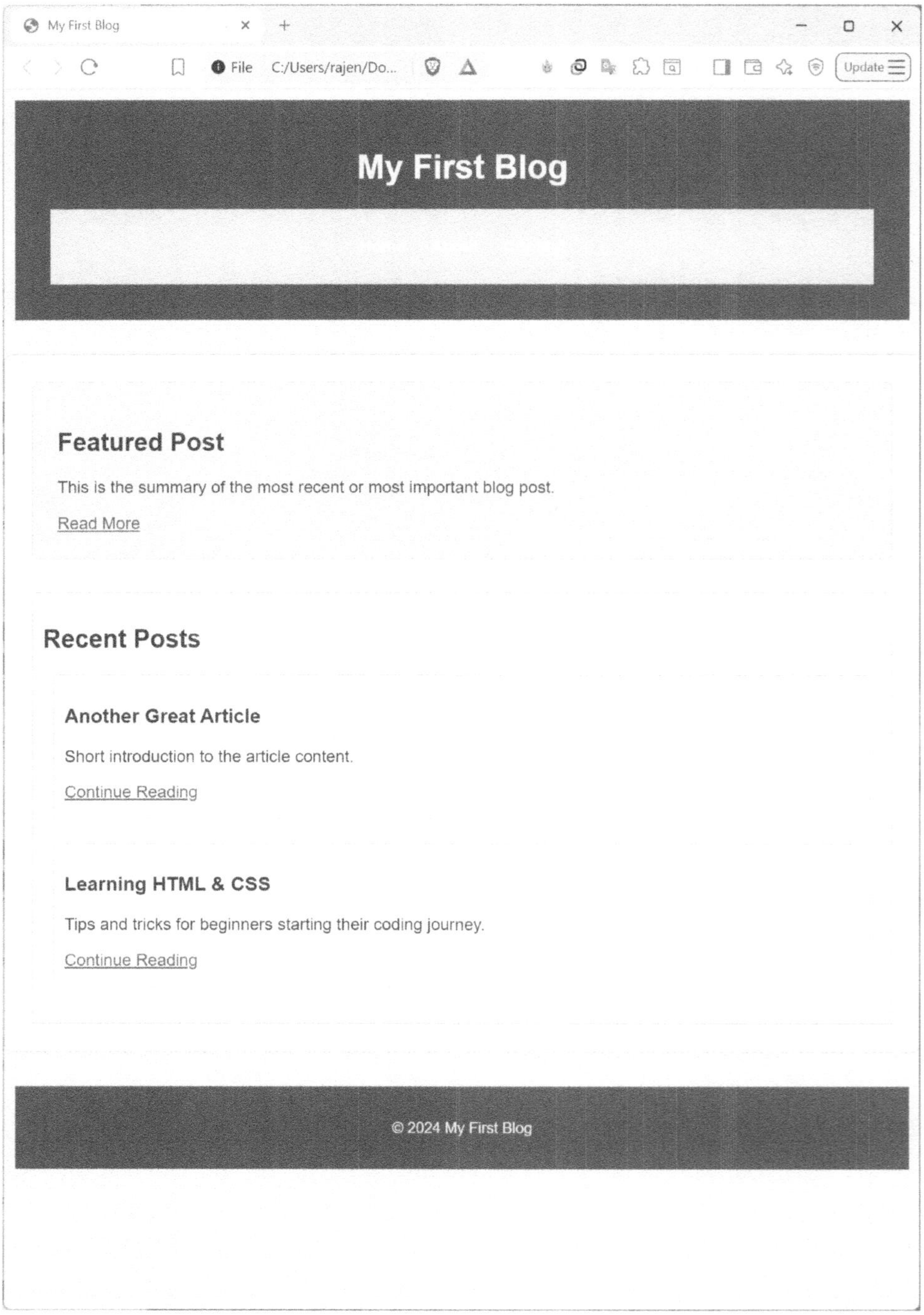

In this section, you've learned the value of thoughtful planning, structured semantic HTML, and responsive CSS for your first complete website. Practice by sketching a layout, writing your HTML, and styling with CSS, then view your site on different devices to see responsive design in action. Each step you take builds confidence and real-world coding skills.

Up next, you'll apply these planning and coding techniques to create your own personal portfolio website, where you can showcase your projects and skills using advanced layouts and creative styling. Get ready to take your web development journey to the next level!

15.2. Building a Personal Portfolio Website

Building your own personal portfolio website is one of the most rewarding ways to put your HTML and CSS skills into practice. A portfolio site not only showcases your coding abilities but also serves as a digital resume to impress potential employers or clients. In this section, you'll create a fully responsive personal portfolio page, using advanced CSS techniques to ensure your site looks great on any device.

Planning the Structure

Before you write any code, it's important to visualize what your portfolio will include. Most professional portfolios feature:

- A clear header with your name and a short tagline
- An About section introducing yourself
- A Projects section with examples of your work
- Contact information or a simple contact form

Let's start with the basic HTML structure.

Step-by-Step: HTML Markup for Your Portfolio

Begin by setting up your HTML file. Open your code editor and create a new file named `index.html`.

Add the basic HTML5 structure:

```
<!DOCTYPE html>
<html lang= "en">
<head>
  <meta charset= "UTF-8">
  <meta name= "viewport"content= "width=device-width, initial-
scale=1.0">
  <title>Your Name | Portfolio</title>
  <link rel= "stylesheet"href= "styles.css">
</head>
<body>
  <header>
    <h1>Your Name</h1>
    <p>Web Developer & Designer</p>
  </header>
  <section id= "about">
    <h2>About Me</h2>
    <p>Hello! I'm a passionate beginner web developer, eager to build
```

```
modern, user-friendly websites.</p>
  </section>
  <section id= "projects">
    <h2>Projects</h2>
    <div class= "project-card">
      <h3>Responsive Gallery</h3>
      <p>A photo gallery that adapts beautifully to mobile and desktop
screens.</p>
      <a href= "#">View Project</a>
    </div>
    <div class= "project-card">
      <h3>Simple E-commerce Demo</h3>
      <p>A mock online store with a clean layout and intuitive
navigation.</p>
      <a href= "#">View Project</a>
    </div>
  </section>
  <footer>
    <p>Contact me: <a href=
"mailto:your.email@example.com">your.email@example.com</a></p>
  </footer>
</body>
</html>
```

This markup provides a clear, semantic structure for your site. Each section is wrapped in meaningful tags like `<header>`, `<section>`, and `<footer>` to help with both screen reader accessibility and search engine optimization.

Styling Your Portfolio with Responsive CSS

Next, let's make your portfolio visually striking and fully responsive using advanced CSS. Create a new file named `styles.css` in the same folder.

Add a modern, clean base style and set up a responsive layout with CSS Grid and Flexbox.

```
/* Reset and base styles */* {
  box-sizing: border-box;
  margin: 0;
  padding: 0;
}
body {
  font-family: 'Segoe UI', Arial, sans-serif;
  background: #f8f9fa;
  color: #222;
  line-height: 1.6;
  min-height: 100vh;
  display: flex;
  flex-direction: column;
}

/* Header styles */header {
  background: linear-gradient(90deg, #67b26f, #4ca2cd);
  color: #fff;
```

```
  text-align: center;
  padding: 2rem 1rem;
}

/* About section */#about {
  max-width: 700px;
  margin: 2rem auto;
  background: #fff;
  padding: 1.5rem;
  border-radius: 8px;
  box-shadow: 0 2px 8px rgba(0,0,0,0.06);
}

/* Projects grid */#projects {
  max-width: 900px;
  margin: 2rem auto;
  display: grid;
  grid-template-columns: repeat(auto-fit, minmax(260px, 1fr));
  gap: 2rem;
}

.project-card {
  background: #fff;
  border-radius: 8px;
  box-shadow: 0 1px 4px rgba(0,0,0,0.08);
  padding: 1.2rem;
  text-align: center;
  transition: transform 0.2s;
}
.project-card:hover {
  transform: translateY(-6px) scale(1.03);
}
.project-card h3 {
  color: #67b26f;
}

/* Footer styles */footer {
  text-align: center;
  padding: 1rem 0;
  background: #e9ecef;
  margin-top: auto;
}
footer a {
  color: #4ca2cd;
  text-decoration: underline;
}

/* Responsive adjustments */@media (max-width: 600px) {
  header, #about, #projects, footer {
    padding-left: 1rem;
    padding-right: 1rem;
  }
  #about {
    margin: 1.2rem auto;
  }
}
```

This CSS leverages responsive techniques such as **CSS Grid** for the projects section, smart use of **media queries**, and subtle **hover effects** to make your portfolio feel interactive and modern. Try adjusting your browser width and see how the project cards reflow on smaller screens for a mobile-friendly experience.

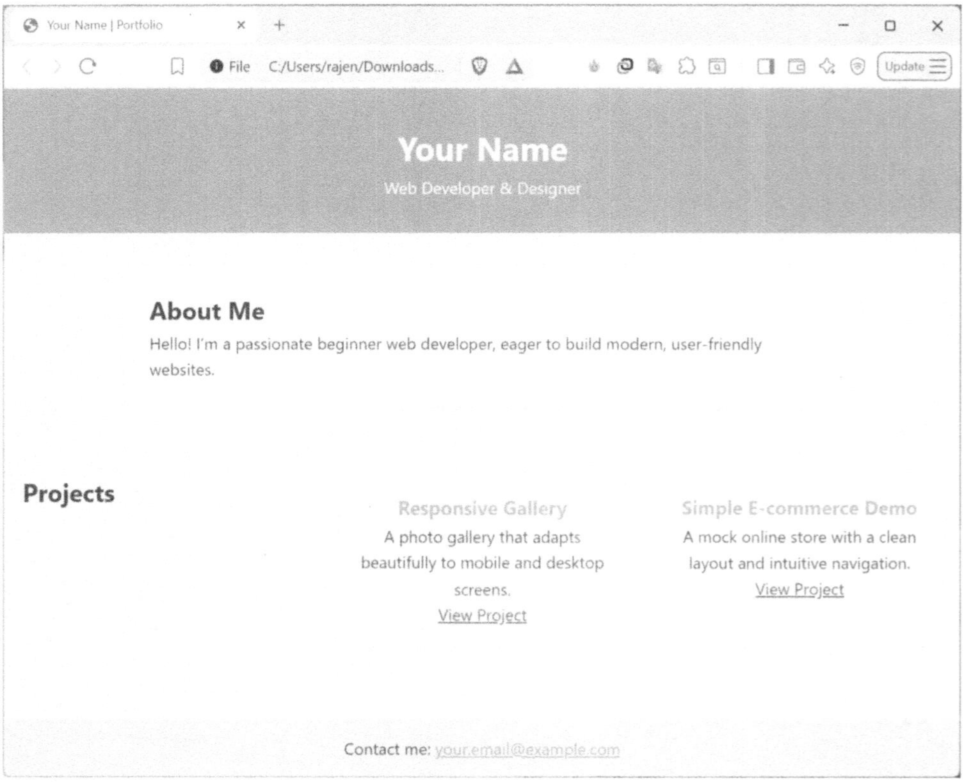

Real-World Inspiration

Visit portfolio sites like dev.to or [Behance](https://www.behance.net/) to see how professional developers and designers structure their profiles. Notice the emphasis on clear sections, visual hierarchy, and mobile accessibility. These are the same principles you're practicing here.

Hands-On Exercise

Customize your portfolio by:

Adding your own photo or avatar in the About section
Expanding the Projects section with real or sample work
Experimenting with color schemes and font choices to reflect your personal style

Test your site on both desktop and mobile devices to confirm its responsiveness.

Coding Summary

You've now built a personal portfolio website from scratch, employing semantic HTML and responsive CSS techniques like Grid, Flexbox, and media queries. This project showcases your growing skills and can be expanded as you learn more.

Next, you'll step up your web development game by building a simple blog or news site, where you'll organize posts, explore new layout strategies, and continue mastering HTML and CSS for content-rich websites. Get ready to create your own digital publication!

15.3. Creating a Simple Blog or News Site

Building a simple blog or news site is a fantastic way to put your HTML and CSS skills into action while learning how real-world websites are structured and styled. Blogs and news sites are everywhere, from global publications like BBC News to personal developer blogs. They offer a dynamic format where content is organized into posts, typically displayed in a list or grid with clear headings, publication dates, and summaries. In this section, you'll create a basic blog homepage that adapts to any device, showcasing articles in a clean, readable layout.

Planning Your Blog Structure

Before writing any code, let's outline the main components of your blog site:

- **Header**: Contains the site title and navigation.
- **Main Content**: Displays a list of blog articles.
- **Sidebar (optional)**: Can include categories or recent posts.
- **Footer**: Displays copyright or extra links.

For this hands-on project, we'll focus on a streamlined version with a header, main content area, and footer.

Writing the HTML Markup

Let's start by building the HTML structure for the blog homepage. Open a new file called `index.html` and follow these steps:

- Create the basic HTML5 document structure.
- Add a header with your blog title and navigation links.
- Create a main section to hold multiple blog articles.
- Each article should include a heading, date, summary, and a "Read More" link.
- Finish with a simple footer.

Here's a sample markup:

```
<!DOCTYPE html>
<html lang= "en">
<head>
  <meta charset= "UTF-8">
  <title>My First Blog</title>
  <link rel= "stylesheet"href= "styles.css">
</head>
<body>
  <header>
    <h1>My First Blog</h1>
    <nav>
      <a href= "#">Home</a>
      <a href= "#">About</a>
      <a href= "#">Contact</a>
    </nav>
  </header>
  <main>
    <article>
      <h2>Welcome to My Blog</h2>
      <time datetime= "2024-06-10">June 10, 2024</time>
      <p>This is my very first post! Here, I'll share coding tips and
web development news to help other beginners get started.</p>
      <a href= "#">Read More</a>
    </article>
    <article>
      <h2>Getting Started with CSS Layouts</h2>
      <time datetime= "2024-06-08">June 8, 2024</time>
      <p>Learn how to create flexible layouts with CSS for your own
sites. Discover the basics of arranging content and making your pages
shine.</p>
      <a href= "#">Read More</a>
    </article>
  </main>
  <footer>
    <p>&copy; 2024 My First Blog</p>
  </footer>
</body>
</html>
```

Styling the Blog with CSS

Now, let's make your blog visually appealing and responsive. Open a new file named `styles.css` and add the following styles:

Use a modern, clean font for readability.
Add spacing and background colors for sections.
Style the header and navigation.
Create a responsive layout using media queries so your blog looks great on desktops, tablets, and phones.

```
body {
  font-family: 'Segoe UI', Arial, sans-serif;
```

```
  margin: 0;
  background: #f7f7f7;
  color: #222;
}

header {
  background: #2c3e50;
  color: #fff;
  padding: 1.5rem 1rem;
  text-align: center;
}

header h1 {
  margin: 0 0 0.5rem 0;
  font-size: 2.5rem;
}

nav a {
  color: #ecf0f1;
  margin: 0 0.75rem;
  text-decoration: none;
  font-weight: 500;
}

main {
  max-width: 700px;
  margin: 2rem auto;
  padding: 0 1rem;
}

article {
  background: #fff;
  margin-bottom: 1.5rem;
  padding: 1.25rem;
  border-radius: 7px;
  box-shadow: 0 2px 8px rgba(44, 62, 80, 0.08);
}

article h2 {
  margin: 0 0 0.5rem 0;
  font-size: 1.5rem;
}

article time {
  color: #888;
  font-size: 0.92rem;
}

article a {
  color: #2980b9;
  font-weight: 600;
  text-decoration: underline;
}

footer {
  background: #2c3e50;
  color: #ecf0f1;
  text-align: center;
```

```
  padding: 1rem 0;
  position: fixed;
  width: 100%;
  bottom: 0;
}

/* Responsive Design */@media (max-width: 600px) {
  main {
    padding: 0 0.5rem;
  }
  header h1 {
    font-size: 2rem;
  }
  article {
    padding: 1rem;
  }
}
```

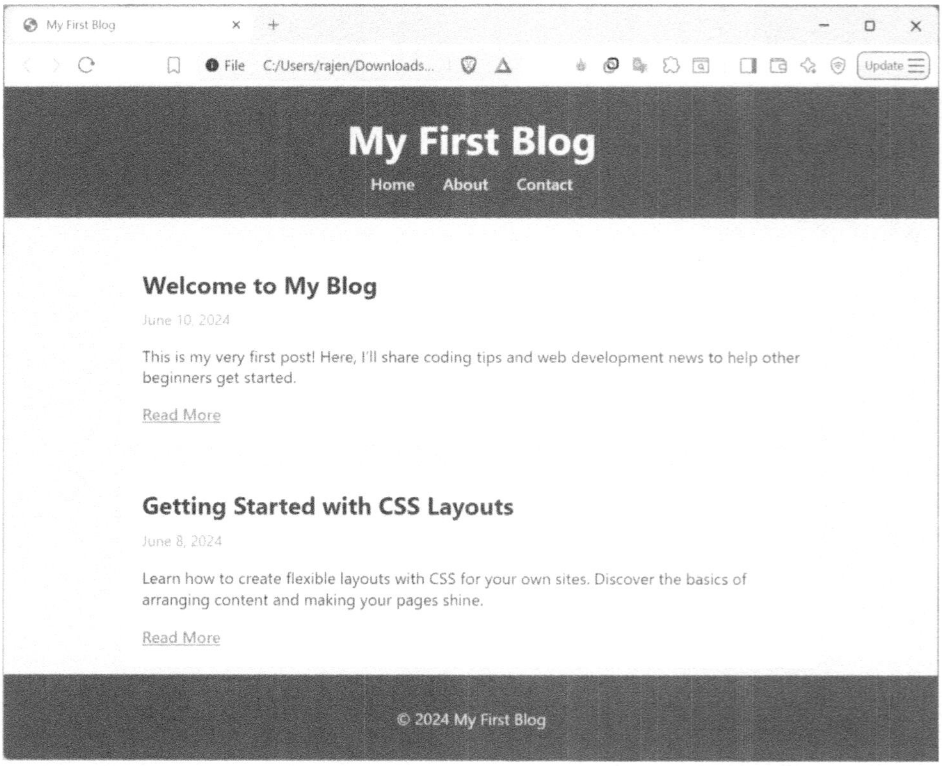

Practical Exercise: Enhance Your Blog

Add another article to your blog with a unique title and summary.
Personalize the navigation with links relevant to your interests.
Experiment with colors or fonts to make your blog stand out.

You've built a functional, responsive blog homepage using semantic HTML and modern CSS techniques. By organizing content with elements like **article**, **header**, and

footer, and styling them for readability and mobile-friendliness, you've taken a big step toward professional web development. Keep practicing by tweaking your blog and exploring other real-world sites for inspiration.

Next, get ready to advance your skills by designing a small e-commerce product page. You'll learn how to present products attractively, use new HTML elements, and apply CSS techniques that make online stores user-friendly and visually engaging!

15.4. Designing a Small E-Commerce Product Page

Building a small e-commerce product page is a fantastic way to apply your knowledge of HTML and CSS to a real-world scenario. E-commerce sites like Etsy and Shopify are built around clear presentation, engaging visuals, and easy navigation. In this hands-on project, you'll craft a responsive product card for a fictional online store. You'll learn to structure product information, display images, highlight price and features, and style a mobile-friendly "Add to Cart" button. This practical exercise introduces advanced HTML elements and responsive CSS techniques that are widely used in professional e-commerce websites.

Structuring Your Product Card in HTML

Begin by creating a semantic layout for your product card. Semantic HTML improves accessibility and makes your code easier to maintain.

Open your code editor and create a new HTML file.
Inside the `<body>`, add a `<section>` to contain your product card.
Use a combination of `<figure>`, `<img>`, `<figcaption>`, and structured text elements for product details.

Here's a sample markup for a single product card:

```
<section class= "product-card">
  <figure>
    <img src= "images/sneaker.jpg"alt= "Stylish Running Sneaker"class=
"product-image">
    <figcaption>Comfort Fit Sneaker</figcaption>
  </figure>
  <div class= "product-info">
    <h2>Comfort Fit Sneaker</h2>
    <p class= "product-desc">Experience all-day comfort with our
lightweight, breathable sneakers. Perfect for running, walking, or
everyday wear.</p>
    <ul class= "features">
      <li>Sizes: 6-12</li>
      <li>Colors: Black, White, Blue</li>
      <li>Free shipping</li>
    </ul>
    <div class= "price-section">
```

```
      <span class= "price">$49.99</span>
      <button class= "add-to-cart">Add to Cart</button>
    </div>
  </div>
</section>
```

This structure separates the image and product information, making it easier to style and adapt for different screen sizes.

Styling the Product Page with Responsive CSS

Now bring your product card to life with CSS. Focus on creating a visually appealing layout that adapts smoothly to mobile and desktop screens.

```
body {
  font-family: 'Segoe UI', Arial, sans-serif;
  background: #f4f4f4;
  margin: 0;
  padding: 40px;
}

.product-card {
  max-width: 350px;
  margin: auto;
  background: #fff;
  border-radius: 12px;
  box-shadow: 0 2px 8px rgba(0,0,0,0.08);
  overflow: hidden;
  transition: box-shadow 0.2s;
}

.product-card:hover {
  box-shadow: 0 8px 24px rgba(0,0,0,0.16);
}

.product-image {
  width: 100%;
  display: block;
  object-fit: cover;
  height: 230px;
}

figcaption {
  text-align: center;
  font-weight: bold;
  padding: 10px 0 0 0;
  color: #222;
}

.product-info {
  padding: 18px;
}

.product-desc {
  color: #555;
  font-size: 1em;
```

```
  margin: 10px 0 16px 0;
}

.features {
  list-style: disc inside;
  margin: 0 0 16px 0;
  padding: 0;
  color: #444;
}

.price-section {
  display: flex;
  align-items: center;
  justify-content: space-between;
}

.price {
  font-size: 1.4em;
  color: #e67e22;
  font-weight: bold;
}

.add-to-cart {
  padding: 10px 20px;
  background: #1abc9c;
  color: white;
  border: none;
  border-radius: 6px;
  cursor: pointer;
  font-size: 1em;
  transition: background 0.2s;
}

.add-to-cart:hover {
  background: #16a085;
}
```

To ensure your product card looks great on all devices, add a responsive media query:

```
@media (max-width: 500px) {
  .product-card {
    max-width: 95%;
    padding: 0;
  }
  .product-info {
    padding: 12px;
  }
}
```

This ensures your card resizes smoothly for mobile users, a vital feature for modern e-commerce.

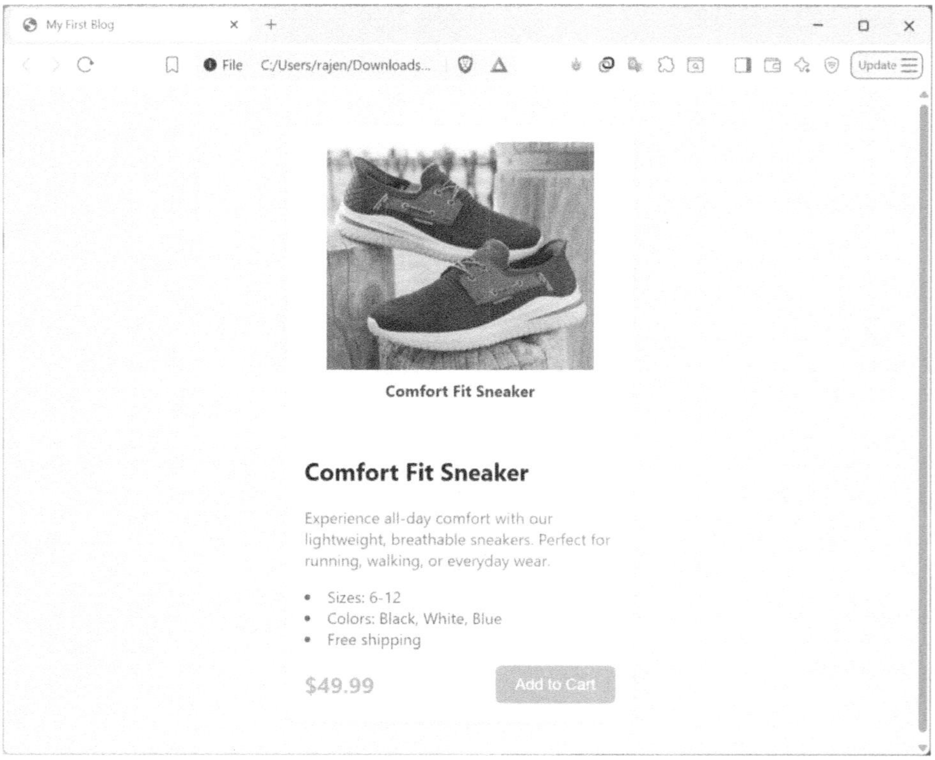

Real-World Scenario: Multiple Products

To simulate a real online shop, you can duplicate your card structure for multiple products inside a flex or grid layout. For example, Etsy and Amazon display responsive product grids that adapt to different screen sizes.

```
<div class= "product-list">
  <!-- Repeat .product-card structure for each product --></div>
```

And style with:

```
.product-list {
  display: flex;
  flex-wrap: wrap;
  gap: 32px;
  justify-content: center;
}
```

Hands-On Coding Summary

You've successfully designed a visually engaging, mobile-friendly product page using semantic HTML and advanced, responsive CSS techniques. Practicing these skills helps you build confidence for real-world projects, such as online shops and digital portfolios. Try customizing the product information, experimenting with different layouts, or adding interactivity with CSS effects.

Next, get ready to expand your design skills as we dive into 'Exploring Global Web Design Trends', where you'll discover how worldwide styles and modern techniques can inspire your own web projects and make your sites stand out.

15.5. Exploring Global Web Design Trends

The web is a vibrant, ever-evolving canvas shaped by cultures, innovations, and users from around the globe. As you level up your web development skills, understanding and applying global web design trends will help your websites feel modern, engaging, and accessible to a wide audience. Let's explore some exciting trends—like dark mode, micro-interactions, and card-based layouts—through hands-on coding exercises that you can immediately try in your own projects.

Embracing Dark Mode: A Modern Aesthetic

Dark mode has become a staple in web design, offering users a visually comfortable experience, especially in low-light environments. Implementing dark mode also shows an understanding of accessibility and user preferences.

Let's create a simple toggle for dark mode using HTML and CSS only. You can extend this later with JavaScript for advanced behavior, but here we'll focus on what you can do with pure CSS.

Start with a basic HTML structure for a personal portfolio header:

```
<!DOCTYPE html>
<html lang= "en">
<head>
  <meta charset= "UTF-8">
  <title>My Global Portfolio</title>
  <link rel= "stylesheet"href= "styles.css">
</head>
<body>
  <header>
    <h1>Maria Gomez - Web Developer</h1>
    <label class= "switch">
      <input type= "checkbox"id= "darkToggle">
      <span class= "slider"></span>
      <span class= "mode-label">Dark Mode</span>
    </label>
  </header>
</body>
</html>
```

Add CSS for the dark mode toggle and the dark theme styles:

```
body {
  background: #f6f6f6;
  color: #222;
  transition: background 0.4s, color 0.4s;
  font-family: Arial, sans-serif;
}

header {
  display: flex;
  justify-content: space-between;
  align-items: center;
  padding: 2rem;
  max-width: 800px;
  margin: 0 auto;
}

/* Toggle Switch Styles */.switch {
  display: flex;
  align-items: center;
  cursor: pointer;
}
.switch input {
  display: none;
}
.slider {
  width: 40px;
  height: 22px;
  background: #ccc;
  border-radius: 12px;
  position: relative;
  margin-right: 8px;
  transition: background 0.4s;
}
.slider::before {
  content: "";
  position: absolute;
  width: 18px;
  height: 18px;
  left: 2px;
  top: 2px;
  background: #fff;
  border-radius: 50%;
  transition: transform 0.4s;
}
input:checked + .slider {
  background: #222;
}
input:checked + .slider::before {
  transform: translateX(18px);
}
input:checked ~ .mode-label {
  color: #222;
}

/* Dark mode styles */input:checked ~ body,
input:checked ~ header {
  background: #222;
  color: #f6f6f6;
```

```
}
```

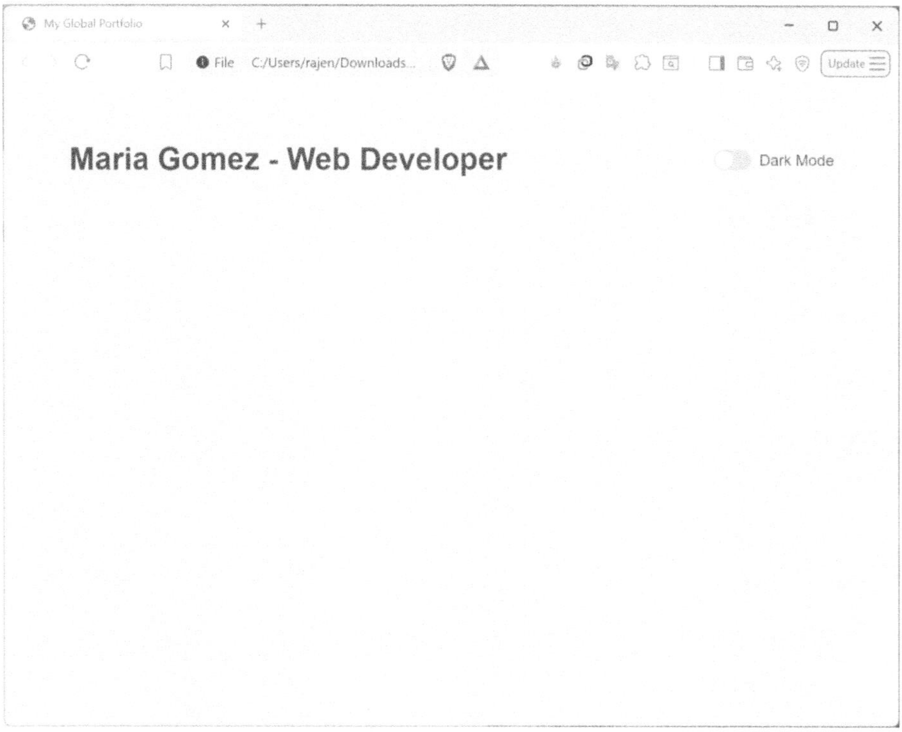

Practice Challenge: Try styling more page elements (like links or cards) to adapt to dark mode, and research how to use CSS variables for even smoother theme switching.

Card-Based Layouts: Flexible, Global Design

A card-based layout is a versatile trend seen in blogs, news sites, and ecommerce platforms worldwide. Cards help organize content visually, making it easier for users to scan and interact.

Let's build a responsive card grid for a simple blog homepage.

Add the HTML for a blog card grid:

```
<main>
  <section class= "card-grid">
    <article class= "card">
      <h2>Exploring Japan</h2>
      <p>Discover the beauty of Japanese gardens and temples in this
photo journey.</p>
    </article>
    <article class= "card">
      <h2>French Cuisine</h2>
      <p>Taste your way through Paris with these authentic French
recipes.</p>
```

```
    </article>
    <article class= "card">
      <h2>Tech in Brazil</h2>
      <p>How Brazilian startups are shaping the future of Latin
American innovation.</p>
    </article>
  </section>
</main>
```

Style the card grid with modern CSS:

```
.card-grid {
  display: grid;
  grid-template-columns: repeat(auto-fit, minmax(260px, 1fr));
  gap: 2rem;
  max-width: 900px;
  margin: 2rem auto;
  padding: 0 1rem;
}
.card {
  background: #fff;
  border-radius: 12px;
  box-shadow: 0 4px 12px rgba(0,0,0,0.09);
  padding: 1.5rem;
  transition: transform 0.2s, box-shadow 0.2s;
}
.card:hover {
  transform: translateY(-6px) scale(1.03);
  box-shadow: 0 8px 32px rgba(0,0,0,0.16);
}
.card h2 {
  margin-top: 0;
  color: #0057b7;
}
```

Hands-On Exercise: Customize your cards with images, buttons, or links. Try adjusting the grid for mobile devices by changing the `minmax` value or adding media queries.

You've just coded two of today's most popular global web design trends: dark mode and card-based layouts. These techniques help your sites look fresh and work well for users worldwide. Keep experimenting by applying these trends to your own projects and exploring other trends like micro-interactions or bold typography.

Next, get ready to take your site live. In the following section, we'll guide you step-by-step through publishing your website on the web, so your work can be seen by the world.

15.6. Publishing Your Site on the Web

Publishing your website is the moment your hard work comes to life for the world to see. Whether you've built a personal portfolio, a blog to share your thoughts, or a mini e-commerce site for a small business, putting your site online is a crucial milestone. Let's walk through the practical steps, from preparing your files to using beginner-friendly publishing tools, and see your site live on the web.

Preparing Your Website Files

Before you upload your site, double-check that your HTML and CSS files are ready:

All linked assets (images, stylesheets, scripts) are inside your project folder.
File names use lowercase letters and no spaces (e.g., `index.html`, `styles.css`).
You have a homepage named `index.html` (this is the default file web servers look for).

Suppose you've built a simple portfolio homepage. Your project folder might look like this:

```
portfolio-site/
│
├── index.html
├── styles.css
└── images/
    └── profile.jpg
```

Choosing a Free Hosting Platform

As a beginner, using a free, reliable hosting service is a smart way to share your site. Two popular choices are **GitHub Pages** and **Netlify**. Both support static sites and require no complex setup.

Let's publish your site using GitHub Pages, which is perfect for HTML and CSS projects.

Step-by-Step: Publishing with GitHub Pages

Create a GitHub Account

Sign up at github.com if you don't have an account.

Set Up Your Repository

1. Click the + icon and select **New repository**.

2. Name your repository (e.g., `portfolio-site`).

3. Make it **public** and click **Create repository**.

Upload Your Website Files

1. On your new repository page, click **Add file > Upload files**.

2. Drag and drop your site files (`index.html`, `styles.css`, and `images/` folder).

3. Click **Commit changes** to upload.

Enable GitHub Pages

1. Go to the **Settings** tab of your repository.

2. Scroll to **Pages** in the sidebar.

3. Under **Source**, select the `main` branch and click **Save**.

GitHub will provide a link like `https://yourusername.github.io/portfolio-site/` where your site is now live.

Real-World Scenario:

Imagine you've built a blog homepage:

```
<!-- index.html --><!DOCTYPE html>
<html lang= "en">
<head>
  <meta charset= "UTF-8">
  <title>Jane's Travel Blog</title>
  <link rel= "stylesheet"href= "styles.css">
</head>
<body>
  <header>
    <h1>Jane's Travel Blog</h1>
    <nav>
      <a href= "#about">About</a>
      <a href= "#stories">Stories</a>
      <a href= "#contact">Contact</a>
    </nav>
```

```
  </header>
  <main>
    <section id= "about">
      <h2>Welcome!</h2>
      <p>Explore my adventures across the globe.</p>
    </section>
    <section id= "stories">
      <h2>Latest Stories</h2>
      <ul>
        <li>Hiking in Peru</li>
        <li>Café hopping in Paris</li>
      </ul>
    </section>
  </main>
</body>
</html>
```

styles.css could include:

```
body {
  font-family: Arial, sans-serif;
  margin: 0;
  padding: 0;
}

header {
  background-color: #0b3d91;
  color: #fff;
  padding: 2rem;
  text-align: center;
}

nav a {
  color: #ffe082;
  margin: 0 1rem;
  text-decoration: none;
  font-weight: bold;
}

main {
  padding: 2rem;
}
```

After uploading, your blog is instantly available to anyone with the link.

Testing and Sharing Your Site

Once live, open your site's URL in different browsers and devices to check its appearance. Ask friends or family to visit and give feedback. If you spot any issues, update your files and re-upload them to GitHub.

Hands-On Summary

You've now taken your first website from a local project to a live site available worldwide. Publishing not only makes your work visible but also gives you valuable experience using professional tools and real-world workflows. Practice by updating your site or creating another mini project, like a single-page product showcase, and publish it using the same steps.

Next, you'll discover how to keep growing your web development skills, explore more advanced HTML and CSS concepts, and take on new challenges. Get ready to continue your web development journey and unlock even more creative possibilities!

15.7. Continuing Your Web Development Journey

As you gain confidence with the fundamentals of HTML and CSS, it's time to stretch your skills further by building real-world, responsive web projects. Applying what you've learned to practical scenarios deepens your understanding and prepares you for the ever-evolving landscape of web development. Let's dive into hands-on practice that will help you master advanced techniques, such as creating layouts that automatically adapt to any device and integrating modern design trends.

Building a Responsive Portfolio Section

A personal portfolio site is a classic beginner project and a valuable tool for showcasing your work. In this exercise, you'll create a sleek, responsive portfolio section that looks great on both desktop and mobile devices.

Start with the HTML structure:

Create a new file named `portfolio.html` and add the following markup inside the `<body>`:

```
<section class="portfolio">
  <h2>My Latest Projects</h2>
  <div class="projects">
    <article class="project-card">
      <img src="project1.jpg" alt="Screenshot of Project 1">
      <h3>Project One</h3>
      <p>A modern, responsive landing page built with HTML &
CSS.</p>
    </article>
    <article class="project-card">
      <img src="project2.jpg" alt="Screenshot of Project 2">
      <h3>Project Two</h3>
      <p>An interactive photo gallery designed for all screen
sizes.</p>
    </article>
  </div>
</section>
```

This layout uses semantic tags like `<section>`, `<article>`, and headings for clarity and accessibility.

Add responsive CSS styling:

Next, style the portfolio to ensure it adapts beautifully to different device widths. Open your CSS file and include:

```
.portfolio {
  max-width: 900px;
  margin: 40px auto;
  padding: 24px;
  background: #f9f9f9;
  border-radius: 12px;
  box-shadow: 0 2px 8px rgba(0,0,0,0.05);
}

.portfolio h2 {
  text-align: center;
  margin-bottom: 28px;
  color: #333;
  font-size: 2rem;
}

.projects {
  display: flex;
  flex-wrap: wrap;
  gap: 24px;
  justify-content: center;
}

.project-card {
  background: white;
  border-radius: 8px;
  box-shadow: 0 2px 4px rgba(0,0,0,0.04);
  padding: 16px;
  width: 340px;
  transition: transform 0.2s;
}

.project-card:hover {
  transform: translateY(-6px) scale(1.03);
}

.project-card img {
  width: 100%;
  border-radius: 5px;
  margin-bottom: 14px;
}

/* Responsive adjustments */
@media (max-width: 768px) {
  .projects {
    flex-direction: column;
    align-items: center;
  }
```

```
    .project-card {
      width: 100%;
      max-width: 400px;
    }
  }
```

Here, you're using **flexbox** for layout, **media queries** for responsiveness, and subtle **hover effects** to enhance interactivity.

Real-World Scenario: Creating a Mobile-Friendly Blog Article

Responsive web design is essential for modern blogs. Here's how to craft a blog article section that adapts smoothly to any screen:

HTML for the blog article:

```
<article class="blog-post">
  <header>
    <h1>Exploring CSS Grid: A Beginner's Guide</h1>
    <p class="meta">Posted on July 7, 2024 by Jane Doe</p>
  </header>
  <img src="css-grid-example.jpg" alt="CSS Grid example layout">
  <section class="content">
    <p>CSS Grid is a powerful tool for creating two-dimensional
layouts. Let's explore its basics and see how we can use it to build
complex designs with simple code.</p>
  </section>
</article>
```

Responsive CSS for the blog article:

```
.blog-post {
  max-width: 650px;
  margin: 32px auto;
  padding: 20px;
  background: #fff;
  border-radius: 10px;
  box-shadow: 0 1px 6px rgba(0,0,0,0.06);
}
.blog-post header h1 {
  font-size: 2.1rem;
  margin-bottom: 6px;
}
.blog-post .meta {
  color: #888;
  font-size: 0.9rem;
  margin-bottom: 12px;
}
.blog-post img {
  width: 100%;
  height: auto;
  border-radius: 6px;
  margin: 16px 0;
}
.blog-post .content p {
  font-size: 1.1rem;
```

```
    line-height: 1.7;
    color: #333;
  }
  @media (max-width: 480px) {
    .blog-post {
      padding: 12px;
    }
    .blog-post header h1 {
      font-size: 1.3rem;
    }
  }
```

By practicing these examples, you'll see how responsive design and modern CSS can make your sites look professional and user-friendly on any device.

You've just created two practical, real-world web components—a portfolio section and a blog article—that highlight how to use semantic HTML, flexbox, media queries, and stylish layouts. Try customizing these examples with your own content, images, and color schemes. Experiment further by adding more projects or articles, and tweak the responsiveness to fit your style.

GLOSSARY OF KEY TERMS

Accessibility

Designing websites and applications so that all users, including those with disabilities, can perceive, understand, navigate, and interact with them. Accessibility involves using semantic HTML, proper color contrast, keyboard navigation, and assistive technologies to make content usable by everyone.

Alt Attribute

An attribute used in the HTML `<img>` tag to provide alternative text for an image. The alt attribute describes the content or function of the image, which helps screen readers convey information to users who cannot see the image.

Anchor Tag (`<a>`)

An HTML element used to create hyperlinks, allowing users to navigate from one page or resource to another. The anchor tag uses the `href` attribute to specify the destination URL and can also link to sections within the same page.

Attribute

A modifier added to HTML elements that provides additional information or functionality. For example, the `src` attribute in an `<img>` tag specifies the image file's location, while the `class` attribute assigns a CSS class for styling.

Block-Level Element

An HTML element that occupies the full available width of its parent container and starts on a new line. Examples include `<div>`, `<h1>`, `<p>`, and `<section>`. Block-level elements are used to structure the main layout of a webpage.

Box Model

A fundamental concept in CSS that describes how elements are rendered on the page. Each element is represented as a box with content, padding, borders, and margins. Understanding the box model helps control spacing and layout in web design.

Breakpoint

A specific screen width defined in CSS where the layout or design of a webpage changes to ensure optimal viewing on different devices. Breakpoints are a key part of responsive design, allowing websites to adapt to mobile, tablet, and desktop screens.

Class Selector

A CSS selector that targets HTML elements with a specific `class` attribute. Classes allow you to apply styles to multiple elements at once and are written in CSS with a period before the class name, such as `.button`.

Color Contrast

The difference in brightness and color between text and its background. Good color contrast is essential for readability and accessibility, ensuring that content is easy to see for all users, including those with visual impairments.

Comment

A line or block of text in HTML or CSS that is ignored by browsers and not displayed on the page. Comments are used to explain code or leave notes for other developers. In HTML, comments are written with `<!-- -->`; in CSS, with `/ /`.

Container

An HTML element, often a `<div>`, used to group and organize other elements within a webpage. Containers help structure content and are commonly used for layout purposes, such as centering content or creating columns.

CSS (Cascading Style Sheets)

A language used to control the appearance and layout of HTML elements on a webpage. CSS allows you to set colors, fonts, spacing, positioning, and more, separating content from design for easier maintenance and flexibility.

CSS Grid

A powerful layout system in CSS that enables you to create complex, responsive grid-based designs. CSS Grid allows you to define rows and columns, place items precisely, and build both simple and advanced page structures.

Declaration Block

A section of CSS code enclosed in curly braces `{}` that contains one or more property-value pairs. Each property defines a specific style, such as color or font-size, and each value sets the desired appearance for that property.

DOCTYPE

A declaration at the very top of an HTML document that tells the browser which version of HTML is being used. The most common is `<!DOCTYPE html>`, which signals that the page uses HTML5. This ensures consistent rendering across browsers.

Element

A basic building block of HTML, represented by tags like `<p>`, `<h2>`, or `<img>`. Elements define the structure and content of a webpage, and can have attributes to specify additional information or behavior.

Em Unit

A relative unit in CSS used for sizing fonts, margins, and other properties. One em equals the current font size of the element. Using em units makes designs more flexible and scalable across different devices.

External Style Sheet

A separate CSS file linked to an HTML document using the `<link>` tag. External style sheets help organize code, enable consistent styling across multiple pages, and make maintaining large websites easier.

Flexbox

A CSS layout model that makes it easier to align, distribute, and order items within a container, even when their size is unknown. Flexbox is ideal for creating flexible, responsive layouts that adjust smoothly to different screen sizes.

Font Family

A CSS property that specifies the typeface used for text on a webpage. You can list multiple fonts as fallbacks, ensuring your text displays correctly even if the preferred font isn't available on a user's device.

Form

An HTML element (`<form>`) that collects user input through fields like text boxes, radio buttons, checkboxes, and submit buttons. Forms are used for tasks like sign-ups, searches, and surveys on websites.

Header

An HTML element (`<header>`) or section at the top of a webpage or section. It typically contains the site logo, navigation links, and introductory content, helping users understand the structure and purpose of the page.

Hex Code

A six-digit code used in CSS to define colors, preceded by a `#` symbol (for example, `#ff5733`). Each pair of digits represents the red, green, and blue (RGB) values, allowing precise color selection for backgrounds, text, and borders.

Hierarchy

The arrangement and organization of elements on a webpage, showing their order of importance. Hierarchy is established through headings, font size, color, and spacing, guiding users through the content in a logical way.

HTML (HyperText Markup Language)

The standard language for creating webpages. HTML uses tags to structure content, such as headings, paragraphs, images, and links, forming the backbone of every website on the internet.

Hyperlink

A clickable element, usually created with the `<a>` tag, that takes users to another webpage or location. Hyperlinks are essential for website navigation, connecting related information across the web.

ID Selector

A CSS selector that targets a single HTML element with a specific `id` attribute. IDs are unique within a page and are referenced in CSS with a `#` symbol, like `#main-header`. They are useful for applying styles or scripting to one element only.

Inline Element

An HTML element that flows alongside other content and only takes up as much width as necessary. Examples include `<span>`, `<a>`, and `<strong>`. Inline elements are typically used to style parts of text within a block-level element.

Internal Style Sheet

A block of CSS code placed within the `<style>` tag in the `<head>` section of an HTML document. Internal style sheets apply styles only to that page, making them useful for small projects or single-page websites.

JavaScript

A programming language commonly used in web development to add interactivity, dynamic content, and logic to webpages. JavaScript works alongside HTML and CSS to create responsive menus, animations, and user-driven features.

Layout

The arrangement of elements on a webpage, including navigation, content sections, sidebars, and footers. Good layout improves usability and aesthetics, making it easier for visitors to find information and interact with the site.

Margin

A CSS property that creates space outside an element's border, separating it from other elements. Margins help control the distance between elements and prevent content from appearing crowded.

Media Query

A CSS technique used to apply styles based on device characteristics, such as screen width or resolution. Media queries enable responsive design by allowing different styles for mobile, tablet, and desktop devices.

Navigation Bar

A section of a webpage, usually at the top or side, containing links to important pages or sections. Navigation bars help users move around the site and find content easily, often using lists or menus.

Padding

A CSS property that adds space inside an element, between its content and border. Padding helps improve readability and visual separation within buttons, boxes, or other components.

Position Property

A CSS property that determines how an element is positioned on the page. Values include static, relative, absolute, fixed, and sticky, each providing different ways to control the layout and stacking of elements.

Pseudo-Class

A keyword added to a CSS selector that applies styles to elements in a specific state, such as when hovered over or visited. Common pseudo-classes include `:hover`, `:active`, and `:focus`, enabling interactive design effects.

Responsive Design

An approach to web design that ensures websites look and function well on all devices and screen sizes. Responsive design uses flexible layouts, images, and media queries to adapt the site's appearance for desktops, tablets, and smartphones.

Selector

A pattern in CSS used to target and apply styles to specific HTML elements. Selectors can reference tags, classes, IDs, attributes, and more, allowing precise control over which elements are styled.

Semantic HTML

Using HTML elements that clearly describe their meaning and role in a webpage, such as `<article>`, `<nav>`, `<footer>`, and `<header>`. Semantic HTML improves accessibility and makes code easier to understand and maintain.

Span

An inline HTML element (`<span>`) used to group and style small sections of text or other inline elements. Spans are useful for applying CSS styles to part of a paragraph or sentence without affecting the entire block.

Style Attribute

An HTML attribute that allows you to apply CSS styles directly to a single element. While quick and convenient for small changes, using the style attribute is less efficient than external or internal style sheets for larger projects.

Tag

A keyword surrounded by angle brackets, such as `<p>` or `<img>`, used in HTML to create elements. Tags often come in pairs, with an opening tag and a closing tag, and are the building blocks of webpages.

Typography

The art and technique of arranging type on a webpage, including font selection, size, spacing, and alignment. Good typography improves readability, accessibility, and the overall look and feel of a website.

Unordered List (`<ul>`)

An HTML element that displays a list of items with bullet points. Each item is placed within a `<li>` (list item) tag, making unordered lists useful for navigation menus and grouping related content.

User Experience (UX)

The overall feeling and ease of use that a visitor has when interacting with a website or app. UX involves design, usability, accessibility, and performance, aiming to make digital experiences enjoyable and effective.

User Interface (UI)

The visual and interactive elements of a website, such as buttons, forms, menus, and icons. UI design focuses on making websites attractive and easy to navigate, contributing to a positive user experience.

Viewport

The visible area of a web page on a user's device. Responsive design uses the viewport to adapt content and layout, ensuring websites look good on different screens. The viewport is controlled in HTML with the `<meta name="viewport">` tag.

Web Hosting

A service that stores website files and makes them accessible on the internet. Web hosting providers offer servers, bandwidth, and management tools so users around the world can visit your site.

Web Server

A computer system or software that delivers web pages to users over the internet. When someone visits your website, the web server processes the request and sends the appropriate files to the browser for display.

Whitespace

The empty space between elements, text, or images on a webpage. Whitespace improves readability, creates balance, and helps guide users through the content by preventing clutter.

Wireframe

A basic visual guide used in the early planning stages of web design to outline the structure and layout of a webpage. Wireframes focus on placement of elements without detailed design, helping developers and clients agree on the site's organization.

Z-Index

A CSS property that controls the stacking order of overlapping elements. Elements with a higher z-index appear in front of those with a lower value, which is especially useful in complex layouts with modals, dropdowns, or layered content.

ANNEXURE A: ESSENTIAL HTML REFERENCE TABLE

A solid understanding of HTML tags is crucial for every aspiring web developer. This annexure provides a concise, beginner-friendly reference table for the most commonly used HTML elements, including their purpose, basic syntax, and example usage. Keep this table handy as you build your projects—it's your quick guide to structuring web content with confidence.

Using the Reference Table

Whenever you're unsure which tag to use or need a refresher on syntax, consult this table. It covers the foundational tags you'll use in nearly every project, from creating headings and paragraphs to embedding images and links.

Core HTML Elements

TAG	*PURPOSE*	*BASIC SYNTAX / EXAMPLE*
<!DOCTYPE>	Declares the document type	<!DOCTYPE html>
<html>	Root element of a web page	<html> ... </html>
<head>	Container for metadata	<head> ... </head>
<title>	Sets the page title (in browser tab)	<title>My Page</title>
<meta>	Defines metadata (charset, viewport, etc.)	<meta charset="UTF-8">
<link>	Links external resources (CSS, favicon)	<link rel="stylesheet" href="style.css">
<style>	Internal CSS styling	<style>body {color:red;}</style>
<script>	Embeds or links JavaScript	<script src="app.js"></script>
<body>	Main content of the page	<body> ... </body>
<h1>–<h6>	Headings (largest to smallest)	<h1>Main Heading</h1>
<p>	Paragraph of text	<p>Hello, world!</p>
<a>	Hyperlink to another page/site	<a href="https://example.com">Visit</a>
<img>	Embeds an image	<img src="cat.jpg" alt="A cute cat">
<figure>	Groups media with a caption	<figure><img src="pic.jpg"><figcaption>Caption</figcaption></figure>

<figcaption>	Caption for <figure>	<figcaption>Image description</figcaption>
<ul>	Unordered (bulleted) list	<ul><li>Item</li></ul>
<ol>	Ordered (numbered) list	<ol><li>First</li></ol>
<li>	List item (inside <ul> or <ol>)	<li>List item</li>
<dl>	Description list	<dl><dt>HTML</dt><dd>Markup Language</dd></dl>
<dt>	Term in description list	<dt>HTML</dt>
<dd>	Description in description list	<dd>Markup Language</dd>
<div>	Block-level container	<div>Section</div>
<span>	Inline container for text	<span>Highlight</span>
* *	Line break	Hello World
<hr>	Horizontal line	<hr>
<strong>	Strong importance (bold)	<strong>Important</strong>
<em>	Emphasis (italic)	<em>Note</em>
<b>	Bold text (styling only)	<b>Bold</b>
<i>	Italic text (styling only)	<i>Italic</i>
<u>	Underlined text	<u>Underline</u>
<small>	Smaller text	<small>Fine print</small>
<mark>	Highlighted text	<mark>Highlight</mark>
<code>	Inline code snippet	<code>console.log("Hi")</code>
<pre>	Preformatted text / code block	<pre>line 1\nline 2</pre>
<blockquote>	Quotation block	<blockquote>Quote here</blockquote>
<cite>	Citation	<cite>Author</cite>
<abbr>	Abbreviation	<abbr title="HyperText Markup Language">HTML</abbr>
<address>	Contact information	<address>Email: me@example.com</address>
<time>	Defines a time/date	<time datetime="2025-08-16">Aug 16, 2025</time>
<form>	Form container	<form> ... </form>
<input>	User input field	<input type="text" placeholder="Name">
<textarea>	Multi-line text input	<textarea rows="4"></textarea>
<button>	Clickable button	<button>Click Me</button>

<select>	Dropdown menu	<select><option>A</option><option>B</option></select>
<option>	Dropdown option	<option>Choice 1</option>
<label>	Label for form element	<label for="name">Name:</label>
<fieldset>	Group related form controls	<fieldset><legend>Info</legend>...</fieldset>
<legend>	Caption for <fieldset>	<legend>Personal Info</legend>
<table>	Table container	<table> ... </table>
<tr>	Table row	<tr> ... </tr>
<td>	Table data cell	<td>Cell</td>
<th>	Table header cell	<th>Header</th>
<thead>	Table head section	<thead> ... </thead>
<tbody>	Table body section	<tbody> ... </tbody>
<tfoot>	Table footer section	<tfoot> ... </tfoot>
<colgroup>	Groups table columns	<colgroup><col span="2"></colgroup>
<col>	Column formatting in tables	<col style="background:yellow">
<audio>	Embeds audio file	<audio controls src="song.mp3"></audio>
<video>	Embeds video file	<video controls src="movie.mp4"></video>
<source>	Defines media resource (audio/video)	<source src="video.mp4" type="video/mp4">
<track>	Subtitles for video/audio	<track src="subs.vtt" kind="subtitles">
<iframe>	Inline frame (embed another page)	<iframe src="https://example.com"></iframe>
<canvas>	Drawing graphics via JS	<canvas id="myCanvas"></canvas>
<svg>	Inline vector graphics	<svg><circle cx="50" cy="50" r="40"/></svg>
<section>	Thematic section of content	<section>...</section>
<article>	Independent content (blog/news post)	<article>...</article>
<aside>	Sidebar or complementary content	<aside>...</aside>
<header>	Page or section header	<header>...</header>
<footer>	Page or section footer	<footer>...</footer>
<nav>	Navigation links	<nav><a href="#">Home</a></nav>
<main>	Main page content	<main>...</main>

<details>	Expandable disclosure widget	<details><summary>Info</summary><p>Hidden</p></details>
<summary>	Summary/label for <details>	<summary>Click to expand</summary>
<progress>	Progress indicator	<progress value="50" max="100"></progress>
<meter>	Scalar measurement (like a gauge)	<meter value="0.7">70%</meter>
<template>	Template content (hidden until used in JS)	<template><p>Clone me</p></template>
<slot>	Placeholder in web components	<slot></slot>
<dialog>	Dialog / modal window	<dialog open>Hi!</dialog>

Example Application: Structuring a Simple Webpage

Here's how some of these tags come together to create a basic HTML page:

```
<!DOCTYPE html>
<html>
  <head>
    <title>My First Webpage</title>
  </head>
  <body>
    <h1>Welcome to My Portfolio</h1>
    <p>Hello! I am learning web development.</p>
    <ul>
      <li>About Me</li>
      <li>Projects</li>
      <li>Contact</li>
    </ul>
    <img src= "profile.jpg"alt= "My Profile Photo">
    <a href= "mailto:me@example.com">Email Me</a>
  </body>
</html>
```

Practical Coding Exercise

Try creating a simple "About Me" page using at least five different HTML tags from this table. Add a heading, a paragraph, a list of hobbies, a profile picture, and a link to your favorite website.

This reference table is designed to speed up your workflow and help you recall essential HTML tags as you build new web pages. Refer back to it as your projects grow in complexity.

ANNEXURE B: CSS PROPERTY CHEAT SHEET

Styling your HTML content is where your creativity shines. CSS (Cascading Style Sheets) allows you to control the appearance of your website, from colors and fonts to layout and spacing. This cheat sheet lists the most useful CSS properties for beginners, with examples to help you quickly style your web pages.

How to Use This Cheat Sheet

When you want to style an element, look up the relevant property in this sheet. Each property includes its purpose, a sample usage, and a quick tip for beginners.

Common CSS Properties

Property	***Purpose***	***Example Usage***	***Quick Tip***
color	Sets text color	color: blue;	Use color names, hex, rgb(), or hsl()
background	Sets background color/image	background: #f0f0f0;	Combine with background-image
background-color	Background color only	background-color: lightblue;	Useful for highlights
background-image	Sets background image	background-image: url("bg.jpg");	Use background-size: cover for full coverage
background-size	Controls background scaling	background-size: cover;	Use contain or cover for responsiveness
background-repeat	Controls background repetition	background-repeat: no-repeat;	Avoid tiling when not needed
background-position	Positions background image	background-position: center;	Combine with cover for hero images
font-size	Changes text size	font-size: 20px;	Use em/rem for scalable text
font-family	Sets the font	font-family: Arial, sans-serif;	Always provide a fallback
font-weight	Boldness of text	font-weight: bold;	Values: 100–900
font-style	Italics or normal	font-style: italic;	Common values: normal, italic
text-align	Horizontal text alignment	text-align: center;	left, right, justify too
text-decoration	Adds underline/line-through	text-decoration: underline;	Also supports none, overline
text-transform	Changes case	text-transform: uppercase;	Options: uppercase, lowercase, capitalize
line-height	Space between lines	line-height: 1.5;	Improves readability
letter-spacing	Adjusts space between letters	letter-spacing: 2px;	Useful for headings

word-spacing	Adjusts space between words	word-spacing: 5px;	Use carefully
margin	Space outside an element	margin: 10px 20px;	Shorthand for all sides
margin-top etc.	Individual margins	margin-top: 20px;	Use when only one side needs spacing
padding	Space inside an element	padding: 10px;	Increases clickable area for buttons
padding-top etc.	Individual paddings	padding-left: 15px;	Useful for fine-tuning
border	Border around elements	border: 1px solid #333;	Shorthand for width, style, color
border-style	Defines border style	border-style: dashed;	Options: solid, dotted, double
border-width	Border thickness	border-width: 3px;	Can set per side
border-color	Border color	border-color: red;	Can set multiple values
border-radius	Rounds element corners	border-radius: 8px;	Use 50% for circles
outline	Outer highlight	outline: 2px solid blue;	Doesn't affect layout (unlike border)
width / height	Sets size of an element	width: 100px; height: 50px;	% for responsive design
max-width	Restrains element width	max-width: 1200px;	Essential for responsive layouts
min-height	Sets minimum height	min-height: 300px;	Prevents shrinking
display	Layout behavior	display: flex;	Options: block, inline, grid
visibility	Hides element but keeps space	visibility: hidden;	Use display:none to remove completely
opacity	Transparency of element	opacity: 0.5;	Use with hover effects
flex	Flexible sizing in flex layouts	flex: 1;	Use inside display: flex
flex-direction	Direction of flex items	flex-direction: column;	Row (default) or column
flex-wrap	Wraps flex items	flex-wrap: wrap;	Helps with responsiveness
justify-content	Aligns items horizontally (flex)	justify-content: space-between;	Other values: center, flex-end
align-items	Aligns items vertically (flex)	align-items: center;	Use with flexbox
align-content	Controls multiple flex lines	align-content: space-around;	Works when wrapping
grid-template	Defines grid rows and columns	grid-template: 1fr 1fr / 200px auto;	Use with display: grid
grid-gap / gap	Space between grid/flex items	gap: 10px;	Replaces grid-gap, row-gap
position	Controls element positioning	position: absolute;	Options: relative, fixed, sticky
top, left, etc.	Positioning offsets	top: 10px; left: 20px;	Use with position

float	Floats element left/right	float: right;	Old method (use flex/grid instead)
clear	Prevents floating overlap	clear: both;	Often used after floats
overflow	Handles content overflow	overflow: auto;	Values: hidden, scroll
z-index	Layering of elements	z-index: 10;	Higher numbers on top
cursor	Mouse cursor style	cursor: pointer;	Useful for buttons/links
box-shadow	Adds shadow around elements	box-shadow: 2px 2px 5px #888;	Subtle shadows add depth
text-shadow	Adds shadow to text	text-shadow: 1px 1px 3px gray;	Good for headers
transform	Transforms (rotate, scale, skew)	transform: rotate(45deg);	Combine multiple transforms
transition	Smoothly animate property changes	transition: all 0.3s ease;	Great for hover effects
animation	Defines animation shorthand	animation: bounce 2s infinite;	Combine with @keyframes
clip-path	Defines a clipping region	clip-path: circle(50%);	Great for creative shapes
object-fit	How media (img/video) fits container	object-fit: cover;	Perfect for responsive images
object-position	Position of replaced element content	object-position: center top;	Use with object-fit
filter	Applies graphical effects	filter: blur(5px) brightness(0.8);	CSS image effects without Photoshop
resize	Allows user to resize element	resize: both;	Works with overflow
scroll-behavior	Smooth scrolling	scroll-behavior: smooth;	Great for single-page sites

Example Application: Styling a Button

Let's style a button using several properties from the cheat sheet:

```
.button {
  background: #007bff;
  color: #fff;
  padding: 12px 24px;
  border: none;
  border-radius: 6px;
  font-size: 16px;
  cursor: pointer;
  box-shadow: 1px 1px 4px #bbb;
  transition: background 0.2s;
}
.button:hover {
  background: #0056b3;
}
```

This creates a modern, interactive button perfect for landing pages or e-commerce sites.

Practical Coding Exercise

Pick an HTML element (such as a `<div>`, `<h1>`, or `<button>`) and apply at least five different CSS properties from this cheat sheet. Experiment with colors, spacing, and fonts until you achieve a design you like.

Coding Summary

Use this CSS cheat sheet as your styling companion. As you become more comfortable, you'll find yourself memorizing these properties and combining them to create beautiful, responsive web pages.

ANNEXURE C: BEGINNER PROJECT TEMPLATES

Jumpstarting your web development journey is easier when you have ready-to-use templates. This annexure provides practical starter templates for three classic projects: a personal portfolio, a simple blog, and a mini e-commerce landing page. These templates align with the book's hands-on approach and can be customized as you learn new skills.

Getting Started with Templates

Each template includes essential HTML structure and basic CSS, giving you a foundation to build on. Copy and adapt these templates as you progress, adding your own content and styles.

1. Personal Portfolio Homepage

Showcase your skills and personality with a clean portfolio homepage. This template includes sections for your introduction, featured projects, and contact details. You can also find this code in our book github repository:

https://github.com/JambaAcademy/HTML_CSS

```
<script type="text/javascript">
        var gk_isXlsx = false;
        var gk_xlsxFileLookup = {};
        var gk_fileData = {};
        function filledCell(cell) {
          return cell !== '' && cell != null;
        }
        function loadFileData(filename) {
        if (gk_isXlsx && gk_xlsxFileLookup[filename]) {
            try {
                var workbook = XLSX.read(gk_fileData[filename], { type: 'base64'
});
                var firstSheetName = workbook.SheetNames[0];
                var worksheet = workbook.Sheets[firstSheetName];
                // Convert sheet to JSON to filter blank rows
                var jsonData = XLSX.utils.sheet_to_json(worksheet, { header: 1,
blankrows: false, defval: '' });
                // Filter out blank rows (rows where all cells are empty, null,
or undefined)
                var filteredData = jsonData.filter(row => row.some(filledCell));
                // Heuristic to find the header row by ignoring rows with fewer
filled cells than the next row
                var headerRowIndex = filteredData.findIndex((row, index) =>
                  row.filter(filledCell).length >= filteredData[index +
1]?.filter(filledCell).length
                );
                // Fallback
                if (headerRowIndex === -1 || headerRowIndex > 25) {
                  headerRowIndex = 0;
                }
                // Convert filtered JSON back to CSV
```

```
                var csv =
XLSX.utils.aoa_to_sheet(filteredData.slice(headerRowIndex)); // Create a new
sheet from filtered array of arrays
                csv = XLSX.utils.sheet_to_csv(csv, { header: 1 });
                return csv;
            } catch (e) {
                console.error(e);
                return "";
            }
        }
        return gk_fileData[filename] || "";
        }
        </script><!DOCTYPE html>
<html lang="en">
<head>
    <meta charset="UTF-8">
    <meta name="viewport" content="width=device-width, initial-scale=1.0">
    <title>Personal Portfolio</title>
    <style>
        * {
            margin: 0;
            padding: 0;
            box-sizing: border-box;
            font-family: 'Arial', sans-serif;
        }
        body {
            background-color: #f5f5f5;
            color: #333;
            line-height: 1.6;
        }
        header {
            background: linear-gradient(135deg, #2c3e50, #3498db);
            color: white;
            text-align: center;
            padding: 4rem 2rem;
            position: relative;
        }
        header img {
            width: 150px;
            height: 150px;
            border-radius: 50%;
            border: 4px solid white;
            margin-bottom: 1rem;
            object-fit: cover;
        }
        header h1 {
            font-size: 2.5rem;
            margin-bottom: 0.5rem;
        }
        header p {
            font-size: 1.2rem;
            opacity: 0.9;
        }
        nav {
            background-color: #fff;
            box-shadow: 0 2px 5px rgba(0, 0, 0, 0.1);
            position: sticky;
            top: 0;
            z-index: 100;
        }
        nav ul {
            display: flex;
            justify-content: center;
            list-style: none;
            padding: 1rem 0;
```

```
}
nav ul li {
    margin: 0 1.5rem;
}
nav ul li a {
    text-decoration: none;
    color: #2c3e50;
    font-weight: bold;
    font-size: 1.1rem;
    transition: color 0.3s;
}
nav ul li a:hover {
    color: #3498db;
}
section {
    max-width: 1000px;
    margin: 3rem auto;
    padding: 0 2rem;
}
section h2 {
    font-size: 2rem;
    color: #2c3e50;
    margin-bottom: 1.5rem;
    text-align: center;
}
.card {
    background-color: white;
    padding: 2rem;
    border-radius: 10px;
    box-shadow: 0 4px 10px rgba(0, 0, 0, 0.1);
    margin-bottom: 2rem;
}
.card h3 {
    font-size: 1.5rem;
    color: #3498db;
    margin-bottom: 1rem;
}
.card p {
    margin-bottom: 0.8rem;
}
.timeline {
    position: relative;
    padding-left: 2rem;
}
.timeline::before {
    content: '';
    position: absolute;
    left: 0.5rem;
    top: 0;
    bottom: 0;
    width: 4px;
    background-color: #3498db;
}
.timeline-item {
    margin-bottom: 2rem;
    position: relative;
}
.timeline-item::before {
    content: '';
    position: absolute;
    left: -1.8rem;
    top: 0.5rem;
    width: 12px;
    height: 12px;
    background-color: white;
```

```
        border: 2px solid #3498db;
        border-radius: 50%;
    }
    .skills-grid {
        display: grid;
        grid-template-columns: repeat(auto-fit, minmax(150px, 1fr));
        gap: 1rem;
    }
    .skill-item {
        text-align: center;
        padding: 1rem;
        background-color: #f0f4f8;
        border-radius: 8px;
        transition: transform 0.3s;
    }
    .skill-item:hover {
        transform: translateY(-5px);
    }
    .projects-grid {
        display: grid;
        grid-template-columns: repeat(auto-fit, minmax(250px, 1fr));
        gap: 1.5rem;
    }
    .project-item img {
        width: 100%;
        height: 150px;
        object-fit: cover;
        border-radius: 8px;
        margin-bottom: 1rem;
    }
    .contact-form {
        display: flex;
        flex-direction: column;
        gap: 1rem;
        max-width: 600px;
        margin: 0 auto;
    }
    .contact-form input,
    .contact-form textarea {
        padding: 0.8rem;
        border: 1px solid #ddd;
        border-radius: 5px;
        font-size: 1rem;
    }
    .contact-form textarea {
        resize: vertical;
        min-height: 100px;
    }
    .contact-form button {
        background-color: #3498db;
        color: white;
        padding: 0.8rem;
        border: none;
        border-radius: 5px;
        cursor: pointer;
        font-size: 1rem;
        transition: background-color 0.3s;
    }
    .contact-form button:hover {
        background-color: #2c3e50;
    }
    footer {
        background-color: #2c3e50;
        color: white;
        text-align: center;
```

```
            padding: 2rem;
            margin-top: 3rem;
        }
        footer a {
            color: #3498db;
            text-decoration: none;
        }
        footer a:hover {
            text-decoration: underline;
        }
        @media (max-width: 768px) {
            header h1 {
                font-size: 2rem;
            }
            nav ul {
                flex-direction: column;
                gap: 0.5rem;
            }
            section {
                padding: 0 1rem;
            }
        }
    </style>
</head>
<body>
    <header>
        <img src="images/profile.jpg" alt="Profile Picture">
        <h1>Jane Doe</h1>
        <p>Web Developer | Designer | Creative Problem Solver</p>
    </header>
    <nav>
        <ul>
            <li><a href="#about">About</a></li>
            <li><a href="#experience">Experience</a></li>
            <li><a href="#education">Education</a></li>
            <li><a href="#projects">Projects</a></li>
            <li><a href="#skills">Skills</a></li>
            <li><a href="#research">Research</a></li>
            <li><a href="#contact">Contact</a></li>
        </ul>
    </nav>
    <section id="about">
        <h2>About Me</h2>
        <div class="card">
            <p>I'm Jane Doe, a passionate web developer with a knack for
creating user-friendly, responsive websites. With a background in design and
technology, I specialize in crafting clean, modern interfaces using HTML and
CSS. My goal is to build digital experiences that are both functional and
visually stunning, helping clients bring their visions to life. When I'm not
coding, you can find me exploring new design trends or mentoring aspiring
developers.</p>
        </div>
    </section>
    <section id="experience">
        <h2>Work Experience</h2>
        <div class="card timeline">
            <div class="timeline-item">
                <h3>Web Developer - TechTrend Innovations</h3>
                <p>June 2023 - Present</p>
                <p>Designed and developed responsive websites for clients across
industries, focusing on clean code and modern design principles. Collaborated
with UX designers to optimize user experiences.</p>
            </div>
            <div class="timeline-item">
                <h3>Junior Developer - CreativeWeb Solutions</h3>
```

```
                <p>Jan 2021 - May 2023</p>
                <p>Built and maintained static websites using HTML and CSS,
improving site performance by 20%. Assisted in implementing responsive layouts
for mobile-first designs.</p>
            </div>
        </div>
    </section>
    <section id="education">
        <h2>Education</h2>
        <div class="card timeline">
            <div class="timeline-item">
                <h3>B.S. in Computer Science - University of Tech</h3>
                <p>2017 - 2021</p>
                <p>Graduated with honors, specializing in web development and
user interface design. Completed a capstone project on responsive web
frameworks.</p>
            </div>
            <div class="timeline-item">
                <h3>Web Development Bootcamp - CodeAcademy</h3>
                <p>2020</p>
                <p>Intensive 12-week program focused on HTML, CSS, and
responsive design techniques.</p>
            </div>
        </div>
    </section>
    <section id="projects">
        <h2>Projects</h2>
        <div class="projects-grid">
            <div class="card project-item">
                <img src="images/portfolio-project.jpg" alt="Portfolio Website">
                <h3>Personal Portfolio</h3>
                <p>A sleek, responsive portfolio showcasing my skills and
projects, built with HTML and CSS.</p>
            </div>
            <div class="card project-item">
                <img src="images/blog-project.jpg" alt="Blog Website">
                <h3>Tech Blog</h3>
                <p>A modern blog platform with custom CSS styling and responsive
layouts for tech enthusiasts.</p>
            </div>
            <div class="card project-item">
                <img src="images/ecommerce-project.jpg" alt="E-commerce
Website">
                <h3>E-commerce Store</h3>
                <p>A product page with a clean, user-friendly design, optimized
for mobile and desktop.</p>
            </div>
        </div>
    </section>
    <section id="skills">
        <h2>Skills</h2>
        <div class="skills-grid">
            <div class="skill-item">HTML5</div>
            <div class="skill-item">CSS3</div>
            <div class="skill-item">Responsive Design</div>
            <div class="skill-item">Flexbox</div>
            <div class="skill-item">CSS Grid</div>
            <div class="skill-item">UI/UX Design</div>
        </div>
    </section>
    <section id="research">
        <h2>Research Papers</h2>
        <div class="card">
            <h3>"Responsive Design Trends in Modern Web Development"</h3>
            <p>Published in WebTech Journal, 2024</p>
```

```
            <p>Explored the evolution of responsive design techniques and their impact on user engagement.</p>
        </div>
    </section>
    <section id="contact">
        <h2>Contact Me</h2>
        <div class="card">
            <form class="contact-form">
                <input type="text" placeholder="Your Name" required>
                <input type="email" placeholder="Your Email" required>
                <textarea placeholder="Your Message" required></textarea>
                <button type="submit">Send Message</button>
            </form>
        </div>
    </section>
    <footer>
        <p>&copy; 2025 Jane Doe. All rights reserved. | <a href="mailto:jane.doe@example.com">jane.doe@example.com</a> | <a href="https://github.com/JambaAcademy/HTML_CSS">GitHub</a></p>
    </footer>
</body>
</html>
```

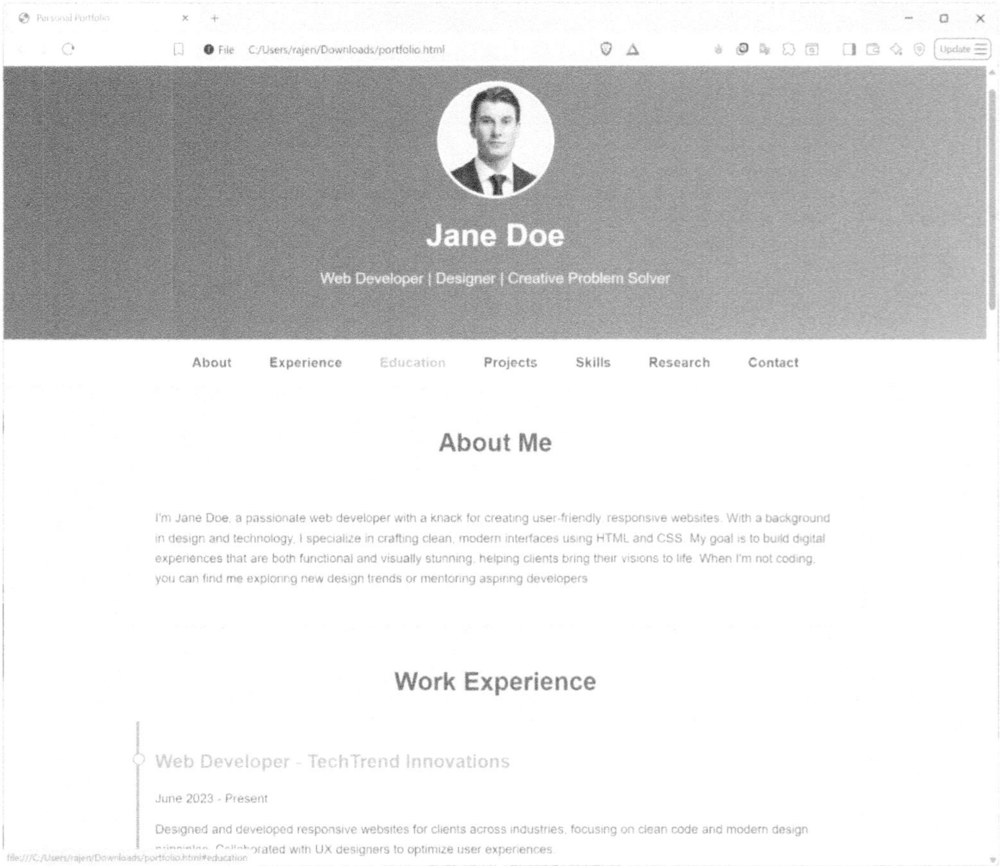

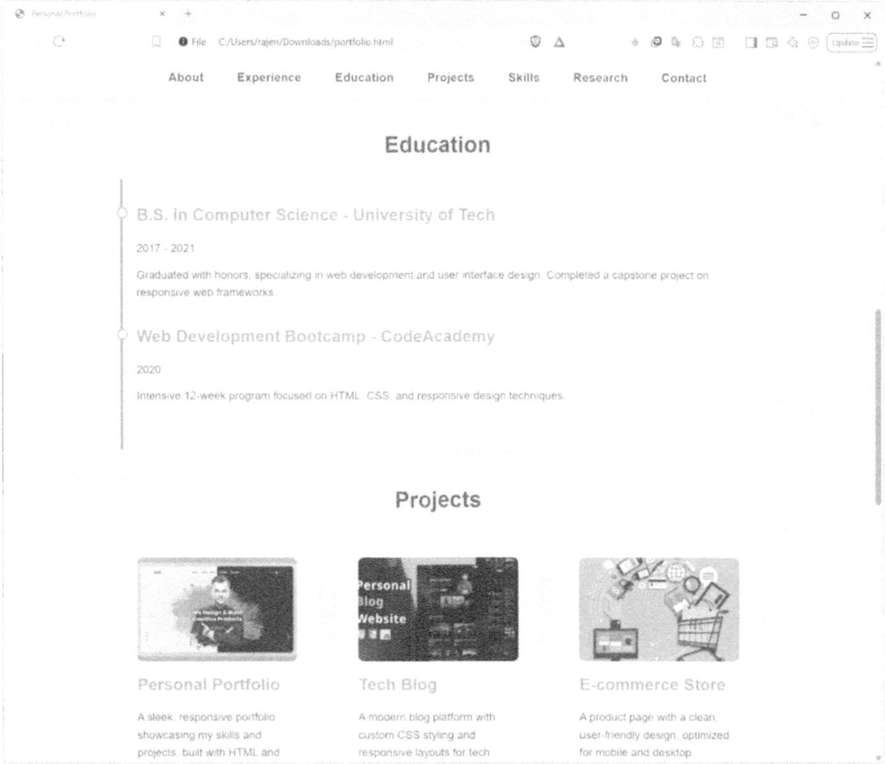
About
Experience
Education
Projects
Skills
Research
Contact
Education
B.S. in Computer Science - University of Tech
2017 - 2021
Graduated with honors, specializing in web development and user interface design. Completed a capstone project on responsive web frameworks.
Web Development Bootcamp - CodeAcademy
2020
Intensive 12-week program focused on HTML, CSS, and responsive design techniques.
Projects
Personal Portfolio
A sleek, responsive portfolio showcasing my skills and projects, built with HTML and
Tech Blog
A modern blog platform with custom CSS styling and responsive layouts for tech
E-commerce Store
A product page with a clean, user-friendly design, optimized for mobile and desktop

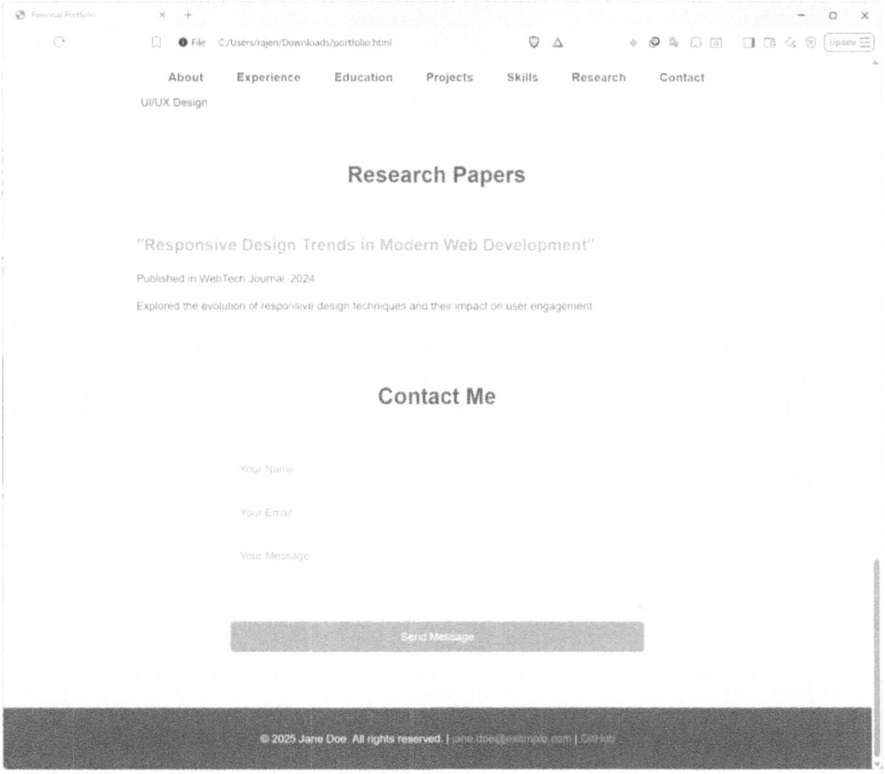
About
Experience
Education
Projects
Skills
Research
Contact
UI/UX Design
Research Papers
"Responsive Design Trends in Modern Web Development"
Published in WebTech Journal, 2024
Explored the evolution of responsive design techniques and their impact on user engagement.
Contact Me
Your Name
Your Email
Your Message
Send Message
© 2025 Jane Doe. All rights reserved. | jane.doe@example.com | GitHub

2. Simple Blog Post

A basic blog template helps you practice content layout, headings, and text formatting. You can also find this code in our book github repository:

https://github.com/JambaAcademy/HTML_CSS

```
<script type="text/javascript">
        var gk_isXlsx = false;
        var gk_xlsxFileLookup = {};
        var gk_fileData = {};
        function filledCell(cell) {
          return cell !== '' && cell != null;
        }
        function loadFileData(filename) {
        if (gk_isXlsx && gk_xlsxFileLookup[filename]) {
            try {
                var workbook = XLSX.read(gk_fileData[filename], { type: 'base64'
});
                var firstSheetName = workbook.SheetNames[0];
                var worksheet = workbook.Sheets[firstSheetName];

                // Convert sheet to JSON to filter blank rows
                var jsonData = XLSX.utils.sheet_to_json(worksheet, { header: 1,
blankrows: false, defval: '' });
                // Filter out blank rows (rows where all cells are empty, null,
or undefined)
                var filteredData = jsonData.filter(row => row.some(filledCell));

                // Heuristic to find the header row by ignoring rows with fewer
filled cells than the next row
                var headerRowIndex = filteredData.findIndex((row, index) =>
                  row.filter(filledCell).length >= filteredData[index +
1]?.filter(filledCell).length
                );
                // Fallback
                if (headerRowIndex === -1 || headerRowIndex > 25) {
                  headerRowIndex = 0;
                }

                // Convert filtered JSON back to CSV
                var csv =
XLSX.utils.aoa_to_sheet(filteredData.slice(headerRowIndex)); // Create a new
sheet from filtered array of arrays
                csv = XLSX.utils.sheet_to_csv(csv, { header: 1 });
                return csv;
            } catch (e) {
                console.error(e);
                return "";
            }
        }
        return gk_fileData[filename] || "";
        }
        </script><!DOCTYPE html>
<html lang="en">
<head>
    <meta charset="UTF-8">
    <meta name="viewport" content="width=device-width, initial-scale=1.0">
    <title>Modern Blog Template</title>
    <style>
        * {
```

```
        margin: 0;
        padding: 0;
        box-sizing: border-box;
        font-family: 'Arial', sans-serif;
    }

    body {
        background-color: #f8f9fa;
        color: #333;
        line-height: 1.6;
    }

    header {
        background: linear-gradient(135deg, #1e3a8a, #3b82f6);
        color: white;
        text-align: center;
        padding: 4rem 2rem;
        position: relative;
    }

    header h1 {
        font-size: 3rem;
        margin-bottom: 0.5rem;
        text-shadow: 1px 1px 2px rgba(0, 0, 0, 0.2);
    }

    header p {
        font-size: 1.2rem;
        opacity: 0.9;
    }

    nav {
        background-color: #fff;
        box-shadow: 0 2px 5px rgba(0, 0, 0, 0.1);
        position: sticky;
        top: 0;
        z-index: 100;
    }

    nav ul {
        display: flex;
        justify-content: center;
        list-style: none;
        padding: 1rem 0;
    }

    nav ul li {
        margin: 0 1.5rem;
    }

    nav ul li a {
        text-decoration: none;
        color: #1e3a8a;
        font-weight: bold;
        font-size: 1.1rem;
        transition: color 0.3s;
    }

    nav ul li a:hover {
        color: #3b82f6;
    }

    .container {
        max-width: 1200px;
        margin: 3rem auto;
```

```
    padding: 0 2rem;
    display: grid;
    grid-template-columns: 2fr 1fr;
    gap: 2rem;
}

.main-content {
    display: flex;
    flex-direction: column;
    gap: 2rem;
}

.blog-post {
    background-color: white;
    border-radius: 10px;
    box-shadow: 0 4px 10px rgba(0, 0, 0, 0.1);
    overflow: hidden;
    transition: transform 0.3s;
}

.blog-post:hover {
    transform: translateY(-5px);
}

.blog-post img {
    width: 100%;
    height: 250px;
    object-fit: cover;
}

.post-content {
    padding: 1.5rem;
}

.post-content h2 {
    font-size: 1.8rem;
    color: #1e3a8a;
    margin-bottom: 0.5rem;
}

.post-meta {
    font-size: 0.9rem;
    color: #6b7280;
    margin-bottom: 1rem;
}

.post-content p {
    margin-bottom: 1rem;
}

.read-more {
    display: inline-block;
    text-decoration: none;
    color: #3b82f6;
    font-weight: bold;
    transition: color 0.3s;
}

.read-more:hover {
    color: #1e3a8a;
}

.sidebar {
    display: flex;
    flex-direction: column;
```

```
        gap: 2rem;
    }

    .sidebar-section {
        background-color: white;
        padding: 1.5rem;
        border-radius: 10px;
        box-shadow: 0 4px 10px rgba(0, 0, 0, 0.1);
    }

    .sidebar-section h3 {
        font-size: 1.5rem;
        color: #1e3a8a;
        margin-bottom: 1rem;
    }

    .sidebar-section ul {
        list-style: none;
    }

    .sidebar-section ul li {
        margin-bottom: 0.8rem;
    }

    .sidebar-section ul li a {
        text-decoration: none;
        color: #333;
        transition: color 0.3s;
    }

    .sidebar-section ul li a:hover {
        color: #3b82f6;
    }

    .search-bar {
        display: flex;
        gap: 0.5rem;
    }

    .search-bar input {
        flex: 1;
        padding: 0.8rem;
        border: 1px solid #ddd;
        border-radius: 5px;
        font-size: 1rem;
    }

    .search-bar button {
        background-color: #3b82f6;
        color: white;
        border: none;
        padding: 0.8rem;
        border-radius: 5px;
        cursor: pointer;
        transition: background-color 0.3s;
    }

    .search-bar button:hover {
        background-color: #1e3a8a;
    }

    footer {
        background-color: #1e3a8a;
        color: white;
        text-align: center;
```

```
            padding: 2rem;
            margin-top: 3rem;
        }

        footer a {
            color: #3b82f6;
            text-decoration: none;
        }

        footer a:hover {
            text-decoration: underline;
        }

        @media (max-width: 768px) {
            .container {
                grid-template-columns: 1fr;
            }

            header h1 {
                font-size: 2rem;
            }

            nav ul {
                flex-direction: column;
                gap: 0.5rem;
            }

            .blog-post img {
                height: 200px;
            }
        }
    </style>
</head>
<body>
    <header>
        <h1>Tech Insights</h1>
        <p>Your Source for Web Development Trends and Tips</p>
    </header>

    <nav>
        <ul>
            <li><a href="#home">Home</a></li>
            <li><a href="#tutorials">Tutorials</a></li>
            <li><a href="#trends">Trends</a></li>
            <li><a href="#about">About</a></li>
            <li><a href="#contact">Contact</a></li>
        </ul>
    </nav>

    <div class="container">
        <div class="main-content">
            <div class="blog-post">
                <img src="images/post1.jpg" alt="Web Development Trends">
                <div class="post-content">
                    <h2>Top Web Development Trends for 2025</h2>
                    <div class="post-meta">Posted on August 18, 2025 by Jane
Doe</div>
                    <p>Explore the latest trends shaping the future of web
development, from AI-driven design tools to progressive web apps. This article
dives into how modern frameworks and responsive design are transforming user
experiences.</p>
                    <a href="#" class="read-more">Read More</a>
                </div>
            </div>
            <div class="blog-post">
```

```
                <img src="images/post2.jpg" alt="CSS Grid Tutorial">
                <div class="post-content">
                    <h2>Mastering CSS Grid: A Step-by-Step Guide</h2>
                    <div class="post-meta">Posted on August 10, 2025 by John
Smith</div>
                    <p>Learn how to create flexible, responsive layouts using
CSS Grid. This tutorial walks you through practical examples, from basic grid
setups to complex multi-column designs.</p>
                    <a href="#" class="read-more">Read More</a>
                </div>
            </div>
            <div class="blog-post">
                <img src="images/post3.jpg" alt="HTML Semantics">
                <div class="post-content">
                    <h2>Why HTML Semantics Matter for Accessibility</h2>
                    <div class="post-meta">Posted on August 5, 2025 by Jane
Doe</div>
                    <p>Discover the importance of semantic HTML in building
accessible websites. This article covers best practices for using tags like
<article> and <section> to enhance user experience.</p>
                    <a href="#" class="read-more">Read More</a>
                </div>
            </div>
        </div>

        <div class="sidebar">
            <div class="sidebar-section">
                <h3>Search</h3>
                <div class="search-bar">
                    <input type="text" placeholder="Search posts...">
                    <button type="submit">Search</button>
                </div>
            </div>
            <div class="sidebar-section">
                <h3>Recent Posts</h3>
                <ul>
                    <li><a href="#">Top Web Development Trends for 2025</a></li>
                    <li><a href="#">Mastering CSS Grid: A Step-by-Step
Guide</a></li>
                    <li><a href="#">Why HTML Semantics Matter</a></li>
                </ul>
            </div>
            <div class="sidebar-section">
                <h3>Categories</h3>
                <ul>
                    <li><a href="#">HTML</a></li>
                    <li><a href="#">CSS</a></li>
                    <li><a href="#">Responsive Design</a></li>
                    <li><a href="#">Web Accessibility</a></li>
                </ul>
            </div>
            <div class="sidebar-section">
                <h3>About the Author</h3>
                <p>Jane Doe is a web developer with a passion for sharing
knowledge on modern web technologies. Follow her for the latest insights!</p>
            </div>
        </div>
    </div>

    <footer>
        <p>&copy; 2025 Tech Insights. All rights reserved. | <a
href="mailto:contact@techinsights.com">contact@techinsights.com</a> | <a
href="https://github.com/JambaAcademy/HTML_CSS">GitHub</a></p>
    </footer>
</body>
```

```
</html>
```

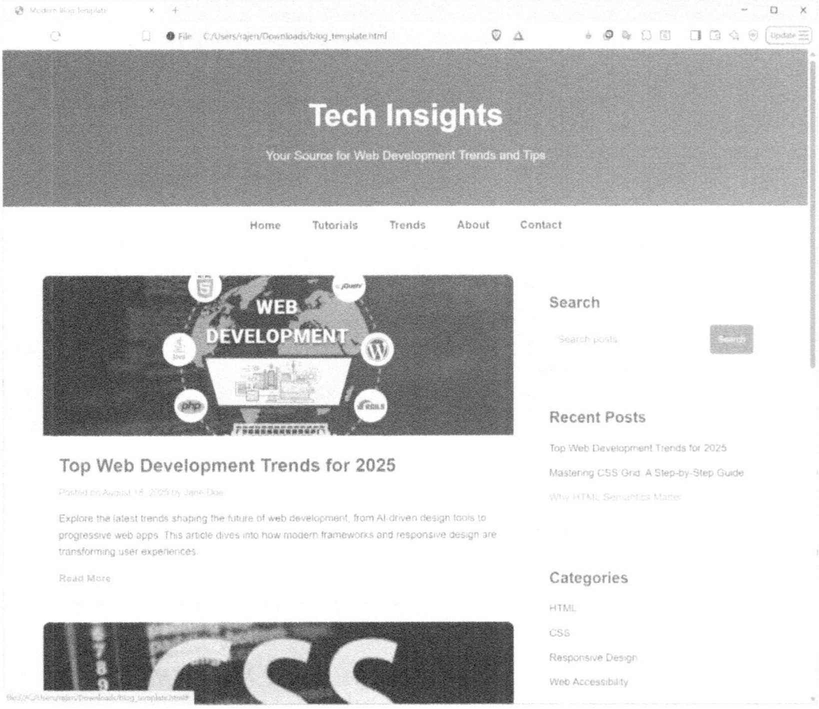
Tech Insights
Your Source for Web Development Trends and Tips
Home
Tutorials
Trends
About
Contact
WEB DEVELOPMENT
Top Web Development Trends for 2025
Explore the latest trends shaping the future of web development, from AI-driven design tools to progressive web apps. This article dives into how modern frameworks and responsive design are transforming user experiences.
Read More
Search
Recent Posts
Top Web Development Trends for 2025
Mastering CSS Grid: A Step-by-Step Guide
Categories
HTML
CSS
Responsive Design
Web Accessibility

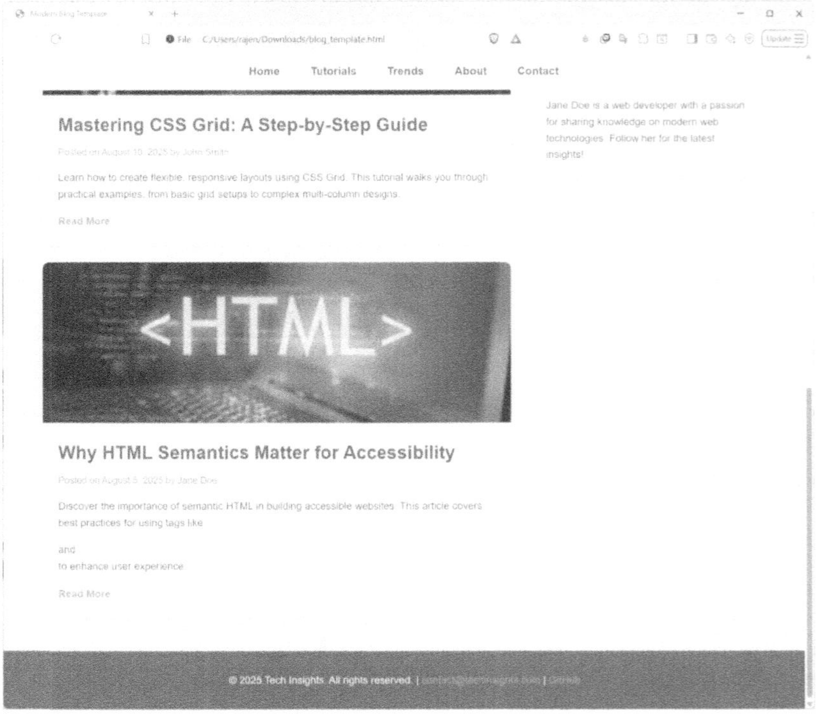
Home
Tutorials
Trends
About
Contact
Mastering CSS Grid: A Step-by-Step Guide
Learn how to create flexible, responsive layouts using CSS Grid. This tutorial walks you through practical examples, from basic grid setups to complex multi-column designs.
Read More
Jane Doe is a web developer with a passion for sharing knowledge on modern web technologies. Follow her for the latest insights!
<HTML>
Why HTML Semantics Matter for Accessibility
Discover the importance of semantic HTML in building accessible websites. This article covers best practices for using tags like
and
to enhance user experience
Read More
© 2025 Tech Insights. All rights reserved.

3. Mini E-Commerce Landing Page

This template introduces layout and visual hierarchy, perfect for building a product showcase. You can also find this code in our book github repository:

https://github.com/JambaAcademy/HTML_CSS

```
<script type="text/javascript">
        var gk_isXlsx = false;
        var gk_xlsxFileLookup = {};
        var gk_fileData = {};
        function filledCell(cell) {
          return cell !== '' && cell != null;
        }
        function loadFileData(filename) {
        if (gk_isXlsx && gk_xlsxFileLookup[filename]) {
            try {
                var workbook = XLSX.read(gk_fileData[filename], { type: 'base64' });
                var firstSheetName = workbook.SheetNames[0];
                var worksheet = workbook.Sheets[firstSheetName];

                // Convert sheet to JSON to filter blank rows
                var jsonData = XLSX.utils.sheet_to_json(worksheet, { header: 1, blankrows:
false, defval: '' });
                // Filter out blank rows (rows where all cells are empty, null, or
undefined)
                var filteredData = jsonData.filter(row => row.some(filledCell));

                // Heuristic to find the header row by ignoring rows with fewer filled cells
than the next row
                var headerRowIndex = filteredData.findIndex((row, index) =>
                  row.filter(filledCell).length >= filteredData[index +
1]?.filter(filledCell).length
                );
                // Fallback
                if (headerRowIndex === -1 || headerRowIndex > 25) {
                  headerRowIndex = 0;
                }

                // Convert filtered JSON back to CSV
                var csv = XLSX.utils.aoa_to_sheet(filteredData.slice(headerRowIndex)); //
Create a new sheet from filtered array of arrays
                csv = XLSX.utils.sheet_to_csv(csv, { header: 1 });
                return csv;
            } catch (e) {
                console.error(e);
                return "";
            }
        }
        return gk_fileData[filename] || "";
        }
        </script><!DOCTYPE html>
<html lang="en">
<head>
    <meta charset="UTF-8">
    <meta name="viewport" content="width=device-width, initial-scale=1.0">
    <title>Digital Products E-Commerce Landing Page</title>
    <style>
        * {
            margin: 0;
            padding: 0;
            box-sizing: border-box;
            font-family: 'Arial', sans-serif;
        }

        body {
            background-color: #f9fafb;
            color: #1f2937;
            line-height: 1.6;
        }
```

```
header {
    background: linear-gradient(135deg, #4f46e5, #7c3aed);
    color: white;
    text-align: center;
    padding: 5rem 2rem;
    position: relative;
}

header h1 {
    font-size: 3rem;
    margin-bottom: 1rem;
    text-shadow: 1px 1px 3px rgba(0, 0, 0, 0.2);
}

header p {
    font-size: 1.3rem;
    max-width: 600px;
    margin: 0 auto 2rem;
    opacity: 0.9;
}

.cta-button {
    display: inline-block;
    background-color: #ffffff;
    color: #4f46e5;
    padding: 1rem 2rem;
    text-decoration: none;
    font-weight: bold;
    border-radius: 5px;
    transition: background-color 0.3s, color 0.3s;
}

.cta-button:hover {
    background-color: #4f46e5;
    color: white;
    border: 1px solid white;
}

nav {
    background-color: #fff;
    box-shadow: 0 2px 5px rgba(0, 0, 0, 0.1);
    position: sticky;
    top: 0;
    z-index: 100;
}

nav ul {
    display: flex;
    justify-content: center;
    list-style: none;
    padding: 1rem 0;
}

nav ul li {
    margin: 0 1.5rem;
}

nav ul li a {
    text-decoration: none;
    color: #1f2937;
    font-weight: bold;
    font-size: 1.1rem;
    transition: color 0.3s;
}

nav ul li a:hover {
    color: #4f46e5;
}

.container {
    max-width: 1200px;
    margin: 3rem auto;
    padding: 0 2rem;
}
```

```
.featured-products {
    text-align: center;
    margin-bottom: 3rem;
}

.featured-products h2 {
    font-size: 2.5rem;
    color: #1f2937;
    margin-bottom: 2rem;
}

.products-grid {
    display: grid;
    grid-template-columns: repeat(auto-fit, minmax(250px, 1fr));
    gap: 2rem;
}

.product-card {
    background-color: white;
    border-radius: 10px;
    box-shadow: 0 4px 10px rgba(0, 0, 0, 0.1);
    overflow: hidden;
    transition: transform 0.3s;
}

.product-card:hover {
    transform: translateY(-5px);
}

.product-card img {
    width: 100%;
    height: 200px;
    object-fit: cover;
}

.product-content {
    padding: 1.5rem;
}

.product-content h3 {
    font-size: 1.5rem;
    color: #1f2937;
    margin-bottom: 0.5rem;
}

.product-content p {
    font-size: 1rem;
    color: #6b7280;
    margin-bottom: 1rem;
}

.product-content .price {
    font-size: 1.2rem;
    font-weight: bold;
    color: #4f46e5;
    margin-bottom: 1rem;
}

.product-content a {
    display: inline-block;
    background-color: #4f46e5;
    color: white;
    padding: 0.8rem 1.5rem;
    text-decoration: none;
    border-radius: 5px;
    font-weight: bold;
    transition: background-color 0.3s;
}

.product-content a:hover {
    background-color: #7c3aed;
}

.testimonials {
    background-color: #f3f4f6;
```

```
        padding: 3rem 2rem;
        text-align: center;
    }

    .testimonials h2 {
        font-size: 2.5rem;
        margin-bottom: 2rem;
    }

    .testimonial-grid {
        display: grid;
        grid-template-columns: repeat(auto-fit, minmax(300px, 1fr));
        gap: 2rem;
    }

    .testimonial-card {
        background-color: white;
        padding: 2rem;
        border-radius: 10px;
        box-shadow: 0 4px 10px rgba(0, 0, 0, 0.1);
    }

    .testimonial-card p {
        font-style: italic;
        margin-bottom: 1rem;
    }

    .testimonial-card .author {
        font-weight: bold;
        color: #4f46e5;
    }

    .features {
        text-align: center;
        margin: 3rem 0;
    }

    .features h2 {
        font-size: 2.5rem;
        margin-bottom: 2rem;
    }

    .features-grid {
        display: grid;
        grid-template-columns: repeat(auto-fit, minmax(200px, 1fr));
        gap: 2rem;
    }

    .feature-item {
        padding: 1.5rem;
    }

    .feature-item img {
        width: 60px;
        height: 60px;
        margin-bottom: 1rem;
    }

    .feature-item h3 {
        font-size: 1.3rem;
        color: #1f2937;
        margin-bottom: 0.5rem;
    }

    footer {
        background-color: #1f2937;
        color: white;
        text-align: center;
        padding: 2rem;
        margin-top: 3rem;
    }

    footer a {
        color: #4f46e5;
        text-decoration: none;
```

```
        }

        footer a:hover {
            text-decoration: underline;
        }

        @media (max-width: 768px) {
            header h1 {
                font-size: 2rem;
            }

            header p {
                font-size: 1.1rem;
            }

            nav ul {
                flex-direction: column;
                gap: 0.5rem;
            }

            .product-card img {
                height: 150px;
            }

            .container {
                padding: 0 1rem;
            }
        }
    </style>
</head>
<body>
    <header>
        <h1>Discover Premium Digital Products</h1>
        <p>Unlock your potential with our curated collection of eBooks, templates, and
software tools designed for creators and professionals.</p>
        <a href="#products" class="cta-button">Shop Now</a>
    </header>

    <nav>
        <ul>
            <li><a href="#home">Home</a></li>
            <li><a href="#products">Products</a></li>
            <li><a href="#features">Features</a></li>
            <li><a href="#testimonials">Testimonials</a></li>
            <li><a href="#contact">Contact</a></li>
        </ul>
    </nav>

    <div class="container">
        <section id="products" class="featured-products">
            <h2>Featured Products</h2>
            <div class="products-grid">
                <div class="product-card">
                    <img src="images/ebook.jpg" alt="eBook Cover">
                    <div class="product-content">
                        <h3>Master Web Design eBook</h3>
                        <p>A comprehensive guide to creating stunning websites with HTML and
CSS.</p>
                        <div class="price">$19.99</div>
                        <a href="#">Add to Cart</a>
                    </div>
                </div>
                <div class="product-card">
                    <img src="images/template.jpg" alt="Website Template">
                    <div class="product-content">
                        <h3>Portfolio Template</h3>
                        <p>A modern, responsive template to showcase your creative work.</p>
                        <div class="price">$29.99</div>
                        <a href="#">Add to Cart</a>
                    </div>
                </div>
                <div class="product-card">
                    <img src="images/software.jpg" alt="Software Tool">
                    <div class="product-content">
                        <h3>DesignPro Software</h3>
```

```
                    <p>A powerful tool for creating professional-grade UI/UX
designs.</p>
                    <div class="price">$49.99</div>
                    <a href="#">Add to Cart</a>
                </div>
            </div>
        </div>
    </section>

    <section id="features" class="features">
        <h2>Why Choose Us?</h2>
        <div class="features-grid">
            <div class="feature-item">
                <img src="images/icon-quality.png" alt="Quality Icon">
                <h3>High Quality</h3>
                <p>Curated products designed to meet professional standards.</p>
            </div>
            <div class="feature-item">
                <img src="images/icon-support.png" alt="Support Icon">
                <h3>24/7 Support</h3>
                <p>Access expert support anytime you need assistance.</p>
            </div>
            <div class="feature-item">
                <img src="images/icon-update.png" alt="Update Icon">
                <h3>Regular Updates</h3>
                <p>Receive free updates to keep your tools current.</p>
            </div>
        </div>
    </section>

    <section id="testimonials" class="testimonials">
        <h2>What Our Customers Say</h2>
        <div class="testimonial-grid">
            <div class="testimonial-card">
                <p>"The eBook transformed my approach to web design. It's clear,
practical, and inspiring!"</p>
                <div class="author">Sarah L.</div>
            </div>
            <div class="testimonial-card">
                <p>"The portfolio template saved me hours of work. It's sleek and
professional!"</p>
                <div class="author">Michael T.</div>
            </div>
            <div class="testimonial-card">
                <p>"DesignPro is a game-changer for my projects. Highly recommend!"</p>
                <div class="author">Emily R.</div>
            </div>
        </div>
    </section>
</div>

<footer>
    <p>&copy; 2025 Digital Creators Hub. All rights reserved. | <a
href="mailto:support@digitalcreators.com">support@digitalcreators.com</a> | <a
href="https://github.com/JambaAcademy/HTML_CSS">GitHub</a></p>
</footer>
</body>
</html>
```

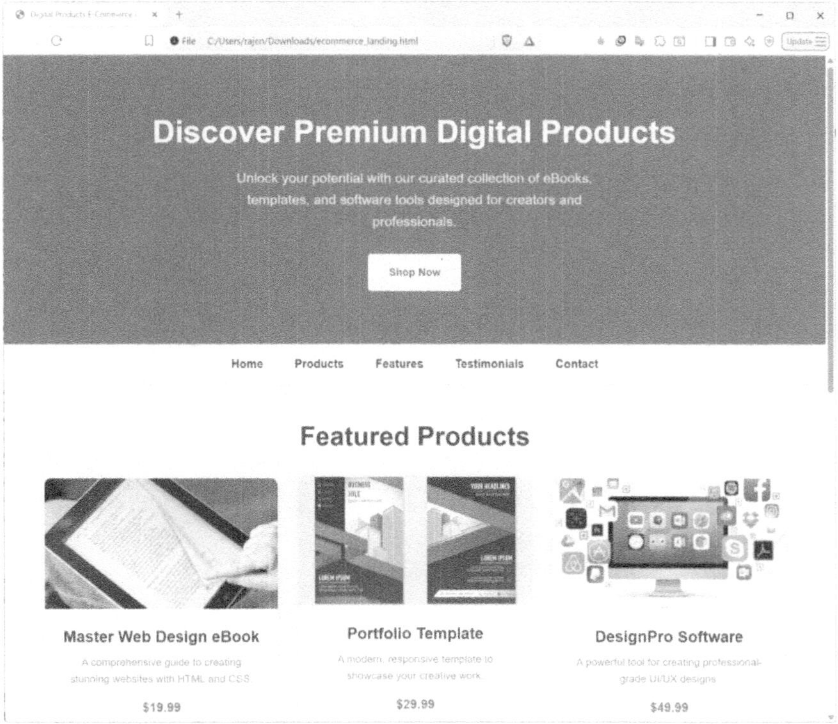
C:/Users/rajen/Downloads/ecommerce_landing.html
Discover Premium Digital Products
Unlock your potential with our curated collection of eBooks, templates, and software tools designed for creators and professionals.
Shop Now
Home
Products
Features
Testimonials
Contact
Featured Products
Master Web Design eBook
A comprehensive guide to creating stunning websites with HTML and CSS
$19.99
Portfolio Template
A modern, responsive template to showcase your creative work
$29.99
DesignPro Software
A powerful tool for creating professional-grade UI/UX designs
$49.99

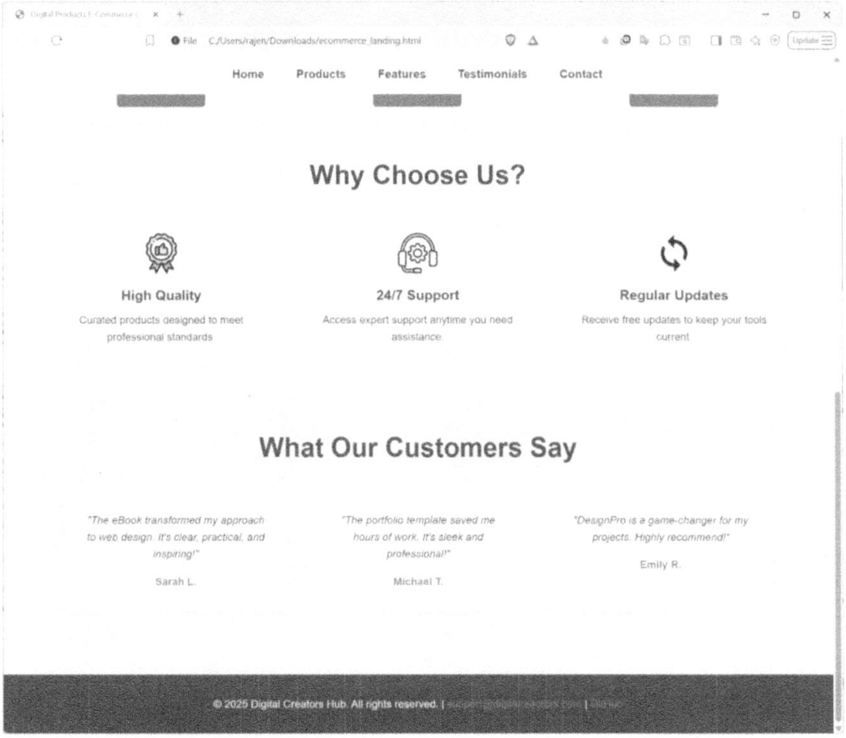
C:/Users/rajen/Downloads/ecommerce_landing.html
Home
Products
Features
Testimonials
Contact
Why Choose Us?
High Quality
Curated products designed to meet professional standards
24/7 Support
Access expert support anytime you need assistance
Regular Updates
Receive free updates to keep your tools current
What Our Customers Say
"The eBook transformed my approach to web design. It's clear, practical, and inspiring!"
Sarah L.
"The portfolio template saved me hours of work. It's sleek and professional!"
Michael T.
"DesignPro is a game-changer for my projects. Highly recommend!"
Emily R.
© 2025 Digital Creators Hub. All rights reserved. |

Example Application: Customizing Your Template

Start with the portfolio template. Change the name, add your own project details, and modify the colors in the CSS. Experiment with layout by adding or removing elements. This process helps you understand how HTML and CSS work together.

Coding Summary

Templates are your launchpad into real-world web development. Adapt these examples as you practice, and gradually add complexity as your skills grow.

ANNEXURE D: WEB DEVELOPMENT RESOURCE LIST

Learning web development is a continuous journey. The following curated list of resources will guide your growth, answer your questions, and keep you updated on best practices and industry trends. Whether you seek documentation, tutorials, communities, or inspiration, these links and recommendations connect you to the global web development community.

Core Documentation & References

MDN Web Docs (https://developer.mozilla.org)

The definitive resource for HTML, CSS, and JavaScript documentation. Clear explanations, interactive examples, and up-to-date browser compatibility tables.

W3Schools (https://www.w3schools.com)

Beginner-friendly tutorials and live editors for practicing HTML and CSS basics.

Interactive Coding & Practice Platforms

freeCodeCamp (https://www.freecodecamp.org)

Offers a full curriculum with hands-on coding challenges. Great for building real projects and earning certificates.

Codecademy (https://www.codecademy.com)

Interactive, guided lessons on HTML, CSS, and modern web development tools.

CodePen (https://codepen.io)

A playground for experimenting with HTML, CSS, and JavaScript in real-time. Browse, share, and fork community projects.

Design Inspiration & Trends

Awwwards (https://www.awwwards.com)

Showcases cutting-edge web design from around the world. See what's possible with HTML and CSS, and get inspired by global trends.

Dribbble (https://dribbble.com)

A creative community for designers. Explore website mockups, UI ideas, and color schemes.

Browser Compatibility Tools

Can I use (https://caniuse.com)

Instantly check if HTML/CSS features work across different browsers and devices.

Community & Support

Stack Overflow (https://stackoverflow.com)

Ask questions, share code, and learn from millions of developers. Search for answers to common HTML and CSS challenges.

r/webdev (https://www.reddit.com/r/webdev/)

A supportive community for web development tips, trends, and feedback.

CSS-Tricks (https://css-tricks.com)

Articles, guides, and code snippets focused on practical web design and advanced CSS techniques.

Example Application: Staying Up-to-Date

Suppose you want to use a new CSS property, such as `grid`, to create advanced layouts. Use "MDN Web Docs" to read detailed guides and syntax examples, then check "Can I use" to ensure it's supported in the browsers your users are likely to use. Visit "CSS-Tricks" for practical tutorials and design patterns, and share your project on "CodePen" for feedback.

Coding Summary

Bookmark these resources and revisit them as you learn and build new projects. The web development community is vast and welcoming—use these tools to deepen your skills, solve problems, and connect with others on your journey to becoming a confident web developer.

BIBLIOGRAPHY

A List Apart. (n.d.). **A List Apart: For people who make websites**. https://alistapart.com/

Atwood, J., Spolsky, J., & Stack Overflow Community. (n.d.). **Stack Overflow**. https://stackoverflow.com/

Beaird, J., & George, J. (2020). **The principles of beautiful web design** (4th ed.). SitePoint.

Berry, J. (2021). **HTML and CSS: Visual quickstart guide** (9th ed.). Peachpit Press.

Bray, T. (Ed.). (2018). **Extensible Markup Language (XML) 1.0 (Fifth Edition)**. W3C. https://www.w3.org/TR/xml/

Brown, J. (2022). **CSS in depth**. Manning Publications.

Budd, T. (2021). **Learning web design: A beginner's guide to HTML, CSS, JavaScript, and web graphics** (5th ed.). O'Reilly Media.

Burns, K. (2023). **Responsive web design with HTML5 and CSS** (4th ed.). Packt Publishing.

Codecademy. (n.d.). **Learn HTML**. https://www.codecademy.com/learn/learn-html

Codecademy. (n.d.). **Learn CSS**. https://www.codecademy.com/learn/learn-css

Coyle, S. (2021). **Practical web development: A beginner's guide to HTML, CSS, JavaScript, and web graphics**. Apress.

Duckett, J. (2014). **HTML and CSS: Design and build websites**. Wiley.

Annotation: Duckett's book is widely praised for its approachable and visually engaging format, making it a favorite among beginners. Its clear explanations, color-coded examples, and hands-on exercises align closely with the hands-on, beginner-friendly philosophy of this guide, helping learners quickly move from theory to practice.

Edwards, B. (2021). **HTML5 and CSS3 all-in-one for dummies** (4th ed.). For Dummies.

Frain, B. (2021). **Responsive web design with HTML5 and CSS** (4th ed.). Packt Publishing.

Frost, B. (2016). **Atomic design**. Brad Frost Web.

GitHub, Inc. (n.d.). **GitHub**. https://github.com/

Goldstein, J. (2023). **Modern CSS: Master the latest web standards to create stunning, responsive websites**. No Starch Press.

Google Developers. (n.d.). **Web Fundamentals**. https://developers.google.com/web/fundamentals

Annotation: Google's Web Fundamentals offers up-to-date, practical guides and hands-on tutorials that cover modern web development essentials, including accessibility, responsive design, and performance. Its authoritative approach and wide-ranging topics make it invaluable for beginners aiming for industry best practices.

Griffiths, D., & Griffiths, D. (2018). **Head First HTML and CSS** (2nd ed.). O'Reilly Media.

Annotation: The "Head First" series is known for its engaging, interactive approach to learning. This book uses puzzles, quizzes, and real-world projects to reinforce concepts, perfectly complementing a hands-on, practice-driven learning approach for novice web developers.

MDN Web Docs. (n.d.). **HTML: HyperText Markup Language**. https://developer.mozilla.org/en-US/docs/Web/HTML

Annotation: MDN Web Docs is a definitive source for web standards and modern coding practices. Its beginner-friendly documentation, examples, and live code editors provide foundational support for hands-on learning and self-paced experimentation.

MDN Web Docs. (n.d.). **CSS: Cascading Style Sheets**. https://developer.mozilla.org/en-US/docs/Web/CSS

Annotation: The CSS resources on MDN are essential for anyone learning modern styling techniques. Its comprehensive guides, property references, and interactive tools are tailored to both beginners and experienced coders, supporting step-by-step learning and experimentation.

Meyer, E. A. (2018). **CSS: The definitive guide** (4th ed.). O'Reilly Media.

Mozilla. (n.d.). **MDN Web Docs: Learn web development**. https://developer.mozilla.org/en-US/docs/Learn

Annotation: Mozilla's "Learn web development" section is specifically designed for beginners, covering everything from HTML basics to advanced CSS and responsive design. Its structured learning paths and hands-on exercises make it a top resource for foundational web skills.

Niemeyer, R. (2022). **Learning web design: A beginner's guide to HTML, CSS, JavaScript, and web graphics** (5th ed.). O'Reilly Media.

Resig, J., & Bibeault, B. (2013). **Secrets of the JavaScript ninja** (2nd ed.). Manning Publications.

Robbins, J. N. (2018). **Learning web design: A beginner's guide to HTML, CSS, JavaScript, and web graphics** (5th ed.). O'Reilly Media.

Annotation: Robbins' book is a cornerstone for new web developers, combining clear, step-by-step instruction with practical coding exercises. Its focus on foundational skills and project-based learning underpins a smarter, more effective learning process.

Shay Howe. (2014). **Learn to code HTML & CSS: Develop & style websites**. New Riders.

Annotation: Howe's book is structured around real-world projects, introducing concepts in a logical order and reinforcing them through hands-on practice. It is particularly well-suited for readers with no prior coding experience, guiding them from basic to advanced web development techniques.

Smashing Magazine. (n.d.). **Smashing Magazine: Web design & development**. https://www.smashingmagazine.com/

Stack Overflow. (n.d.). **Stack Overflow**. https://stackoverflow.com/

Thompson, T. (2021). **HTML5 and CSS3: Building responsive websites**. Packt Publishing.

W3C. (n.d.). **HTML Living Standard**. https://html.spec.whatwg.org/

W3C. (n.d.). **Cascading Style Sheets (CSS) Working Group**. https://www.w3.org/Style/CSS/

W3Schools. (n.d.). **W3Schools Online Web Tutorials**. https://www.w3schools.com/

Annotation: W3Schools is a popular, beginner-friendly platform offering interactive tutorials and code editors for HTML, CSS, and many other web technologies. Its hands-on exercises and practical examples make it a go-to resource for novices eager to practice as they learn.

WebAIM. (n.d.). **WebAIM: Web accessibility in mind**. https://webaim.org/

World Wide Web Consortium (W3C). (n.d.). **Web Content Accessibility Guidelines (WCAG) Overview**. https://www.w3.org/WAI/standards-guidelines/wcag/

Zeldman, J. (2017). **Designing with web standards** (3rd ed.). New Riders.

Annotated Sources Overview

Duckett, J. (2014). HTML and CSS: Design and build websites.

A visually rich, step-by-step guide ideal for absolute beginners. Duckett's approach demystifies HTML and CSS through practical examples and real projects, supporting hands-on learning.

Google Developers. Web Fundamentals.

Offers current best practices in web development, including responsive design and performance optimization. Its practical tutorials and guides ensure readers learn industry-relevant, modern coding techniques.

Griffiths, D., & Griffiths, D. (2018). Head First HTML and CSS.

Uses interactive learning techniques such as quizzes, puzzles, and projects. Perfect for those who learn best by doing and want an engaging, beginner-friendly introduction.

MDN Web Docs. HTML: HyperText Markup Language.

The authoritative resource for HTML standards and best practices, with clear explanations, code samples, and live editors for experimentation.

MDN Web Docs. CSS: Cascading Style Sheets.

Comprehensive, up-to-date, and approachable documentation for all CSS concepts and properties, supporting self-paced, hands-on practice.

Mozilla. MDN Web Docs: Learn web development.

A structured, beginner-focused path covering HTML, CSS, and more, with interactive exercises and project-based learning.

Robbins, J. N. (2018). Learning web design.

Balances foundational theory with hands-on tutorials, ensuring readers can apply what they learn by building real websites.

Shay Howe. (2014). Learn to code HTML & CSS.

Focuses on practical projects and logical concept sequencing, making it ideal for total beginners who want to build their skills through real-world examples.

W3Schools Online Web Tutorials.

A widely used, interactive platform for practicing HTML and CSS in-browser, providing instant feedback and practical coding challenges.

All sources listed are relevant to current HTML and CSS standards, practical web development skills, and the hands-on, accessible learning style emphasized in The Smarter Way to Learn HTML & CSS.

RECOMMENDED RESOURCES FOR HTML & CSS MASTERY

1. MDN Web Docs (Mozilla Developer Network)

A gold standard for web development documentation, MDN Web Docs offers comprehensive, up-to-date references and tutorials on HTML, CSS, and JavaScript. Its beginner-friendly guides feature code samples, live demos, and in-depth explanations, making it an ideal supplement to this book's hands-on approach. MDN also covers advanced topics, web APIs, accessibility, and responsive design, supporting learners as they progress from foundational skills to real-world projects. Bookmark MDN as your go-to reference for troubleshooting and deepening your understanding of web standards.

2. W3C HTML & CSS Specifications

The World Wide Web Consortium (W3C) sets the official standards for HTML and CSS. Their specifications are the definitive source for understanding how HTML elements and CSS properties should work in browsers. While more technical, referencing W3C docs helps you build websites that align with global best practices, ensuring compatibility and future-proofing your skills. Use the W3C as an authoritative reference as you advance beyond beginner projects.

3. freeCodeCamp

freeCodeCamp is a widely respected, free platform offering interactive coding challenges, full curriculum paths, and real-world projects. Its HTML and CSS sections guide absolute beginners through hands-on lessons that reinforce what you learn in this book. As you progress, freeCodeCamp's later modules introduce JavaScript, responsive design, accessibility, and even full stack development, making it a great resource for expanding your web development toolkit.

4. Codecademy: HTML & CSS Courses

Codecademy's interactive courses are perfect for beginners seeking guided practice. The platform's HTML and CSS tracks feature bite-sized exercises, quizzes, and projects that mirror the step-by-step learning style of this book. Codecademy also offers paths to JavaScript and web frameworks, supporting your transition from beginner to job-ready developer. Its instant feedback system is especially helpful for self-learners.

5. A Book Apart: "HTML5 for Web Designers" by Jeremy Keith

This concise, approachable book demystifies the HTML5 standard, focusing on practical uses and how new elements improve web structure and semantics. It complements this guide's foundational approach by introducing best practices for clean, modern HTML. "HTML5 for Web Designers" is an excellent next step for learners ready to deepen their understanding of the language's evolution.

6. CSS-Tricks

CSS-Tricks is a popular blog and resource hub covering everything CSS, from beginner tips to advanced techniques like Flexbox, Grid, and responsive design. Its "Complete Guide to Flexbox" and "Complete Guide to Grid" are industry favorites. Use CSS-Tricks to solve styling challenges, learn about browser quirks, and stay updated on the latest design trends. The friendly, example-driven articles align perfectly with the book's hands-on philosophy.

7. Smashing Magazine

Smashing Magazine publishes practical articles, tutorials, and case studies on web design and development. The site regularly features guides on HTML, CSS, UI/UX, and accessibility, making it a valuable resource for aspiring developers. Smashing Magazine's emphasis on real-world solutions and best practices will help you move from beginner projects to professional-quality websites.

8. "Don't Make Me Think" by Steve Krug

A classic on web usability, this book introduces principles of user-centered design, navigation, and accessibility. While not a coding manual, it's essential reading for anyone who wants to build intuitive, user-friendly sites. "Don't Make Me Think" complements your coding skills with insights into how real users interact with websites, preparing you for freelance and client-based projects.

9. WebAIM (Web Accessibility In Mind)

WebAIM provides clear, actionable guides to making websites accessible to all users, including those with disabilities. Their checklists, articles, and tools (like the WAVE accessibility evaluation tool) are essential for learning about semantic HTML, ARIA roles, and inclusive design. Integrate WebAIM's recommendations into your projects to build sites that meet global accessibility standards.

10. Google Web Fundamentals

Google's Web Fundamentals offers practical guides on modern web development, with a strong focus on responsive design, performance, and best practices. It covers topics like mobile-first design, optimizing load times, and using cutting-edge CSS features. This resource is perfect for advancing your skills and ensuring your sites meet the expectations of today's users.

11. Responsive Web Design Basics (Google Developers)

This concise, visual tutorial introduces the core principles of responsive web design. Learn how to use media queries, flexible layouts, and fluid images to create sites that look great on any device. The guide includes code examples and best practices, making it an ideal companion for the responsive design section of this book.

12. Web.dev by Google

Web.dev is Google's hub for modern web development, featuring interactive tutorials, tools, and articles on performance, accessibility, PWA (Progressive Web Apps), and more. The platform's "Learn CSS" and "Learn HTML" sections provide structured, hands-on instruction to reinforce your foundational knowledge and introduce you to new web technologies as you're ready.

13. Stack Overflow

Stack Overflow is the world's largest online community for programmers. As you encounter challenges while coding, searching or asking questions on Stack Overflow can help you troubleshoot errors, learn best practices, and connect with other developers. Reading solutions to real-world coding problems will accelerate your learning and give you insight into how professionals solve challenges.

14. A11y Project

The A11y Project is a community-driven resource focused on digital accessibility. It features checklists, how-to guides, and straightforward explanations of accessibility concepts for web developers. Incorporate the A11y Project's advice into your workflow to ensure your projects are inclusive and meet legal accessibility requirements.

15. HTML.com

HTML.com offers beginner-friendly tutorials on HTML, CSS, and web design. Its clear explanations, examples, and glossary make it ideal for absolute beginners looking for an approachable starting point. The site's hands-on style closely aligns with this book's teaching philosophy, helping you reinforce and expand your basic skills.

16. The Net Ninja (YouTube Channel)

The Net Ninja's YouTube tutorials cover HTML, CSS, responsive design, and beyond, with clear, engaging explanations and practical projects. Video content can help you visualize concepts and see how real developers approach building web pages. The channel's playlists on Flexbox, Grid, and full website builds are especially useful as you advance.

17. Kevin Powell (YouTube Channel)

Kevin Powell specializes in CSS education, from basics to advanced layouts and animations. His approachable video lessons break down tricky concepts and showcase modern CSS techniques, complementing the book's hands-on learning style. Kevin's focus on real-world design problems will help you build stylish, responsive sites.

18. Frontend Mentor

Frontend Mentor provides real-world coding challenges where you build professional website layouts from provided designs. This platform bridges the gap between learning and working on actual client projects, giving you a portfolio-ready experience. Use Frontend Mentor to practice your HTML and CSS skills, gain feedback, and prepare for freelance opportunities.

19. "Learning Web Design" by Jennifer Robbins

This comprehensive beginner's book covers HTML, CSS, and web graphics with approachable explanations and exercises. It's an excellent companion to this guide, offering additional context, hands-on practice, and coverage of design principles. "Learning Web Design" is highly recommended for those who prefer structured, print resources alongside interactive learning.

20. Can I use

"Can I use" is an invaluable online tool that lets you check browser support for HTML elements, CSS properties, and web APIs. As you build more advanced sites, this resource helps ensure your designs work reliably across all browsers and devices, a crucial step for professional web development.

21. CSS Zen Garden

CSS Zen Garden demonstrates the power of CSS by showcasing how different style sheets can transform the same HTML structure into visually distinct designs. Browsing these examples will inspire your creativity and show the possibilities of advanced CSS, aligning with the book's emphasis on hands-on experimentation.

22. CSS Reference (by Codrops)

CSS Reference offers a visual, interactive glossary of CSS properties, complete with examples and browser compatibility info. Use it to quickly look up syntax and see demonstrations of how each property works, making it a practical supplement to your daily coding practice.

23. GitHub

GitHub is the leading platform for code collaboration and version control. Even as a beginner, learning to use GitHub to organize and share your projects is invaluable for building a professional portfolio or collaborating with others. Explore open-source HTML/CSS projects, contribute to documentation, and track your coding progress.

24. Webflow University

Webflow University combines video tutorials, interactive lessons, and visual examples to teach both code-based and visual web design concepts. While primarily focused on the no-code Webflow platform, its resources on responsive design, layout, and UI/UX are applicable to all web developers, helping you bridge the gap between design and coding.

25. "Refactoring UI" by Adam Wathan & Steve Schoger

This book and its accompanying website offer practical tips on making web interfaces visually appealing and user-friendly. While not focused on HTML/CSS syntax, "Refactoring UI" teaches design thinking and small tweaks that can dramatically improve your sites. Its actionable advice complements your technical skills with a designer's eye, helping your projects stand out in freelance or professional settings.

These resources, paired with the hands-on projects and foundational knowledge gained from this book, will empower you to continue growing as a web developer, build standout websites, and pursue advanced topics or professional opportunities with confidence.

EPILOGUE: YOUR JOURNEY FROM BEGINNER TO WEB CREATOR

As you reach the final pages of *The Smarter Way to Learn HTML & CSS: Hands-On Guide for Beginners and Future Web Developers*, take a moment to appreciate how far you've come. What began as a curiosity about how websites are made has blossomed into genuine skill: you now possess the foundational knowledge and practical experience to build, style, and launch your own web pages. Along the way, you've discovered not just the mechanics of code, but the creative and professional possibilities that web development unlocks.

This epilogue is a celebration of your progress and a gateway to your next chapter as a web creator. Let's reflect on the essential skills you've acquired, explore the real-world impact of your new abilities, and look ahead to the opportunities that await as you continue to learn and build.

Synthesizing Key HTML & CSS Skills

Throughout this guide, you've learned to wield **HTML** and **CSS**—the twin pillars of the web. With these tools, you can structure content, control layout, and craft visually engaging pages that work beautifully across devices. Let's revisit some of the essential abilities you've now mastered:

Structuring the Web with HTML

You began by learning the core concepts of **HTML (HyperText Markup Language)**:

Elements and Tags: You understand how tags like `<h1>`, `<p>`, `<a>`, `<img>`, and `<div>` form the building blocks of every page.

Semantic HTML: You know how to use semantic tags such as `<header>`, `<nav>`, `<main>`, `<section>`, and `<footer>` to give structure and meaning to your content, improving both accessibility and SEO.

Links and Images: You can add hyperlinks and graphics, connecting your pages and making your sites visually dynamic and interactive.

Lists, Tables, and Forms: You've learned to organize information and collect user input, essential for nearly every website.

Styling with CSS

You then discovered the power of **CSS (Cascading Style Sheets)**:

Selectors and Properties: You can target HTML elements with selectors and style them using properties like `color`, `font-size`, `margin`, and `padding`.

Classes and IDs: You know how to apply unique and reusable styles to specific elements, giving you precise control over your design.

Box Model: You understand how elements are sized and spaced, and how to create layouts with borders, margins, and padding.

Typography and Color: You can choose fonts, set text alignment, and create color schemes that enhance readability and appeal.

Responsive Design: You've learned how to use media queries and flexible layouts so your websites look great on any device, from smartphones to desktop monitors.

Putting It All Together

Most importantly, you've practiced combining these skills to create real web projects. You've built portfolio pages, simple blogs, and even the foundations of e-commerce sites. Each exercise and project has added to your confidence and competence, reinforcing the smarter, hands-on approach to learning that is at the heart of this book.

The Broader Impact: Creativity, Career, and Connection

With these skills, you are now part of a global community of creators, problem-solvers, and innovators. Web development is about more than writing code—it's about shaping the experiences people have online, communicating ideas, and building the tools and platforms that move the world forward.

Real-World Applications: From Practice to Purpose

To see the real power of what you've learned, let's explore a few inspiring examples of how basic HTML and CSS can transform lives, careers, and communities.

1. A Personal Portfolio: Launching a Creative Career

Meet Lila, an aspiring graphic designer who knew she needed an online portfolio to showcase her work. With no prior coding experience, Lila used the very skills you've practiced to build a clean, responsive website. She structured her projects using semantic HTML, styled her gallery with custom CSS grids, and crafted an about page that introduced her personality. The result? Not only did she impress potential clients, but her website also caught the eye of a design agency that offered her a full-time position. Lila's journey shows how even a simple, well-built portfolio can be a launchpad for new opportunities.

2. A Community Blog: Giving Voice and Building Networks

Carlos was passionate about local history and wanted to create a blog to share stories from his neighborhood. Using HTML to structure articles and CSS to create an inviting, readable design, Carlos built a site that quickly became a gathering place for community members. As he added features like comment forms and photo galleries, he learned more advanced web concepts, but it was his mastery of HTML and CSS that gave him the confidence to start. His blog not only brought neighbors together but also attracted collaboration with local schools and museums, amplifying his impact.

3. E-Commerce Foundations: Turning Ideas into Income

Aisha dreamed of selling her handmade jewelry online. While popular platforms existed, she wanted her shop to reflect her unique brand. Using the skills from this guide, she built a custom landing page for her products, styled her listings with CSS Flexbox, and embedded simple forms for contacting her. As her site grew, she explored more advanced features, but her strong foundation in HTML and CSS allowed her to iterate quickly and maintain full creative control. Today, Aisha's site is not only a thriving business but also a portfolio piece that has opened doors to freelance web development work.

These stories are just a few examples of how learning HTML and CSS can empower you to create, connect, and build a future on your own terms.

Reflecting on Your Coding Achievements

If you look back to where you started, the progress you've made is remarkable. Maybe you wrote your very first `<h1>` tag with a mix of excitement and uncertainty. Now, you can structure a complete webpage, style it to match your vision, and make it responsive for any device. You've debugged errors, solved layout puzzles, and learned how to bring ideas to life through code.

This journey is about more than technical proficiency. It's about developing a mindset of problem-solving, creativity, and continuous learning. Every time you experiment with a new layout, tweak a color scheme, or refactor your code for clarity, you're building skills that extend far beyond web development. You're learning to think logically, communicate ideas clearly, and adapt to new challenges—abilities that will serve you well in any career.

The Next Steps: Your Path Forward

Now that you have a solid foundation, the possibilities are limitless. Here's how you can continue growing as a web developer and creative professional:

Keep Practicing: The best way to solidify your skills is by building real projects. Tweak your portfolio, start a personal blog, or redesign a favorite website for practice.

Build a Portfolio: Collect your best projects in a portfolio site. This is your showcase to the world—and to potential employers or clients—demonstrating what you can do.

Collaborate and Share: Join online communities, contribute to open-source projects, or team up with friends on new ideas. The web is built on collaboration and sharing.

Learn Advanced Technologies: Once you're comfortable with HTML and CSS, explore **JavaScript** to add interactivity, or delve into modern frameworks such as **React** or **Vue**. Try **CSS preprocessors** like SASS, or learn about **version control** with Git.

Stay Inspired: Follow design trends, study sites you admire, and keep experimenting. The world of web development is always evolving, and there's always something new to learn.

The Transformative Power of Web Development

The skills you've gained are not just technical—they're transformative. With HTML and CSS, you have the power to:

Express Yourself Creatively: Your websites are your canvas. Every color, layout, and interaction is a reflection of your unique perspective.

Shape the Web: The sites you build can inform, inspire, entertain, and empower others. Whether you're creating resources for your community, launching a business, or sharing your story, you're making a real impact.

Build a Career: Web development is a gateway to a world of professional opportunities. From freelance work to full-time roles in tech companies, your skills are in demand across industries and around the globe.

A Final Word: Your Coding Journey Continues

As you close this book, remember: mastery isn't about knowing everything at once. It's about curiosity, persistence, and the willingness to keep learning. Every great web developer started as a beginner, just like you. The sites you admire most were built one line of code at a time.

The web needs your voice, your ideas, and your creativity. Whether you dream of launching your own startup, designing for a non-profit, or simply creating a beautiful blog to share your passions, you now have the skills to turn those dreams into reality.

So keep coding. Keep building. Keep exploring.

The smarter way to learn is not just about following instructions, it's about making, experimenting, collaborating, and growing. The journey you've started here is only the beginning. The next great website could be yours.

Your story as a web creator has just begun. The future of the web is in your hands. Go build it.

Made in United States
Cleveland, OH
05 November 2026

47668307R00250